The Aspiring Adept

The Aspiring Adept

ROBERT BOYLE AND HIS ALCHEMICAL QUEST

Including Boyle's "Lost" *Dialogue on the Transmutation of Metals*

LAWRENCE M. PRINCIPE

PRINCETON UNIVERSITY PRESS

PRINCETON, NEW JERSEY

Published by Princeton University Press, 41 William Street, Princeton, New Jersey 08540
In the United Kingdom: Princeton University Press, Chichester, West Sussex

---

Library of Congress Cataloging-in-Publication Data

Principe, Lawrence.
The aspiring adept : Robert Boyle and his alchemical quest :
including Boyle's "lost" Dialogue on the transmutation of metals /
Lawrence M. Principe.
p. cm.
Includes bibliographical references (p. –) and index.
ISBN 0-691-01678-X (cl. : alk. paper)
1. Boyle, Robert, 1627–1691—Contributions in alchemy.
2. Transmutation of metals. 3. Chemistry—Great Britain—
History—17th century. I. Boyle, Robert, 1627–1691 Dialogue on
the transmutation and melioration of metals. 1998. II. Title.
QD24.B685P75 1998 540′.1′1202—dc21 [B] 97-41793 cip

This book has been composed in Sabon Typeface

Princeton University Press books are printed on acid-free paper and meet the guidelines for permanence and durability of the Committee on Production Guidelines for Book Longevity of the Council on Library Resources

http://pup.princeton.edu

Printed in the United States of America

1 2 3 4 5 6 7 8 9 10

TO KDK

CONTENTS

ACKNOWLEDGMENTS

MANY HISTORIANS of science have assisted in many ways in bringing this project to completion. The initial form of this book was a Ph.D. dissertation completed in March 1996 for the Institute for the History of Science, Medicine, and Technology at The Johns Hopkins University. My progress through the graduate program was guided by Robert Kargon, who not only showed patience with what was often very slow progress, but also provided insightful comments and encouragement that remain much appreciated. I also thank the readers of the original dissertation—Jerome Bylebyl, Stephen Barker, Howard Egeth, and Gary Posner—for their willingness and stamina in reading a three-volume dissertation. During my first year of study at Hopkins I was supported by a George Owen Fellowship of the Humanities, for which I am grateful. Since 1989, I have been a full-time member of the chemistry faculty, and I would like to express my gratitude to three chairmen who were wonderfully understanding and supportive of me and my work in the history of science: Gary Posner, Craig Townsend, and David Draper.

Two colleagues, William Newman and Michael Hunter, deserve particular mention. Michael Hunter began by generously sharing with me, in 1989, a draft copy of his subsequently published and invaluable *Guide* to the Royal Society Boyle Papers, and since that time he has been a constant source of information, critical comment, and assistance. His continuing contributions to Boyle studies, in terms of both his own scholarly publications and his role as a medium of communication among other Boyle scholars, have advanced the field enormously. William Newman, my colleague in all things alchemical, shared freely of his staggeringly wide knowledge of alchemical literature during years of conversations, has read and commented on most of the document, and occasionally provided much-needed exhortations to bring it to completion. Many other scholars have read all or part of the manuscript and offered their comments, most notably Antonio Clericuzio, Edward B. Davis, John Harwood, Seymour Mauskopf, Bruce Moran, Margaret J. Osler, Rose-Mary Sargent, Kathleen Whalen, and Jan Wojcik.

This study is heavily dependent upon archival materials, for access to which I thank the British Library, the Royal Society of London, the Bodleian Library, the Église Protestante de Londres, the Archives Départmentales du Calvados, and the University of Glasgow. At the Royal Society, the librarians Sandra Cummings, Sheila Edwards, and Mary Sampson deserve special mention for unflagging help and patience. I also thank Mason Barnett for checking on Heinrich Screta at the archives of the Academia naturae curiosorum, Dr. René Specht, Chief Librarian at the Stadtbibliothek Schaffhausen, for his help in tracking local materials on Screta, and David Weston,

Principal Assistant Librarian of the Special Collections at the University of Glasgow, for checking on a rare edition of Boyle's *Anti-Elixir* tract.

I also wish to express my gratitude to the American Philosophical Society, which funded a research trip to French archives to uncover traces of Georges Pierre and the Asterism, as detailed in chapter 4.

Finally, in terms of daily maintenance of my sanity, constant encouragement, proofreading, the explosion of feeble arguments, and a host of other things, I am especially thankful to K. D. Kuntz.

NOTE ON PRIMARY SOURCES

I HAVE consistently used the seventeenth-century editions of all of Boyle's works, as I find them preferable to the eighteenth-century Birch editions for a number of reasons. All of Boyle's works are available on the Wing microfilm collection *Early English Books* and are thus actually more readily available to most readers than Birch. Furthermore, the new edition of Boyle's *Works* now in preparation will fully supersede the Birch edition and will include marginal notes indicating the original pagination. One slight problem in using these early editions is the erratic pagination: in cases of ambiguity (i.e., several sets of page numbers in one volume) I have explicitly cited the name of the essay that I am referencing; in cases of no pagination (i.e., Boyle's lengthy prefaces, notes, and advertisements), rather than sending the reader generally to "preface" or relying on infrequent and cumbersome signature markings, I have merely counted (in lower case Roman numerals) from the first printed page and have supplied these page references in square brackets. Additionally, all translations and transcriptions are mine unless otherwise noted.

ABBREVIATIONS

Alchemy and Chemistry	*Alchemy and Chemistry in the Sixteenth and Seventeenth Centuries: Proceedings of the Warburg Colloquium 1989,* ed. Piyo Rattansi and Antonio Clericuzio (Dordrecht: Kluwer, 1994).
BC	John Ferguson, *Bibliotheca chemica,* 2 vols. (Glasgow, 1906).
BCC	*Bibliotheca chemica curiosa,* ed. J. J. Manget, 2 vols. (Geneva, 1702; reprint, Sala Bolognese: Arnaldo Forni, 1976).
BL	Royal Society Boyle Letters.
BP	Royal Society Boyle Papers.
Boas, *RBSCC*	Marie Boas [Hall], *Robert Boyle and Seventeenth-Century Chemistry* (Cambridge: Cambridge University Press, 1958).
DNB	*Dictionary of National Biography.*
Hunter, *Guide*	Michael Hunter, *Letters and Papers of Robert Boyle: A Guide to the Manuscripts and Microfilm* (Bethesda: University Publications of America, 1992).
Maddison, *Life*	R.E.W. Maddison, *The Life and Works of the Honorable Robert Boyle F.R.S.* (London: Taylor & Francis, 1969).
MH	*Musaeum hermeticum reformatum et amplificatum* (Frankfurt, 1678; reprint, Graz: Akademische Druck, 1970).
Newman, *Gehennical Fire*	William R. Newman, *Gehennical Fire: The Lives of George Starkey, an American Alchemist of Harvard in the Scientific Revolution* (Cambridge: Harvard University Press, 1994).
NRRS	*Notes and Records of the Royal Society of London.*
RBR	*Robert Boyle Reconsidered,* ed. Michael Hunter (Cambridge: Cambridge University Press, 1994).
RSMS	Royal Society Miscellaneous Manuscript Series.
TC	*Theatrum chemicum,* ed. Lazarus Zetzner, 6

	vols. (Strasbourg, 1659–1661; reprint, Turin: Bottega d'Erasmo, 1981).
TCB	*Theatrum chemicum britannicum,* ed. Elias Ashmole (London, 1652; reprint, New York: Johnson Reprint Co., 1967).
Works	*The Works of the Honourable Robert Boyle,* ed. Thomas Birch, 6 vols. (London, 1772).

The Aspiring Adept

INTRODUCTION

THIS IS A BOOK about a well-known figure's involvement in a still poorly known subject. Robert Boyle (1627–1691) is undoubtedly one of the most influential and fascinating characters of the seventeenth century. Recognition of his impact on the development of science is widespread—every student who has taken introductory chemistry reads his name in connection with his air-pump experiments and learns the fundamental gas law that bears his name. Many science textbooks also routinely refer to Boyle as the "Father of Modern Chemistry," citing him as the man who broke once and for all from the irrational, misguided alchemy that preceded him. While it is the task of this volume to refute the double error of that last assertion, there is no denying Robert Boyle's profound impact on the development of experimental science, scientific method, and scientific culture that has earned him an enduring place in history.

Of late there has been a remarkable revival of interest in Boyle's thought, life, and works. The past decade (since the appearance of Steven Shapin's and Simon Schaffer's celebrated *Leviathan and the Air-Pump*) has witnessed the publication of at least six books and a score of scholarly articles on Boyle. Interest in Boyle now occupies a wide range of specialists—historians of science, theology, literature, and rhetoric, as well as some sociologists. Critically for historical studies, the enormous mass of his surviving papers at the Royal Society has at long last been cataloged by Michael Hunter and is now available on microfilm. Presently, a new edition of Boyle's complete works (the first since 1772) is being prepared by the able collaboration of Hunter and Edward B. Davis, while Hunter and Antonio Clericuzio are editing Boyle's complete surviving correspondence for the first time. Both of these major contributions are scheduled to appear before the end of the decade.

In 1991, during several frosty December days, the currently active Boyle scholars gathered at a country hotel near Robert Boyle's (now demolished) hereditary estate of Stalbridge in commemoration of the tercentenary of his death. This symposium, organized by Michael Hunter, showcased the fresh new directions in Boyle studies that will undoubtedly be pursued for many years to come. The papers presented there (gathered into the appropriately titled volume *Robert Boyle Reconsidered*) display the young vitality of Boyle studies and, by extension, a renewal of scholarly interest in seventeenth-century science, a subject that, as recently as fifteen years ago, was widely considered to have been "mined out."

In contrast to the renown of a figure like Boyle, a considerable part of the field of alchemy still remains unexplored or misunderstood. Notwithstanding the important advances in understanding that have been forthcoming from the pens of scholars during the past thirty years, much of the realm of alchemy still presents to historical inquirers as great, as vast, as confusing, and as ill-charted a domain as it did to the aspiring Sons of the Art three centuries ago. The causes for this lack of understanding are many. Lingering prejudices against what has been seen for the past two centuries primarily as irrational and misguided—at worst mere charlatanry and at best pseudoscientific—still persist in some quarters. Alchemy also makes a very convenient (and simplistic) foil against which to set off modern science. Further, much published literature on alchemy is simply *bad*—pitiful translations of primary sources, ahistorical caricatures masquerading as history, and a parade of modern esoterica and assorted fluff presented as if they were directly connected to historical alchemy. Nonetheless, there now seems to be a revival of scholarly interest in alchemy; the careful, considered study of its vast primary literature and its contributions to modern science advocated for so long by pioneers like Allen Debus may finally be creaking into motion. Recently we have seen the editing and publication of several source-works in scholarly format, as well as the revisiting and resolution of a few long-standing biographical and bibliographical puzzles. Yet the light of these labors illuminates not only their immediate topics but also the dim outlines of a vast labyrinth of alchemical thought, as intriguing and potentially valuable as it is unexplored.

In bringing together a grand figure celebrated by modern science and the misty topic of alchemy rejected by that same science, I am of course preceded by the work of the late Betty Jo Teeter Dobbs, whose two books and numerous articles on the alchemical preoccupations of Sir Isaac Newton ushered in a completely new view of the man who is probably the most widely popularly recognized figure in the history of science. The work of Dobbs, Richard S. Westfall, and Karin Figala on Newton's alchemy has made my job considerably easier. The number of those who would bristle at the mere mention of alchemy in the same breath with the invocation of a hero of the Scientific Revolution is now much diminished, and those few well-intentioned but misguided hangers-on to positivist hagiography (often the very same ones who balked at the idea of Newton as alchemist) are now marginalized to irrelevance. That Newton, or Boyle, or Locke, or any other seventeenth-century natural philosopher should have believed, dabbled, labored, or sweated in alchemy should no longer horrify the sensibilities.

We must acknowledge, however, that much of that previous horror was based upon a very poor notion of what alchemy was. Those previous, immediately negative, views of alchemy arose not out of a view of historical alchemy as it was practiced by serious workers, but rather out of an acceptance of the obscuring layers of negative propaganda and esoteric detritus accreted around alchemy during subsequent generations. Preserving the rationality of the architects of modern science from the taint of some putative alchemical irrationality was seen at one time, I suppose, as a noble endeavor. Recent work has softened much of the earlier *Schwarzweissmalerei,* and while a great deal of labor must yet be expended to flesh out the now-rising new framework of the development of science in the seventeenth century, we do see clearly the artificiality of many of the earlier divisions of historical epochs and schools of thought into categories as absolute as naive.

The work I have done on Robert Boyle since 1988 has actually been an extended digression. During the mid-1980s I was concentrating on the works of the mysterious Basil Valentine, a supposed Benedictine monk of Erfurt and author of a large number of very popular and influential alchemical tracts purportedly written in the fifteenth century, but first published only in 1599 and after. In the course of that study, convinced that alchemical symbolism and allegory masked real experimental endeavors and results, I labored on deciphering the allegorical emblems and enigmas contained in Valentine's first-published work *Ein kurtz Summarischer Tractat,* better known as *Die Zwölff Schlüssel* or *The Twelve Keys.* I was fairly confident that one of these figures portrayed the volatilization of gold, a process that fit well with the universal alchemical dictum "to make the fixed volatile" in order to prepare the Philosophers' Stone. I believed that modern chemical information on the volatility of gold compounds would help to illuminate the process and the origins of its metaphorical description, and I found as one of the most extensive studies of this topic in the literature a late-nineteenth-century paper on the volatility of gold chloride. I was surprised to find that author referencing Robert Boyle's *Origine of Formes and Qualities* as the earliest mention of such volatility. Upon reading Boyle I was quite amazed to find the process Boyle describes there to be identical, step-by-step, with the one I had been deciphering from Valentine.

Intrigued by this discovery, I set Valentine aside momentarily and began reading Boyle. The more I read, the more allusions and uncited references to alchemical practices, theories, and beliefs I found. I presented a paper on these early findings at an international symposium on alchemy held at Groningen in April 1989.[1] On the way back from that meeting I visited the library of the Royal Society, and there in Boyle's papers I saw the undeniable magnitude of the traditional alchemical component of Boyle's work. Although I had suspected that a careful study of Boyle's texts would in fact reveal significant alchemical influence and content, the sheer extent of clearly traditional alchemical material—not predominantly transcripts as in the case of Newton, but considerable original material and laboratory accounts—among Boyle's manuscripts was astonishing. The present volume is the work I then knew I would have to write, and as a result I have not (yet) returned to the *sixth* key of Basil Valentine.

One of the aspects of this present work on Boyle distinguishing it from some earlier studies is that it approaches him from behind—that is to say, I had been immersed in the earlier alchemical literature before reading any Boyle. As a result, it was not only easier to recognize alchemical tropes and

[1] Lawrence M. Principe, "The Gold Process: Directions in the Study of Robert Boyle's Alchemy," in *Alchemy Revisited,* ed. Z.R.W.M. von Martels (Leiden: Brill, 1990), 200–205.

allusions than it would otherwise have been, but I was also more strongly disposed toward identifying Boyle's sources—what he learned from whom, what he carried over into his own work unchanged, what he modified—than toward finding in his works the germs of later chemical developments. This approach from a different direction has, I think, enabled the discovery of a side of Boyle that other workers did not (or would not) see.

What I stress throughout this volume is a sense of *context;* we cannot understand Boyle except in his intellectual and social context. While this sentiment may strike the reader as obvious or trite, much of the older standard literature on Boyle (and even some of the most popular recent material) does not so treat him. Clearly, Boyle did not work in isolation; he read the works of contemporaries and maintained an extensive network of colleagues and correspondents in Britain and on the Continent. Boyle may have lived on an island, but he was not himself an island. In terms of alchemy, the problem of context is automatically severe simply because primary source material is so infrequently read. This study shows, however, how Boyle fits into the alchemical context of the seventeenth century, and reveals new (and unexpected) sources for his thought as well as targets of his opposition. Without these contemporaneous sources, it is as if we took one of Boyle's own literary dialogues, expunged the speeches of all the characters save one, and then tried to make sense of the remnant of tattered text.

The first chapter traces out how the image of Robert Boyle as a historical character—particularly as concerns his relationship to alchemy—has been drawn and redrawn over the past three centuries. As such this chapter functions as a slightly idiosyncratic literature review. The second chapter is devoted to perceptions of the *Sceptical Chymist* and the presentation of a new (and more contextual) reading of this difficult text. I more accurately identify the real targets of the *Sceptical Chymist,* thus allowing us to proceed unopposed to a consideration of Boyle's positive involvement in alchemy. The third chapter—the most important in terms of new primary source material—analyzes a significant new addition to Boyle's corpus, his "lost" *Dialogue on the Transmutation of Metals.* The extant text of this work, reconstructed from some twenty surviving fragments, is presented along with full scholarly apparatus as the first appendix to this volume. This chapter also introduces some distinctions within transmutational alchemy important for the understanding of Boyle's alchemy. The fourth chapter continues from the transmutation accounts found in the *Dialogue* to investigate Boyle's own witness of transmutation, his meetings with mysterious traveling adepts, and the continuity of Boyle's accounts with the long-standing tradition of transmutation histories. Here I also chronicle Boyle's less positive interactions with alchemists including the twisted tale of the shady Georges Pierre, his master, the supposed patriarch of Antioch, and their "Asterism" of alchemical masters. In the fifth chapter I recount Boyle's alchemical practice, both literary and experimental. With respect to Boyle's textual alchemy, I show how Boyle read alchemical texts and attempted to penetrate their veil of secrecy while covering his own alchemy with a similar veil. In regard to practice, I trace one of Boyle's lifelong alchemical pursuits by tracking a grand alchemical quarry—the Philosophical Mercury, that precious substance needed in the first operations of preparing the Philosophers' Stone. This hunt leads from Boyle's earliest experiences with experimental laboratory science to his deathbed and follows an Ariadne's thread through a complicated web of collaborators and contacts including George Starkey, John Locke, and Sir Isaac Newton. The sixth chapter explores Boyle's changing reasons for pursuing alchemy. These reasons include the potential value of alchemy as a tool for dissecting nature, as a source for potent medicines, and, most dramatically, as a bridge between Boyle's natural philosophy and theology. Here I explore the spiritual aspects of Boyle's chrysopoeia, those topics that do involve true supernaturalism as opposed to laboratory practice. Here again a new text is presented, a fragmentary dialogue (fully transcribed in the third appendix) that clearly links Boyle's alchemical thoughts to theology and angelic visitations. I conclude with a summary of how this study reforms our historical understanding of Robert Boyle and of alchemy, and their places in early modern science.

ALCHEMY AND CHEMISTRY

A CRUCIAL NOTE ON TERMINOLOGY AND CATEGORIES

A SLIPPERY problem that plagues the historian of alchemy/chemistry is the fluidity of the terminology relating to the very subject he is endeavoring to study. The words *alchemy* and *chemistry* (in all their orthographic and grammatical variants) are used interchangeably in almost all seventeenth-century contexts. The modern distinctions between the two were not made and codified until after Boyle's death. Accordingly, in the study of the *Sceptical Chymist* I must explore Boyle's use of the two words and show how we cannot divide up seventeenth-century activities into "alchemy" and "chemistry" based directly (and naively) upon which of these two terms was used in a given situation. It would, however, be helpful to set down defensible and consistent linguistic conventions so that we can talk accurately and meaningfully about alchemy/chemistry in the seventeenth century. Were we dealing with an earlier or a later epoch, the difficulties would be much diminished, but in that century, when we are able to find topics easily alignable with (modern) chemistry alongside accounts of the Philosophers' Stone, it pays to investigate our usages carefully. We want to settle upon terminology that circumvents the ambiguity and value-laden baggage of the terms *alchemy* and *chemistry.*[1]

There are several ways to address this terminological problem. One option is to do nothing; but if we continue to use terminology without consistent and defensible ground rules, we are too liable to employ terms as they suit our historiographic needs, a Procrustean practice of truncating or stretching history to our beliefs rather than constructing our beliefs in accord with history. This has frequently happened in the past when "alchemy" was attached to one author and "chemistry" to another based not upon the actual operations or goals of the practitioner but rather upon his preconceived degree of credibility or "modernity."[2] To speak, for example, of the "alchemy of Eirenaeus Philalethes" and then the "chemistry of Robert

[1] The reader is advised that the conventions outlined here diverge somewhat from those I suggested in an earlier paper ("Boyle's Alchemical Pursuits," in *RBR*, 91–105, on 92). This development of my views on the matter stems largely from discussions with William Newman (who was concerned with the same issues in his *Gehennical Fire*, xi–xiii). Our concerns over the impact on historical writing from the terminological chemistry/alchemy problem and the anachronistic accretions to alchemy have resulted in a pair of joint papers, "Alchemy vs. Chemistry: The Etymological Origins of a Historiographic Mistake" (*Early Science and Medicine,* forthcoming 1998) and "The Historiography of Alchemy" (*Archimedes,* forthcoming 1998). The reader is referred thither for further discussion on these topics.

[2] Examples are not difficult to find; indeed, inconsistency in what constitutes alchemy is a recurring problem in Boas Hall's *RBSSC;* compare, for example, 48 where Frederick Clodius's "alchemistic tendencies" are bad, with 54 where alchemy for Glauber is equated with "pure chemistry."

Boyle" places these authors in different sets bearing different connotations, in spite of the fact that the two (as we shall see later) worked together in the same place, in the same way, at the same time, on the same thing.[3] Common sense and respect for historical sources require that we be consistent and affix the same label to the activities of both. What, however, is that label?

One label we can attach to both is *chymistry*—where the archaic orthography signals that we mean *the sum total of alchemical/chemical topics as understood in the seventeenth century.* The term *chymistry* is extremely convenient for describing a broad historical domain as it was contemporaneously denominated. It avoids the connections implicit in the use of either *chemistry* or *alchemy.* I do not think, however, that *chymistry* alone is a complete solution; its encompassing nature renders it insufficiently descriptive for some fine-grained historical analyses. We also need some more precise terms to differentiate the range of pursuits encompassed within seventeenth-century chymistry.

Fortunately, there are seventeenth-century words that distinguish various branches of chymistry; I suggest we resuscitate and employ them. Since I am interested particularly in Boyle's pursuit of metallic transmutation, especially involving the Philosophers' Stone—the agent of metallic transmutation—I will often use the word *chrysopoeia. Chrysopoeia* and the related *argyropoeia* are derived from the Greek *chrysos* (gold) or *argyros* (silver) plus *poiein* (to make), and denote specifically the transmutation of base metals into the noble ones. Thus, to speak in the accurate and precise sense I advocate, the present study centers on *Boyle's chrysopoeia*—a more rigorously defined realm than "Boyle's alchemy."

Another valuable term is *spagyria. Spagyria* (purportedly compounded from the Greek *span,* to draw forth or separate, and *ageirein,* to collect together) refers to the process of separating a substance into its Essentials—Mercury, Sulphur, and Salt, for example—purifying these ingredients, and then recombining them into a purified, exalted, regenerated body. Further terms are *iatrochemistry* and its older form *chemiatria* for medical chymistry (both from the Greek *iatros,* physician).[4] These subsets can overlap—spagyria, for instance, overlaps with chrysopoeia (some workers believed the Philosophers' Stone to be a spagyric preparation of gold), as well as with iatrochemistry, for most iatrochemists believed that spagyric preparations were especially potent medicines freed from poisonous qualities.

While I hope that this terminology will increase precision and avoid unwanted connotations, it would be idle to imagine that three hundred years of linguistic development could (or *should*) vanish by fiat. Moreover, there

[3] A fuller exposition of the collaboration between Eirenaeus Philalethes (a.k.a. George Starkey) and Robert Boyle forms the basis of a study currently being pursued by William R. Newman and the present author.

[4] See Wolfgang Schneider, "Chemiatry and Iatrochemistry," in *Science, Medicine, and Society in the Renaissance,* ed. Allen G. Debus (New York: Science History Publications, 1972), 141–50.

are topics pursued by Boyle and other seventeenth-century chymists that do not fall under any of the subsets described above. We need not abandon altogether the words *alchemy* and *chemistry* but only use them cautiously.[5] Since chrysopoeia and spagyria are the chief pursuits of those chymists retrospectively called alchemists, I will also employ the phrase *traditional alchemy,* or occasionally simply *alchemy,* to refer to these two pursuits; I will never use *chemistry* to refer to seventeenth-century activities.

Finally, those who have delved into traditional alchemical texts are well aware of the difficulties arising from equivocal or intentionally ambiguous nomenclature. The names of the Principles, particularly *Mercury,* play heavily into the hands of those writers who strove for obfuscation. To avoid the recurrence of that difficulty here, I will adopt the following convention: The names of Principles will always be capitalized, while the names of the common substances will not be (e.g., *mercury* is quicksilver, while *Mercury* is an alchemical substance not necessarily equivalent to the modern chemist's mercury, the element Hg).

There will frequently be cause for digressions to fill in background material on some aspect of alchemy. In the absence of any comprehensive "primer of alchemy" to which to refer, the historian of alchemy must write not only the history at hand but also the histories of related topics so that his main story makes sense and displays its real significance. Every attempt will be made to prevent such digressions from becoming too Boylean.

[5] Taken to extremes, the use of *chymistry* alone would imply a position of radical nonpersistence of terms—that the practices and knowledge of the seventeenth century are so wholly "other" that our terminology cannot be *at all* mapped out upon them. That way lies historical nihilism.

CHAPTER I

Boyle Spagyricized

After the death of the Honorable Robert Boyle on New Year's Eve 1691, his vast accumulation of personal papers and manuscripts had to be set in order. Boyle had appointed three friends to sort his chymical papers—John Locke, the well-known philosopher; Edmund Dickinson, physician to Charles II and James II; and Daniel Coxe, another physician and longtime friend. Although these three soon gave up their work on the project in despair (more on the reasons later), their abortive project did give rise to a remarkable exchange of letters whose contents have intrigued readers ever since. Less than a month after Boyle's death, Sir Isaac Newton wrote to John Locke, "I understand Mr. Boyle communicated his process about the red earth and mercury to you as well as to me & before his death procured some of that earth for his friends."[1] Two weeks later, Newton explained briefly how he knew that Locke "had the receipt" (Locke seems to have been surprised), and then Newton's original hint about the "earth for his friends" apparently paid off, for Locke sent him some of this mysterious "red earth" before midsummer. Having gotten hold of the earth, Newton also requested that Locke, whom he knew to have access to Boyle's papers, locate for him some particular chymical processes. Accordingly, Locke sent Newton a peculiar process on mercury written in a curious code as it was transcribed from Boyle's papers. Newton replied a week later that "this receipt I take to be that thing for the sake of which Mr. B. procured the repeal of the Act of Parliment against Multipliers," referring to the 1689 repeal of a statute dating from the reign of Henry IV, which seemed to outlaw the transmutation of metals into gold and silver. Newton went on to link this recipe with the curious paper Boyle had published sixteen years earlier (1676) in the *Philosophical Transactions* concerning a special mercury that would grow hot upon mixing with gold.[2] At that time, Newton had fired off a letter which rather inconsistently denied that Boyle's mercury had any special virtue while also recommending that Boyle "preserve high silence" regarding so sensitive and potentially critical a topic.[3] In his final letter to Locke about this matter, Newton was again dismissive in spite of his manifest eagerness to get his hands on these materials that Boyle would not give him during his lifetime.

[1] Newton to Locke, 26 January 1691/92, *The Correspondence of Isaac Newton,* ed. H. W. Turnbull, 7 vols. (Cambridge: Cambridge University Press, 1960), 3:193.

[2] Newton to Locke, and Locke to Newton, 16 February, 7 July, 26 July, and 2 August 1691, ibid., 195, 215–19. Boyle's paper on the incalescent Mercury will be fully discussed in chapter 5.

[3] Newton to Oldenburg, 26 April 1676, ibid., 2:1–3.

The content of this rather cryptic exchange between Newton and Locke undeniably ties Robert Boyle to transmutational alchemy. Newton and Locke both knew of Boyle's alchemical interest and its overlap with their own; in the waning years of the seventeenth century, Boyle's interest was neither surprising nor notably scandalous. But to later historians these well-known references, along with the existence of Boyle's rare and brief tract *An Historical Account of a Degradation of Gold made by an Anti-Elixir* (published anonymously in 1678) represented constant (and generally annoying) reminders of Boyle's alchemical activities. Like the signs of Newton's alchemy circumspectly rumored since the time of his death, grudgingly admitted and profusely apologized over by Sir David Brewster in 1831, but not seriously and exhaustively explored until the recent researches of Richard Westfall, Betty Jo Teeter Dobbs, and Karin Figala, these tidbits about Boyle have always been known but have remained largely unpursued. In actuality, they represent only the tip of the proverbial iceberg. What activities and motivations lie behind that mysterious red earth we hear about only after Boyle's death? How does the *Anti-Elixir* tract fit in with the rest of Boyle's output? In sum, how important overall were traditional alchemical activities to Boyle? It is the goal of this study to map out the contours of Boyle's hitherto little-known alchemical endeavors—to explore their goals, their context both intellectual and social, and their significance to the whole of Boyle's thought as well as to our understanding of chymistry in the seventeenth century.

This study contends that Robert Boyle's traditional alchemical interests were serious and persistent, constituting a significant and influential dimension of his life, thought, and works. A necessary corollary to this thesis is the contention that our current general understanding of Boyle predominantly as a modern—whether in terms of the origins of modern chemistry, science in general, or the scientific social dynamic—is based upon a selective and sometimes incorrect reading of his activities and works. While it will be the task of the succeeding chapters to uncover and to document Boyle's chrysopoetic practices, and then to trace out their importance to the whole of Boyle's thought and to our understanding of seventeenth-century science, it is valuable first to recount how the generally reigning portrait of Boyle has developed over the past three hundred years. How and why did Boyle's chrysopoetic endeavors, well-known to Locke and Newton (and, as I shall show later, to a great many of his contemporaries), become obscured, forgotten, realigned, and explained away? Such a study taken to its furthest extent would touch not only upon the changing evaluation of alchemy during the intervening centuries but also upon the varied apologetic or illustrative purposes to which a prominent but deceased figure and his works can be put by successive generations.[4] Clearly, such an inquiry could easily ex-

[4] The shifting emphases in biographies of Boyle are treated in Michael Hunter, "Robert Boyle and the Dilemma of Biography in the Age of the Scientific Revolution," in *Telling Lives*

pand to fill an entire (and not uninteresting) volume itself. For the present purpose, however, it will suffice to sketch out Boyle's reconfiguration from the time of his death to the present age of professional history of science. I ask the reader not to view the following analysis as a refutation or ridicule (as otiose as unfair) of writers who are long dead or who worked on Boyle well before the present author saw the light of day. Rather, the collection and scrutiny of these writers on Boyle provides a view of the spirit of their times. Indeed, as Jost Weyer has pointed out in terms of the historiographic fortunes of alchemy, the sentiments of many nineteenth- and twentieth-century writers show "less alchemy itself than the *Zeitgeist* of the epoch in which the historian lived."[5] The tracking of that spirit's evolution will be both explanatory and valuable to other studies.

Soon after his death in 1691, Boyle's chrysopoetic interests and influences began either to fade or to be effaced. Wilhelm Gottfried Leibniz, for example, in the context of his correspondence with Samuel Clarke, used Boyle as a contrast to what he saw as the occult, nonmechanical action-at-a-distance implicit in Newton's gravity. "In the time of Mr. Boyle . . . no body would have ventured to publish such chimerical notions . . . Mr. Boyle made it his chief business to inculcate that every thing was done mechanically in natural philosophy."[6] In this and similar ways, Boyle's science was made more universally and rigidly mechanical and comprehensive than Boyle had ever made it, and it became a science from which Helmontian and other alchemical notions such as seminal principles, plastic powers, and occult forces (important to Boyle) had been purged.[7]

The epitomes of Boyle's works that were produced at the beginning of the eighteenth century provided a selective view of his work. The first of these epitomes, four volumes published by Richard Boulton in 1699–1700, already begins to skew Boyle's works away from chymistry. Boulton defers the *Sceptical Chymist* and its 1680 appendix *Producibleness of Chemical Principles* (which defends some important chrysopoetic notions) to the final volume, and reduces them from the original 700 pages to only about 150. Boyle's most clearly alchemical writings, the incalescent mercury paper and the *Anti-Elixir* tract, are entirely absent.[8] By contrast, a full volume and a

in Science: Studies in Scientific Biography, ed. Michael Shortland and Richard Yeo (Cambridge: Cambridge University Press, 1996), 115–37.

[5] Jost Weyer, "The Image of Alchemy in Nineteenth and Twentieth Century Histories of Chemistry," *Ambix* 23 (1976): 65–70, on 65.

[6] *The Leibniz-Clarke Correspondence,* ed. H. G. Alexander (Manchester: Manchester University Press, 1956), 92.

[7] Antonio Clericuzio, "A Redefinition of Boyle's Chemistry and Corpuscular Philosophy," *Annals of Science* 47 (1990): 561–89, on 561–63. We are only now coming to understand that many chrysopoeians themselves used mechanical and corpuscularian frameworks; see William R. Newman, "The Corpuscular Transmutational Theory of Eirenaeus Philalethes," in *Alchemy and Chemistry,* 161–82, and "Boyle's Debt to Corpuscular Alchemy," in *RBR,* 107–18.

[8] Richard Boulton, *The Works of the Honourable Robert Boyle, Esq. Epitomized,* 4 vols. (London, 1699–1700). These volumes include only Boyle's natural philosophy; in 1715

half are devoted to Boyle's work on pneumatics and air. It may also be noted that Boulton's epitomizing often completely misrepresents Boyle's actual intent. For example, in one passage of *Producibleness* Boyle describes a "noble mercury," which he went on to link with chrysopoetic themes. The original reads,

> I had once the opportunity to examine *Hydrostatically* a noble *Mercury*, for the impregnating whereof neither corporall *Gold* nor *Silver* was employ'd, and yet having carefully weighed this *Quicksilver* in water, according to the method I elsewhere teach, in the presence of a famous and very heedfull *Virtuoso*, I found it, as I had foretold, not only manifestly, but very considerably heavier in *specie* (that is *bulke* for *bulke*) than common *Quicksilver*, though that *Mercury* had been severall times distill'd, and by other waies depurated.[9]

Boulton's epitome garbles the passage to meaninglessness, omitting the fact that a special Mercury is being described:

> Mercury distilled or otherwayes depurated, without the Addition of Gold or Silver, being heavier than common Quicksilver; . . .[10]

Such errors are common throughout Boulton and are especially frequent in chymical topics. It would, however, be incorrect to impute such misrepresentation to a desire on Boulton's part to suppress alchemical content; rather, it seems more likely that they arise out of Boulton's own innocence of chymistry and perhaps a goodly measure of sloppiness as well. Nonetheless, anyone using Boulton's epitome would get a very distorted view of Boyle's notions.

A second epitome, and one much more widely used, was produced by the chemist Peter Shaw in 1725. In this case, the apologetics in which Shaw felt obliged to indulge are quite revealing of the early eighteenth century's view of both chrysopoeia and the late Robert Boyle. Shaw's introduction contains a passionate four-page defense of Boyle from the "imputation of credulity" under which he had fallen.

> [Boyle] is particularly cautious . . . to prevent his imposing upon himself by any superficial observation, any fallacious or contingent experiment, whether made in his own person, or deliver'd by others. So much the more afflicting is it to find, that he shou'd himself be charged with having attributed too much to the power of nature and art; and to have easily credited uncertain accounts of things uncommon and extraordinary. In this particular I beg leave to dissent from the general opinion which has crept abroad of Mr. Boyle.[11]

Boulton produced a three-volume epitome of Boyle's theological works. Note that *Anti-Elixir* was never publicly owned by Boyle, and that may explain its absence from Boulton.

[9] Robert Boyle, *The Producibleness of Chymical Principles* (Oxford, 1680), 221–22.

[10] Boulton, *Works*, 4:162.

[11] Peter Shaw, *The Philosophical Works of the Honourable Robert Boyle Esq.*, 3 vols. (London, 1725), 1:ix.

The view of Boyle's gullibility was apparently quite widespread, for it is described as being a "common report." Rising to Boyle's defense, Shaw argues that Boyle saw more unusual things than most people, and so he better recognized the vast potentiality of nature.[12] Nonetheless, Shaw's epitome shows many marks of being "cleansed" of alchemical taints. Shaw prunes down the *Anti-Elixir* tract into less than a single page and silently inserts it into the middle of *Usefulnesse of Experimental Natural Philosophy.* Alchemical reports from the late *Experimenta et Observationes Physicae* are similarly pared down and buried in the same tract.[13] Language that contains manifestly alchemical overtones is modernized throughout. Several clear examples occur in the *Producibleness of Chymical Principles*—for instance, Boyle wrote that the mercury he was accustomed to prepare by heating cinnabar with iron filings was "perhaps somewhat impregnated with a martial vertue," but Shaw deletes the alchemical-sounding planetary nomenclature and the dangerously nonmechanical connotations of "vertue" and writes in much more modern prose that such mercury is "perhaps, impregnated with a ferruginous quality."[14]

Shaw's touchstone of greatness is resemblance to "the happiest philosopher the world ever yet cou'd boast, the great Sir Isaac Newton."[15] Shaw's notes and introductions rarely fail to cite Newton at every opportunity. For the eighteenth century, Newton was the paragon of rational thought and mathematical precision, far removed (so they thought) from suspect alchemical musings. Thus it was important for Shaw to "Newtonize" Boyle—to make his works more systematic and universal by shuffling their order in the epitome, something that Boyle had expressly rejected throughout his life. What Shaw did for Boyle is similar to what Roger Cotes did for Newton in the preface of the second edition of the *Principia*—notions that could be assailed as esoteric or nonmechanical are gently removed or glossed over, even though (we come to know now) these very issues were of great interest and importance to their authors.[16]

Shaw's epitomization eviscerates Boyle's discursive (admittedly verbose) style that allowed him so wide a latitude of expression and belief. While it is unnecessary to dwell upon the inadequacy of such an epitome in representing the views of Boyle, it must be recognized that Shaw's volumes were in fact used as primary source material (sometimes by historians of science) and thus have had a real effect in constructing an anachronistic portrait of Boyle.[17] Shaw's Boyle is more a figure of the eighteenth century, and with

[12] Ibid., ix–x and xii.

[13] Ibid., 78 and 70.

[14] Boyle, *Producibleness,* 207; Shaw, *Works,* 3:410.

[15] Shaw, *Works,* 3:cclx.

[16] Recent scholarship reveals the ironic error of this Enlightenment idealization of Newton; see, for example, Edward B. Davis, "Newton's Rejection of the 'Newtonian World View,'" *Science and Christian Belief* 3 (1991): 103–17.

[17] J. C. Ferdinand Hoefer, *Histoire de la chemie,* 2 vols. (Paris, 1842–1843), for example,

alchemical leanings masked or expunged, now less liable to the "imputation of credulity."

The longest-lived portrait of Robert Boyle is that drawn by Thomas Birch in his 1744 biography.[18] But, as Michael Hunter has shown in his study of biography, the development of Birch's attitude toward Boyle's alchemy is quite intriguing.[19] In 1734, when he wrote Boyle's entry in the enlarged English edition of Pierre Bayle's *General Dictionary,* Birch seemed fascinated by Boyle's transmutational pursuits. He devoted a lengthy note to a synopsis of the *Anti-Elixir* tract, the only Boylean work to merit such consideration. Indeed, Birch's note is about as long as the entire tract as epitomized by Shaw. Another lengthy note describes Boyle's belief in the Philosophers' Stone (the agent of metallic transmutation), his transmutation of gold into silver using a solvent he called *menstruum peracutum* (described in *Origine of Formes and Qualities*), and his *Philosophical Transactions* paper on the incalescent mercury including Isaac Newton's response to it. Finally, a third note reprises Peter Shaw's "vindication of Mr. Boyle" from the imputation of credulity. Birch's *Life of Boyle* of 1744, however, although many times the length of the dictionary entry, contains a very small proportion of such material, including only a brief mention of Boyle's confession of belief in the Stone to Edmund Halley, and a short notice of the *Anti-Elixir* tract. Likewise, the correspondence published in Birch's 1744 *Works of the Honourable Robert Boyle* is carefully chosen, excluding almost everything with alchemical content. Some of this alchemical correspondence not only was excluded but may actually have been destroyed—some missing alchemical items are marked "N[o] W[orth]" or "in a mystic strain" in the inventories made by Birch's assistant Henry Miles. One example is provided by the now-missing letters from one John Matson; the single surviving letter (missed because it was bundled up in a collection of recipes) contains questions regarding the Philosophers' Stone, the preparation of which Matson had underway with Boyle's help.[20]

used Shaw's epitome and thought that Shaw's "Preliminary discourse" was Boyle's: "Ce discours preliminaire est un chef-d'oeuvre de logique qui ne serait déplacé dans aucun traité de science" (2:156).

[18] Thomas Birch, *The Life of the Honourable Robert Boyle* (London, 1744). Problems in producing a Boyle biography prior to Birch's are detailed in Michael Hunter, *Robert Boyle by Himself and His Friends* (London: William Pickering, 1994); see also R.E.W. Maddison, "A Summary of Former Accounts of the Life and Work of Robert Boyle," *Annals of Science* 13 (1957): 90–108.

[19] Hunter, "Dilemma of Biography," 129–30.

[20] Ibid., 115–37. On Birch and Miles, see also Marie Boas Hall, "Henry Miles, F.R.S. (1698–1763) and Thomas Birch (1705–1766)," *NRRS* 18 (1963): 39–44. The surviving Matson letter is BL, 4:42; I shall return to Matson's letter in chapter 5. Boyle's lost letters will be treated in Hunter and Clericuzio's forthcoming edition of Boyle's *Correspondence,* and the large number of Boyle documents extant in the 1740s but now missing will be discussed in Hunter and Principe, "The Lost Papers of Robert Boyle."

The gradual attrition of alchemy from Birch's portrayal of Boyle suggests that Birch the antiquarian found Boyle's chrysopoeia fascinating at first but as a biographer and editor decided (or was convinced) that such interests, unseemly by mid-eighteenth-century standards, were better suppressed. Other "nonrational" details of Boyle's character fared similarly, like his belief that 1 May was an unlucky day for him—a surviving editorial memo suggests that it would better for Boyle's image to omit mention of that eccentricity.[21]

From the middle of the eighteenth century until the beginning of the twentieth, accounts of Robert Boyle's work were to be found only in general histories of chemistry. In that location his role as a revolutionary was codified. Trommsdorf's 1806 history romantically portrays the emergence of chemistry (and early modern science in general) as a grand struggle between light and darkness, with Boyle as the key hero. He states, curiously, that Boyle was himself once among the number of alchemists, but he then realized their "emptiness" (*Nichtigkeit*) and opposed them, striking down their arrogance and showing their impotence.[22]

J. F. Gmelin (1797) sets Boyle as the definer of an epoch, titling his first section of "modern history of chemistry" "The Age of Boyle" (*Boyles Zeitalter*). Here again, the presumed late-seventeenth-century origin of chemistry at the hands of Boyle is seen as the "first rays of light . . . into the mist of alchemical and theosophical error."[23] While Gmelin's text abounds in bibliographical details of seventeenth-century alchemy, his citation of these sources is directed toward the drawing of a sharp contrast between them and Boyle. Thus it is somewhat surprising that K. C. Schmieder's 1832 study of the history of alchemy refers to Boyle as an advocate (*Vertheidiger*) of alchemy. Schmieder states that Boyle was at first a doubter of alchemy but then "came in fully on the side of the alchemists" when he published the *Anti-Elixir* tract in 1678.[24] Hermann Kopp, providing a more scholarly treatment than had previous authors, recognizes some of the usual signs of Boyle's alchemy (the *menstruum peracutum* and the *Anti-Elixir* tract) but, more interestingly, also briefly notes that even in the *Sceptical Chymist* Boyle seemed to allow some of the tenets of traditional alchemy when he

[21] Miles to Birch, British Library Add. MS 4314, fol. 70; see Michael Hunter, "The Conscience of Robert Boyle: Functionalism, Dysfunctionalism, and the Task of Historical Understanding," in *Renaissance and Revolution: Humanists, Scholars, Craftsmen, and Natural Philosophers in Early Modern Europe,* ed. J. V. Field and F.A.J.L. James (Cambridge University Press, 1993), 147–59 on 148–49.

[22] Johann B. Trommsdorf, *Versuch einer allgemeinen Geschichte der Chemie* (Erfurt, 1806; reprint, Leipzig: Zentral-Antiquariat der DDR, 1965), pt. 2, 119–55; on 130–31.

[23] Johann Friedrich Gmelin, *Geschichte der Chemie,* 3 vols. (Göttingen, 1797; reprint, Hildesheim: Georg Olms, 1965), 2:1. The linking of "theosophy" and alchemy is itself a product of the late eighteenth century.

[24] Karl Christoph Schmieder, *Geschichte der Alchemie* (Halle, 1832; reprint, Ulm: Arkana-Verlag, 1958), 455–59, on 456; "trat er . . . ganz auf die Seite der Alchemisten."

cites traditional authorities.[25] Kopp's brief observations, however, seem never to have been picked up or extended by other writers.

A brief pyrotechnic display enkindled in the late nineteenth century serves to illuminate the contemporaneous attitude toward alchemy and Boyle. In 1869, *Notes and Queries* began to compile a "General Index" of alchemy and, on the basis of the *Anti-Elixir* tract, listed Boyle's name among the alchemists. This act provoked a fulminating address to the American Antiquarian Society by Charles O. Thompson who declared that "no greater act of injustice has ever been committed against the fame of a great man." The author then proceeds to misrepresent the contents of *Anti-Elixir* in order to show it to contain foresighted chemistry rather than disreputable alchemy; rather unfortunately, in his anxiety he falters on his chemistry as much as on his history.[26] Thompson's paper is nonetheless valuable for its hagiographic portrayal of the English seventeenth century as seen in the nineteenth century, at least in the Anglophone world: "All modern thought, so far as it is scientific, is largely dependent upon the labors of three men—Isaac Newton, Robert Boyle, and John Locke."[27]

The present century witnessed the rise of the history of science as an autonomous profession; however, this step has not always succeeded in providing a more balanced or historically sensitive account of the development of science. Until the 1960s, the strong influence of positivism on many historians of science manifested itself in their desire to find the germs of modern knowledge in historical figures. While this approach undoubtedly shed light on a number of previously neglected figures and laid the foundations of the profession, it sometimes devolved into a competition to find the earliest adumbration of some modern (read "correct") notion without a concomitant appreciation of historical context. This approach is really no more than a continuation of the preexistent, "preprofessional" history of science carried out by scientists or ex-scientists themselves, whose interest was primarily directed toward the origins of their own ideas. Of course, this positivist or "Whig" approach resulted in a sifting of historical texts, preserving and modernizing the modern-sounding, and rejecting the archaic. Coupled with early formulations of the Scientific Revolution, this outlook led to renewed interest in Robert Boyle. He had long been acclaimed as a founder of modern chemistry, and his life overlapped those of both Galileo and Newton, who were recognized as heroic, revolutionary figures. Consequently, most of the views of Boyle provided by the early professional history of science were hagiographic.

Simultaneously, the same historiography ensured that alchemy did not fare well. Not only was alchemy something "unsuccessful" in the develop-

[25] Hermann Kopp, *Die Alchemie in Älterer und Neuerer Zeit* (Heidelberg, 1886; reprint, Hildesheim: Georg Olms Verlag, 1971), 53–56, on 54.

[26] Charles O. Thompson, "Robert Boyle," *Proceedings of the American Antiquarian Society,* n.s., 2 (1882): 54–79, on 56.

[27] Ibid., 54.

ment of science—having been replaced during the Scientific Revolution—but it also seemed to embody all the nonlogical, mysterious, metaphysical, superstitious, and occult qualities that were anathema to twentieth-century positivism. This view of alchemy—which is not by and large supported by historical texts—was forged by historians laboring at a time when alchemy had been co-opted and thoroughly misrepresented by the occultist revival of the nineteenth and early twentieth centuries. The retrospective linking of alchemy with esotericism resulted in a strongly pejorative view of alchemy as a subject of serious historical inquiry.[28] Alchemy came to be classed with magic, witchcraft, and astrology; still today, the unfortunate survival of George Sarton's indexing scheme for the *Isis Bibliography* continues to class alchemy under the rubric of *pseudo-science.* Accordingly, the notion that respected historical figures dabbled in alchemical explorations was not generally pursued.

In the case of Boyle, there were always those few well-known examples that seemed to indicate at least interest in chrysopoeia if not outright belief; as a result, early historians of science were at pains either to explain them away or to make them seem as rational and as sensible as possible—protochemistry rather than pseudoscience. A paradigmatic example of this view of the history of science in general and Boyle in particular is supplied by George Sarton, who saw Boyle as "one of the best prototypes of the modern man of science" and the alchemists as "fools or knaves, or more often a combination of both in various proportions." In a 1950 paper, he cites the *menstruum peracutum* and Boyle's transmutation of gold into silver as evidence that Boyle did believe in transmutation, which he, like many positivist apologists, sees as "justified by the very progress of modern chemistry"—presumably referring to transmutation by atomic decay, a concept related only in name to anything in alchemical thought. But he also inexplicably confounds Boyle's *menstruum peracutum* not only with the alkahest (the universal solvent) but with the Philosophers' Stone (the universal transmuting agent) as well—sufficient testimony to the minimal understanding then available of even the most rudimentary of alchemical principles. At any rate, after this measured admission, Sarton emphasizes that Boyle had nothing whatsoever to do with the "occult implications" of alchemy, and thus concludes that "Boyle was not an alchemist, although he accepted the postulate that the transmutation of metals was possible."[29]

At about the same time, Thomas Kuhn took a similar view of Boyle's thought. Citing the known instances of Boyle's alchemical interests, Kuhn strove to portray them merely as chemical studies to ascertain the nature of an element. Such motivations for Boyle's interest in debasing gold, Kuhn concluded, were "more creditable to Boyle than the more usual apologia for

[28] See Principe and Newman, "Historiography of Alchemy."

[29] George Sarton, "Boyle and Bayle, the Sceptical Chemist and the Sceptical Historian," *Chymia* 3 (1950): 155–89, on 160–62.

his excessive concern to discover the 'elixir' [Philosophers' Stone] of the alchemists."[30] The idea that a historian should base his explanations upon what is "creditable" to a historical figure is today sufficiently odd to render any emphasis upon it superfluous.

Rather than framing explanations or apologia for the long-known signs of Boyle's alchemy, Lynn Thorndike rejects them dismissively. Although he recognizes the residuals of sympathetic magic in several of Boyle's medical receipts, alchemy was apparently even less desirable. He denies any chrysopoetic content to the incalescent mercury paper and disputes the authenticity of the *Anti-Elixir* tract, claiming that it "reads more like a parody or mild satire upon alchemical treatises than a serious account of actual experimentation."[31]

Against this fairly consistent backdrop of the denial or dismissal of Boyle's alchemical interests appeared a few contrasting views, remarkable both for their relative isolation and for their dispassionate citation and treatment of some of Boyle's interests in alchemy. The earliest is a 1941 paper by Louis T. More, simply but provocatively entitled "Boyle as Alchemist," which later appeared in slightly reworked form as a chapter in his 1944 biography of Boyle. More notes the usual signs of Boyle's alchemy but points out that "one is apt to overlook what an obsession it was," and clear-sightedly refers the neglect of Boyle's alchemical pursuits to the emphasis laid by biographical writers upon his "institution of modern chemistry."[32] More chronicles the usual instances of Boyle's red earth, incalescent mercury, anti-Elixir, and transmutation of gold into silver by the *menstruum peracutum* quite without any expressions of surprise or distaste. The above-cited writings of Sarton, Kuhn, and Thorndike all assail More's text; but while More's work is indeed not of uniform quality, he did nevertheless view Boyle with less programmatic eyes and was thereby enabled to see Boyle's alchemy much more clearly than they did.

In 1953, Harold Fisch argued that while Boyle's studies were modern, the "theological framework" in which he pursued them was akin to that of the foregoing alchemical tradition, as well as to that of the not unrelated contemporary philosophical systems of Thomas Browne and Henry More. Fisch discriminated between seventeenth-century atomists: some (like Hobbes) adopted Epicurean-style atomism, but Boyle's corpuscularian system was built upon a conception of nature that was closely linked to God—the scientist's actions in nature parallel those of the priest in the temple.[33] In 1961, Muriel West showed much critical discernment in her reading of the

[30] Thomas S. Kuhn, "Robert Boyle and Structural Chemistry in the Seventeenth Century," *Isis* 43 (1952): 12–36, on 21; see also 28–29.

[31] Lynn Thorndike, *A History of Magic and Experimental Science*, 8 vols. (New York: Columbia University Press, 1958), 8:192.

[32] Louis Trenchard More, "Boyle as Alchemist," *Journal of the History of Ideas* 2 (1941): 61–76, and *The Life and Works of the Honourable Robert Boyle* (New York: Oxford University Press, 1944), 214–30.

[33] Harold Fisch, "The Scientist as Priest: A Note on Robert Boyle's Natural Theology," *Isis* 44 (1953): 252–65.

Sceptical Chymist—recognizing, for example, Boyle's "deliberate secretiveness" in using the classically alchemical technique of the dispersion of knowledge, a subject to which we will return in chapter 5. She also noted Boyle's approving use of traditional authors such as Geber, pseudo-Lull, Roger Bacon, and Paracelsus.[34]

While the contributions of More, Fisch, and West conflicted with the predominantly positivist and "Whig" tenor of the times, those of the most influential of Boyle's surveyors, Marie Boas Hall, largely conformed to it. Her writings have been chief in framing standard images of Boyle's natural philosophy and especially his chemistry. Boas Hall claims that Boyle eschewed alchemy immediately, that he had "no use for secrets," that the *Sceptical Chymist* was a "withering blast" against alchemy, and so forth.[35] Not unlike the eighteenth-century epitomizers, she chose for her collection *Robert Boyle on Natural Philosophy* selections from Boyle's massive corpus that specifically addressed the mechanical philosophy and other topics appropriate for the then-ascendant model of the Scientific Revolution which focused on physics as the paradigm of the sciences. Thus although she rightly denies the modernity of Boyle's views on elements as expressed in the *Sceptical Chymist* (against the then-prevalent view), she nonetheless constructs another modernized portrait of Boyle. These results are informed not only by her Whiggish inclinations but also by her aims. In her study of Boyle's chemistry, her goal was to argue for the role of the seventeenth century, Boyle, and England in the development of chemistry, as against the view—proposed by Butterfield—of an eighteenth-century "postponed chemical revolution" at the hands of Lavoisier.[36] Thus she was at pains to organize and rationalize Boyle's chemistry, making it as much like physics and as little like the foregoing chymistry as possible, thereby portraying Boyle as a reformer and revolutionary. The topics from Boyle that she chose to emphasize are indeed similar to those which we associate with Lavoisier. Her word choices are pregnant with intent; we find Boyle "anticipating" Lavoisier or "preparing the way" for him. Laudable characters among Boyle's predecessors exhibit "premature modernity" as if "modern times" were just over a frontier everyone was waiting for Boyle to cross. In particular, the alchemists are dismissed as "mystic" or "mystical."

Boas Hall's curious use of the word *mystical* encapsulates her attitude toward alchemy. For her, *mystical* seems to have a host of meanings, none of which are recognized by modern dictionaries, and all of which are intended to mark out errors of the most noxious sort. *Mystical* seems to be a general word to juxtapose against modern thought; it appears repeatedly in her denial of Newton's alchemy (against Westfall) in 1975 and recurs in her

[34] Muriel West, "Notes on the Importance of Alchemy to Modern Science in the Writings of Francis Bacon and Robert Boyle," *Ambix* 9 (1961): 102–14.

[35] Boas, *RBSCC; Robert Boyle on Natural Philosophy* (Bloomington: Indiana University Press, 1965).

[36] Herbert Butterfield, *The Origins of Modern Science* (New York: Macmillan, 1951), 150–63.

denial of Boyle's alchemy in 1994.[37] Here it certainly does not carry its usual signification of some intangible correspondence of the temporal to the spiritual (for example, she refers rather incomprehensibly to "that great mystic Sir Kenelm Digby"). Rather, in her usage it is a nebulous (mystical?) term meaning *mysterious, unclear, secret,* and *wrong* all at once.[38] For example, in explaining away Boyle's belief in metallic transmutation as nothing more than a logical consequence of his corpuscularian hypothesis, she remarks that "while misguided, [it] does not mean that Boyle was slipping into mystical alchemism."[39] Her meaning seems to be a characteristically positivist one, a notion denigrating whatever is perceived as not rigidly logical and rationalist in the context of modern science, rather like the Vienna Circle's use of the word *metaphysical.* Ironically, we will see later that Boyle did in fact embrace a *real mysticism* closely tied to his alchemy.

In spite of the limitations imposed by Boas Hall's historiographic dispositions, her work was still groundbreaking in the truest sense of the word. We owe a great deal of what we know about Boyle and a great deal of the continued interest in Boyle to her early endeavors. She succeeded admirably in calling attention to Boyle and arguing for his important place in the history of science, especially chemistry. But the price exacted for this feat was the "Enlightenmentization" of Boyle—finer-grained but philosophically analogous to the eighteenth-century portrayal of Boyle—by the magnification of his modern-oriented labors and the concomitant suppression of unwelcome traces of alchemical interest and belief. In Boas Hall's study of Boyle's chemistry, the touchstone of legitimacy is the resemblance to Lavoisier and the "accepted" chemical revolution. What we get is not Boyle as a seventeenth-century natural philosopher but Boyle as a chemical John the Baptist. Boas Hall's successful demonstration of the influence of earlier studies, particularly Boyle's, on Lavoisier only transfers "revolution" to an earlier generation and across the Channel, while omitting a broad range of pursuits of great importance to Boyle himself.[40] Such topics (had they not

[37] Marie Boas Hall, "Newton's Voyage in the Strange Seas of Alchemy," in *Reason, Experiment, and Mysticism,* ed. M. L. Righini Bonelli and William R. Shea (New York: Science History Publications, 1975), 239–46; Marie Boas Hall and A. Rupert Hall, "Newton's Chemical Experiments," *Archives internationales d'histoire des sciences* 11 (1958): 113–52; Marie Boas Hall, book review of *Robert Boyle Reconsidered, Ambix* 41 (1994): 111–12. The review ends with the amusing remark that "Boyle deserves commemoration not for his alchemical interests or for his religious writings but, and chiefly, because he was an important and innovative natural philosopher," as if it were the duty of historians to pen panegyrics rather than to provide accurate historical details and to exercise critical discernment.

[38] Boas, *RBSSC,* 25; while *mystical* does bear the archaic meaning "secret" or "hidden" (as Boyle himself sometimes uses it even to refer to alchemical writers), such an archaism does not seem to explain Boas Hall's usage; the archaic sense certainly does not encompass the opprobrium she inseparably fuses with it. Richard S. Westfall noted that in such usage, mysticism "lacks any precise meaning" whatsoever; "Newton and the Hermetic Tradition," in *Science, Medicine, and Society in the Renaissance,* ed. Allen G. Debus (New York: Science History Publications, 1972), 183–92, on 183.

[39] Ibid., 102. *Alchemism* is another curious word, but let it pass.

[40] The gap of some eighty years between Boyle and Lavoisier has only recently begun to be

been omitted) would not only have provided a truer portrait of Boyle but could also have answered the question that titled Allen Debus's brief but penetrating essay review of Boas Hall's book thirty years ago—"And Boyle stood on the shoulders of whom?"[41]

The question of Boyle's dependence upon earlier writers was in fact approached in the mid-1960s. Allen Debus and Charles Webster both called attention to the dependence of Boyle's *Sceptical Chymist* upon the writings of the Flemish chymist Jan Baptista van Helmont.[42] Debus outlined the role of fire-analysis among Paracelsians in the seventeenth century, van Helmont's rejection of fire as an accurate means of dividing the principles of mixed bodies, and the recapitulation of these arguments by Boyle in the *Sceptical Chymist.* Webster followed van Helmont's famous willow tree experiment (arguing for the primacy of water as the elementary principle) through its various reenactments and variations in England, to its incorporation into the *Sceptical Chymist.* Such treatments began to reveal the "overmodernization" of Boyle that was contemporaneously pointed out by Robert Kargon.[43] The overmodernization of Boyle in particular, and of much of the seventeenth century in general, was elegantly critiqued by J. E. McGuire in 1972. McGuire rightly noted that the problem of overmodernization "results, primarily, from interpreting seventeenth-century developments in the light of eighteenth-century history."[44] McGuire showed the important resonances of Boyle's way of thinking about the world with preceding traditions, casting him appropriately in a seventeenth-century context, rather than distorting him by reading history backward through the lenses of subsequent ages.

The history of science as a discipline has largely moved beyond the positivist (and internalist) historiography exemplified by Sarton and Boas Hall. But this earlier historiography has been supplanted in some quarters by a new historiography that, unfortunately, is no less programmatic and obscuring of historical reality. The earlier historiography wrested the "facts" pre-

---

filled; see, for example, Frederick L. Holmes, *Eighteenth Century Chemistry as an Investigative Enterprise* (Berkeley and Los Angeles: University of California Press, 1989).

[41] Allen G. Debus, "And Boyle Stood on the Shoulders of Whom?" (essay review of Boas's *Robert Boyle on Natural Philosophy*), *Isis* 57 (1966): 125–26; see also "The Significance of the History of Early Chemistry," *Cahiers d'Histoire Mondiale* 9 (1965): 3–58, on 53–54.

[42] Allen G. Debus, "Fire Analysis and the Elements in the Sixteenth and the Seventeenth Centuries," *Annals of Science* 23 (1967): 128–47; see also "The Chemical Debates of the Seventeenth Century: The Reaction to Robert Fludd and Jean Baptiste van Helmont," in *Reason, Experiment, and Mysticism,* ed. M. L. Righini Bonelli and William R. Shea (New York: Science History Publications, 1975), 19–48; Charles Webster, "Water as the Ultimate Principle of Nature: The Background to Boyle's Sceptical Chymist," *Ambix* 13 (1966): 96–107; Michael T. Walton, "Boyle and Newton on the Transmutation of Water and Air, from the Root of Helmont's Tree," *Ambix* 27 (1980): 11–18.

[43] Robert H. Kargon, *Atomism in England from Hariot to Newton* (Oxford: Clarendon Press, 1966), 93.

[44] J. E. McGuire, "Boyle's Conception of Nature," *Journal of the History of Ideas* 33 (1972): 523–42, on 524.

sented by a given historical figure from their historical matrix and judged that figure's importance by their relative proximity to modern scientific notions. The resultant history thus failed to acknowledge that historical characters exist within a complex web of intellectual, social, institutional, cultural, and other "nonscientific" interactions. The now popular sociological approach emphasizes the social networks that earlier studies wrongly ignored, but in some of its stronger applications often ignores the fact that our historical characters believed themselves to be more than the manipulators or the manipulated of a social system. Such applications rather too glibly deemphasize the role of the natural world in directing the thoughts and activities of those who commit themselves to studying it. The "facts" that were once the criteria of historical eminence become not merely unimportant but nonexistent, the products of social interaction and "public validation." Fortunately, historians need not choose between these extremes; a wide and fertile ground for historians stretches between the deserts of positivism and the swamps of social constructivism.

Boyle has been set up as the centerpiece of sociological programs by Steven Shapin, Simon Schaffer, Jan Golinski, and others.[45] While this approach has been valuable in bringing forth aspects of Boyle's enterprise that had earlier been overlooked or taken for granted, it is not without problems. Rose-Mary Sargent has rightly questioned the validity of some philosophical assumptions and the "collectivist" approach of *Leviathan and the Air-Pump,* whereby the opinions of Boyle's contemporaries are unproblematically transferred to Boyle, and Boyle's views similarly extended to the Royal Society.[46] Using a formidable array of sensitively interpreted archival material, Michael Hunter has revealed a complex and deeply troubled Boyle who little resembles the calculating, strategizing Boyle of the sociological school.[47] Jan Wojcik has questioned the "dogmatic Boyle" presented by sociological studies, and she notes that such portrayals owe much to the long tradition of "re-creating" Boyle set forth in this chapter. Wojcik's observation points out that although the sociological is the most recent and apparently innovative model, it is curiously retrogressive, for it has taken "at face value the portrait of Boyle drawn by previous generations of Boyle scholars," a portrait that is far from complete.[48] Indeed, this retrogressive nature extends somewhat further, for the sociological program also assumes an older view of the devel-

[45] Steven Shapin and Simon Schaffer, *Leviathan and the Air-Pump* (Princeton: Princeton University Press, 1985); Steven Shapin, "The House of Experiment in Seventeenth Century England," *Isis* 79 (1988): 373–404; "Pump and Circumstance: Robert Boyle's Literary Technology," *Social Studies of Science* 14 (1984): 481–520; Jan V. Golinski, "A Noble Spectacle: Phosphorus and the Public Culture of Science in the Early Royal Society," *Isis* 80 (1989): 11–39.

[46] Rose-Mary Sargent, *The Diffident Naturalist* (Chicago: University of Chicago Press, 1995), 9, 131–35, 211–16.

[47] Hunter, "Conscience of Robert Boyle."

[48] Jan W. Wojcik, *Robert Boyle and the Limits of Reason* (Cambridge: Cambridge University Press, 1997), 217–19; see also Hunter, *Robert Boyle by Himself,* lxiv.

opment of science. It implicitly assumes, as did earlier historiography, that the origins of modern science lie in a seventeenth-century revolution with the Royal Society of the Restoration at its center. Having (perhaps unconsciously) made this assumption, when adherents of the sociological approach seek for sociological factors responsible for the emergence or "reification" of unique aspects of modern science, they focus on that time and place and subsequently on its chief figure—Boyle. But if, as more broadly based studies show, the Royal Society is not the sole (or perhaps even chief) template of modern scientific practice, then the assumption that its social space and strategies are somehow straightforwardly explanatory of our own, or of the rise of modern scientific culture and practice, is undermined. Perhaps more critically, such studies selectively delineate the Royal Society itself, centering on its "modern" and successful affairs—much in the tradition of the Halls. I might add that finding sociological explanations too frequently becomes, as Karl Popper might say, like finding Marxist or psychoanalytical ones: they are almost always discoverable when sought, but rarely verifiable as causes. Rather more troubling is the apparent assertion by some supporters of social constructivism that the historical background is relatively unimportant; one supportive reviewer writes in regard to *Leviathan and the Air-Pump* that "it mattered little whether the portraits of Hobbes or, especially, Boyle were historically accurate."[49] The troubling ramifications of such an utterance need not be insisted upon.

In reference specifically to alchemy, the sociological school, by adopting the formulations of an earlier historiography, has further obscured that aspect of Boyle's activities. In *Leviathan and the Air-Pump,* for example, alchemy is routinely and uncomplicatedly set up as a foil to a supposedly modern public culture of science, alchemical obscurantism set against modern openness. The solitary, secretive alchemical adepti contrast with the corporate self-validating Royal Society.[50] These lines of distinction, as we shall see, are by no means clear.

The current renewal of interest in Robert Boyle has provided new information and some fresh, less theory-driven approaches to his chymistry. Michael Hunter, drawing upon largely unpublished manuscript material, has underscored the importance of alchemy to Boyle while lucidly showing the connection of Boyle's alchemy with his concerns about morality, magic, and the spirit realm.[51] In quite another vein, William Newman has pointed out sources of Boyle's corpuscularianism—his chief claim to fame in the stan-

[49] Peter Dear, "Trust Boyle" (essay review of Shapin, *The Social History of Truth*), *British Journal for the History of Science* 28 (1995): 451–54.

[50] Shapin and Schaffer, *Leviathan,* 71–78. The automatic use of alchemy as a representation of "prescientific" thought is still widely prevalent; see Marco Beretta, *The Enlightenment of Matter* (Canton, MA: Science History Publications, 1993). Beretta claims that it is "better to stop regarding alchemy as a scientific or experimental endeavor"; otherwise, we are left to conclude that alchemy was "a pathological state of mind" (331).

[51] Michael Hunter, "Alchemy, Magic and Moralism in the Thought of Robert Boyle," *British Journal of the History of Science* 23 (1990): 387–410.

dard literature—in the particulate theories of matter prevalent in alchemical texts. This corpuscularian view ultimately devolves from the medieval *Summa perfectionis* of Geber.[52] Similarly, Antonio Clericuzio, continuing the study of Boyle's Helmontian connections first noted by Allen Debus and Charles Webster, has outlined the carryover of Helmontian notions by Boyle, their subsequent popularity in England, and their removal from the public perception of Boyle after his death. Clericuzio's reevaluation of Boyle's chymistry is key to the process of resituating Boyle in his context.[53] As the efforts of all three of these scholars are of great importance to the present study, I shall return to their work repeatedly in subsequent chapters.

[52] Newman, "Boyle's Debt to Corpuscular Alchemy"; "The Alchemical Sources of Robert Boyle's Corpuscular Philosophy," *Annals of Science* 53 (1996): 567–85.

[53] Antonio Clericuzio, "Robert Boyle and the English Helmontians," in *Alchemy Revisited*, ed. Z.R.W.M. van Martels (Leiden: Brill, 1990), 192–99; "Redefinition of Boyle's Chemistry"; "Carneades and the Chemists: A Study of the *Sceptical Chymist* and Its Impact on Seventeenth-Century Chemistry," in *RBR*, 79–90; "From van Helmont to Boyle: A Study of the Transmission of Helmontian Chemical and Medical Theories in Seventeenth-Century England," *British Journal for the History of Science* 26 (1993): 303–54.

CHAPTER II

Skeptical of the *Sceptical Chymist*

HAVING REVIEWED the evolving image of Robert Boyle presented in secondary literature, I shall now revisit the work most frequently taken to mark Boyle's rejection of alchemy—the *Sceptical Chymist*. This book has become in retrospect an icon of a historical trend or event and indeed seems to have become rather independent of its author. Some studies make Boyle himself the Sceptical Chymist, inverting the relationship between creator and created. There are few other books that have been so long considered crucial to the development of science yet remain so little read critically and in context. A close examination of the *Sceptical Chymist*, however, reveals that its origins and content are not so straightforward as generally presumed. Much nineteenth- and twentieth-century historiography pitches upon the *Sceptical Chymist* as Boyle's most popular and most influential work, hailing it as the major cusp in the development of modern chemistry. Even among recent chemical histories, William Brock's laudable *History of Chemistry* is organized upon a scheme where the *Sceptical Chymist* constitutes a crucial break between ancient alchemists and modern chemists. Interestingly, this view of the *Sceptical Chymist* is by no means a solely twentieth-century phenomenon; already by the late eighteenth century the *Sceptical Chymist* was viewed as an attack against alchemy. Siegmund Heinrich Güldenfalk, in his 1784 defense of transmutation, equates Boyle's title with antialchemical sentiments, remarking sarcastically that "those 'Sceptical Chymists' will have to wait a good long while" before any Adept Philosopher will deign to visit them with the powder of transmutation.[1]

The *Sceptical Chymist* is Boyle's most cited work in histories of chemistry dating from the first half of this century. Early-twentieth-century historians of chemistry all refer glowingly to it—Thorpe presents it as one of the most important books of the seventeenth century; Holmyard calls it Boyle's "great work"; Stillman, "epoch-making"; and so forth.[2] Modern readers may, however, find it difficult to attach such laudatory sentiments to it. I suspect historians of chemistry have enshrined the *Sceptical Chymist* less owing to its content than to their own anxiety to locate a *Principia* or *De revolutionibus* of chemistry. Praise for the *Sceptical Chymist* centers on the

[1] Siegmund Heinrich Güldenfalk, *Sammlung von mehr als hundert wahrhaftigen Transmutationgeschichten* (Frankfurt and Leipzig, 1784), preface.

[2] T. E. Thorpe, *Essays in Historical Chemistry* (London: Macmillan and Co., 1894), 1–27, esp. 2–3, and 26; E. J. Holmyard, *Chemistry to the Time of Dalton* (London: Oxford University Press, 1925), 63–64; John Maxson Stillman, *The Story of Early Chemistry* (1924; reprint, New York: Dover Publications, 1960), 394–97.

beliefs that it accurately and for the first time defined an element in the modern sense and that it dealt the deathblow to alchemy. When Marie Boas Hall carried out her influential studies of Boyle's chemistry in the 1950s and 1960s, she recognized both that Boyle's definition of "element" was not akin to the modern one, and that earlier writers had exaggerated both the importance and the clarity of the *Sceptical Chymist.* She noted further that "for some reason, [the *Sceptical Chymist*] has in modern times been far more read than it deserves"; nonetheless, she still retained the notion that it constituted a "withering blast" against alchemy.[3] This view of the *Sceptical Chymist* remains widespread. Quite recently, Maurice Crosland, an unflinching supporter of the old positivist dismissal of alchemy from serious consideration, uncritically repeated earlier judgments by asserting that the *Sceptical Chymist* "marks the effective end of alchemy."[4]

TEXTUAL CONFUSION IN THE *SCEPTICAL CHYMIST*

The *Sceptical Chymist* is a very difficult text, perhaps the most difficult of all Boyle's works in terms of readability and coherence. It displays an ample measure of Boyle's notorious prolixity and digressive style, and fully indulges his penchant for equivocation. Some historians have (understandably) found it easier to cite from the briefer and more direct early draft discovered by Boas Hall than from the final published text.[5] The latter is quite unpolished; it contains, for example, references to supposedly foregoing arguments that are not to be found, numerous repetitions, and a totally incongruous plot structure—it is a patchwork text sloppily stitched together. In fact, the textual confusion is hard to imagine for anyone who has encountered only the lucid passages quoted in the secondary literature rather than muddled through the lengthy original itself. The introduction (labeled as "Part of the First Dialogue") begins with five friends meeting in Carneades' garden and chatting about the constituents of mixed bodies—Carneades the host and the skeptic, Philoponus the chymist, Themistius the Aristotelian, Eleutherius the impartial judge, and an unnamed narrator (presumably

[3] Boas, *RBSSC,* 67, 94.

[4] Maurice Crosland, "Chemistry and the Chemical Revolution," in *The Ferment of Knowledge,* ed. G. S. Rousseau and Roy Porter (Cambridge: Cambridge University Press, 1980), 380–416, on 392.

[5] Marie Boas [Hall], "An Early Version of Boyle's *Sceptical Chymist,*" *Isis* 45 (1954): 153–68. William H. Brock (*The Norton History of Chemistry* [New York: W. W. Norton, 1992]) notes the opacity of the published text and chooses to quote almost exclusively from the manuscript, 56 ff.

On the dating of this essay, see also Webster ("Water as the Ultimate Principle of Nature"), who argues for a date of 1659–1660, and Clericuzio ("Carneades and the Chemists," 79–80), who provides convincing arguments regarding the details of the essay's composition, including a date of the mid-1650s.

Boyle himself).[6] Leucippus the atomist, Carneades explains, is unexpectedly out of town. This setting would seem to lend itself to a fairly straightforward presentation of each character's beliefs followed by a lively discussion, but the whole is abruptly derailed after Themistius's opening speech in favor of the four Aristotelian elements and Carneades' "unexpected objection" that the substances separated from a body by fire may not have been preexistent in the unanalyzed body. The reader is then presented with a second full title page that surprisingly intervenes after page 34, and when the following "First Part" begins, Themistius and Philoponus have vanished, and we hear no more of them for almost two hundred pages until (in the "fourth part") Carneades utters the words "if *Themistius* were here . . ." but without offering an explanation of where he has gone.[7] At the beginning of the sixth part ("A Paradoxical Appendix," pp. 347–426) Carneades looks "about him, to discover whether it were Time for him and his Friend to Rejoyne the Rest of the Company," but the reader has no notion of how or whither Carneades and Eleutherius have wandered off, or even if "the Rest of the Company" contains the same characters he met (very briefly) three hundred pages earlier. Near the end, Carneades asks Eleutherius to remember when Philoponus "lately disputed for his Chymists against *Themistius,*" an event that never occurs in the text as we have it.[8] Indeed, interestingly enough, Philoponus, the advocate of the "chymists' principles" under attack, is allowed only a single utterance of about a dozen lines in the entire book.[9] Finally, as Carneades and Eleutherius wander back to the "Company," Eleutherius notes "that *Philoponus,* had he heard you, would scarce have been able in all points to defend the Chymical *Hypothesis,*" but Philoponus never gets the chance, for the dialogue ends after a brief summary.

This patchwork format is probably due to the incorporation of elements from at least two separate works, and a rather hurried plastering over the seams.[10] Boyle himself remarks in the preface that the book is "maim'd and imperfect," and is an agglomeration of several dialogues involving the same speakers.[11] Owing to the convoluted and repetitious (occasionally contradictory) nature of the book, it has been somewhat too easy to find support-

[6] The narrator, who acts as the recording secretary for the discussion, is referred to as if he were Boyle on several occasions (135, 253, 303, 339, 386), but elsewhere "Mr. Boyle" is referred to as an absent third party (78 and 252), and of course, Carneades' experiments are Boyle's own.

[7] Robert Boyle, *Sceptical Chymist* (London, 1661), 202.

[8] Ibid., 402.

[9] Ibid., 27–28.

[10] See the discussion of Boyle's compositional style in the introduction to appendix 1; and especially Michael Hunter and Edward B. Davis, "The Making of Robert Boyle's *Free Enquiry into the Vulgarly Receiv'd Notion of Nature* (1686)," *Early Science and Medicine* 1 (1996): 204–71, esp. 209–25.

[11] Boyle, *Sceptical Chymist,* preface, [i], [ii], and [xii–xiii]. Below I shall suggest the identity of one of these related dialogues.

ing quotations for prevailing historical models by biased selection. Boyle's highly qualified prose has also presented a problem of a general nature for history of science writers, for it is very difficult to find pithy statements in Boyle that do not give a mistaken picture. The qualifiers and endless dependent clauses that crowd Boylean periods, and which tend to be excised in the effort to make quotations wieldy, sometimes subvert the main meaning and sometimes provide important clues to Boyle's subtle thought or to the identity of his adversaries. For this reason I will quote Boyle in extenso on a regular basis and thus ask the reader's forbearance and patience.

I have purposely limited this chapter to a single point: Against whom or what is the *Sceptical Chymist* written? Is it in fact a condemnation of alchemy, that is, chrysopoeia? Several of the important points to be developed throughout this study of Boyle's alchemy will surface in this chapter. In order to concentrate on the *Sceptical Chymist,* I will defer discussions of these issues to subsequent chapters, and their mention here in passing will lay the foundations for later in-depth analyses.

Chymists High and Low

Before advancing to the text itself, we should first understand the title, which in full is *The Sceptical Chymist, or Chymico-physical Doubts & Paradoxes, Touching the Spagyrist's Principles commonly call'd Hypostatical; as they are wont to be Propos'd and Defended by the Generality of Alchymists.* The word *Chymist* to modern ears has a modern ring, and a judgment of this book by its cover is facilitated by the modern-sounding distinction between this "Sceptical *Chymist*" and the "*al*chymists." But I argue that this distinction is apparent rather than real and is spuriously reinforced by the subsequent demise of alchemy and rise of modern chemistry and the prevailing model of the seventeenth-century rise of modern science. It must be asked: What was the distinction in Boyle's English between *chymist* and *alchymist?* Boas Hall claims that the very existence of two words argues that there existed two distinct disciplines.[12] More recently, Alan Rocke usefully examines the emergence and divergence of the words *chymia* and *alchymia;* however, his study treats predominantly Germanic and Gallic sources in either German or Latin, rather than English. He notes the synonymity of the words for much of the seventeenth century, but suggests that "by the time of Robert Boyle and Isaac Newton the words had nearly their modern connotations."[13]

I have not been able to verify this suggestion judging from Boyle's usage. Having made a census of the words *alchemist* and *alchemy* (in all their orthographic variants) in Boyle's writings, I find that although these words are used much less frequently than their more modern counterparts, there is

[12] Boas, *RBSCC,* 48.

[13] Alan J. Rocke, "Agricola, Paracelsus, and 'Chymia,'" *Ambix* 32 (1985): 37–45.

no coherent and consistent discrimination in their use.[14] Nowhere in Boyle's texts have I found *alchemy* and *chemistry* paired as if they referred to separate things. It has been suggested that *chymistry* refers to more technological or medicinal subjects, while *alchemy* is retained for matters relating to the Philosophers' Stone and such like, but I cannot find such a division *using these terms* made by Boyle or by most of his contemporaries; the two words are synonymous in Boyle's usage.[15] Boyle writes, for example, of the enigmatic books of the chymists and of the "supernaturall arcana of the chymists," clearly in reference to what we would now call *alchemy,* certainly not *chemistry.* Occasionally, Boyle uses *alchymist* explicitly in connection with the Philosophers' Stone and gold-making activities, but even more often he uses *chymist* in that same context.[16] The words *alchemy* and *alchemist* are used indiscriminately in conjunction with both opprobrious and laudatory adjectives. Further, I mentioned above that the *Sceptical Chymist* was originally issued with *two* title pages—the second one substitutes the term *spagyrists* for *alchymists.* This second title page was the one reused for the second edition of 1680. In sum, we cannot view the terminology embedded in the title as a sign of "modernity" or as distancing Boyle from the alchemical tradition; indeed, it leaves open the possibility that it actually allies him to that tradition more than a title such as the *Sceptical Natural Philosopher* would have done. To find the *Sceptical Chymist*'s real position vis-à-vis alchemy, we must look further.

While Boyle did not recognize our modern terminological chemist/alchemist dichotomy, he did, in fact, discriminate among chymical practitioners. In the preface to the *Sceptical Chymist* Boyle writes, "believe me when I declare that I distinguish betwixt those Chymists that are either Cheats, or but Laborants, and the true *Adepti.*" By these latter, says Boyle,

[14] The same conclusion regarding seventeenth-century usages has been reached by Robert Halleux, *Les textes alchimiques* (Turnhout, Belgium: Brepols, 1979), 47; and Newman, *Gehennical Fire,* xi.

[15] See my introduction. *Alchemy* is often paired in opposition to *physick,* but not universally; see Boyle, *Producibleness,* 130 ("usefull to considerable purposes both in *Alchymy* and *Physick*") and 225 ("uses, not only in respect of *Alchymy,* but of *Medicine*"). A clear example of synonymity occurs in *Origine of Formes and Qualities* (Oxford, 1666), 302–3, where Boyle writes of a silver compound (silver chloride), noting first that "some Chymists have call'd [it] *Luna Cornea,*" and then repeats himself on the following page, saying that "the name of *Luna Cornea* be already to be met with in the Writings of some Alchymists." The division between alchemy and chemistry is discussed in detail in Newman and Principe, "Etymological Origins."

[16] In the preface to his "Essay on Nitre" (in *Certain Physiological Essays* [London, 1661]), Boyle speaks of "the Elixir, that Alchymists generally hope and toyl for," ([ii]), and similarly in *Origine of Formes and Qualities,* 417, but in many other places he refers to *chymists* in relation to the Elixir; see, for example, "A History of Fluidity and Firmnesse," in *Physiological Essays,* 230; *Sceptical Chymist,* 158, 203; "Of the Mechanical Origine or Production of Fixtness," in *Mechanical Origine of Qualities* (London, 1675), 15. Boyle uses *spagyrist* in the same context in, for example, "Imperfection of the Chymists' Doctrine of Qualities," in *Mechanical Origine,* 34. Boyle uses *alchymist* only four times in the *Sceptical Chymist,* and that only in the compass of a single, two-page digression on 181–82.

he wishes he could be "willingly and thankfully" instructed "concerning the Nature and Generation of metals."[17] Indeed, throughout the *Sceptical Chemist* Boyle distinguishes consistently between one group he calls either adepti, chymical philosophers, or uncommon chymists and another group he calls the common, vulgar, or ordinary chymists. He stresses repeatedly that it is the latter whom he criticizes.[18] In the 1680 second edition, this distinction is drawn even more forcibly. "I make a great difference between the avow'd Cultivators of that Art," writes Boyle, "and look not with the same eyes on the opinions or performances of vulgar *Chymists,* and Chymical Philosophers."[19] Who, then, are these Chymical Philosophers? Fortunately, Boyle is explicit in describing them; they are not mechanical or "New Philosophers," or anything resembling modern chemists. Rather, he recounts of them that

> there lives conceal'd in the world, a sett of *Spagyrists* of a much higher order than those who are wont to write courses of *Chymistry* or other Bookes of that nature; being able to transmute baser Metalls into perfect ones, and do some other things, that the generality of *Chymists* confess to be extreamly difficult; and divers of the more judicious even among the *Spagyrists* themselves have judg'd impossible.[20]

These "higher order" chymists who live "conceal'd in the world" and who can transmute the metals, are clearly traditional alchemists or, more properly speaking, chrysopoeians. Boyle adds that he has been convinced of their existence by the witness of others and, interestingly, by "some more immediate arguments."[21] Boyle notes that if these "adept *Philosophers*" were to show some of their arcana—particularly analyzing menstrua such as the alkahest—they might be able to "induce me, and perhaps the Chymists too, to entertaine other thoughts about the constitution of compounded bodies (as they are wont to be call'd) than either I or they now have."[22] Thus, perhaps surprisingly to many, Boyle does not direct the *Sceptical Chymist* against "alchemical adepts" but rather calls them to be his teachers.

Thus Boyle groups the adepti separately from the common chymists, whose "vulgar" notions these adepti might well be able to refute by virtue of their superior knowledge. Such a division is not unique to Boyle; a very similar distinction among chymists is made by E. R. Arnaud in 1650 and reiterated by Christophle Glaser in 1663. The latter describes various classes of practitioners ranging from the masters of *la haute chymie* (chrys-

[17] Boyle, *Sceptical Chymist,* [xiii].

[18] See, for example, 13, 17, 47, 48, 80, 303, 367, 390, 407, 429.

[19] Boyle, *Producibleness,* preface, [ix–x].

[20] Ibid., [x].

[21] In some contexts (e.g., "Imperfection of the Chymists' Doctrine of Qualities," 48–49) Boyle is somewhat agnostic about the real existence of these secretive adepti, but by 1680 he was sure of their reality; he had met some. Indeed, the "more immediate arguments" are presumably Boyle's meetings with alchemical adepti, some of whom performed transmutation before him. We shall examine these meetings in chapter 4.

[22] Boyle, *Producibleness,* [xii].

opoeians) who write obscurely and have discovered the greatest mysteries and arcana, through those who aspire to such heights and those less capable who possess "little lights" (*petites lumières*) and can do useful but not extraordinary things, like Boyle's "vulgar chymists," down to the "ignorant puffers" (*souffleurs ignorants*), akin to Boyle's "Cheats."[23] This same distinction is almost a commonplace among chrysopoetic writers, some of whom employ very nearly the same terminology as Boyle. For example, the anonymous mid-seventeenth-century *Instructio patris ad filium* teaches "that there is as great a difference between the true Philosophers and the vulgar alchymists as there is between night and day."[24] Upon revisiting the title page of the 1661 edition of the *Sceptical Chymist,* we can now take note of the qualifying "generality of" before "alchymists," which singles out the lower order. The second title page substitutes the synonymous "vulgar spagyrists."

Having identified the "higher order" chymists with chrysopoeian adepti, we must determine who the "vulgar chymists" are. Boyle initially lists two groups: the "cheats" from whom chymistry had earned a bad name, and the "laborants," namely, those distillers, refiners, dyers, apothecaries, and such "technical chymists" whose practice is devoid of any but the barest rudiments of theory and philosophical interest. It has long been recognized that one of the "problems" of chymistry before the eighteenth century was its status as a practical or technical art rather than as a branch of natural philosophy. The low status of chymistry as determined by its use among low technical appliers militated against its acceptance by many natural philosophers. Even among the Hartlib circle of the 1650s, which endorsed and promoted many chymical projects, there occurred some exasperation on this score. When Benjamin Worsley became disillusioned because of the failure of some promising chymists to obtain the "key" to the radical dissolution of minerals and metals (a chief alchemical desideratum), he wrote to Hartlib in early 1653 that "I have laid all considerations in chemistry aside, as things not reaching much above laborants, or strong-water distillers, unless we can arrive at this key." Worsley here longs for the higher arcana known to the adepti in which "both principally and only I conceive [chymical] learning, judgement, or wisdom to consist" in distinction from the common practice of workshop laborants.[25]

One of Boyle's chief contributions to chemistry is in fact his attempt to elevate its status by insisting upon its usefulness to natural philosophy. His apologetic preface to the "Essay on Nitre" is the clearest expression of this

[23] E. R. Arnaud, *Introduction a la chymie* (Lyons, 1650), 8–10; Christophle Glaser, *Traite de la chymie* (Paris, 1663), sig. ai. See Newman and Principe, "Etymological Origins."

[24] *Instructio patris ad filium de arbore solari,* in *TC,* 6:165–94, on 166. "Scias igitur tantam inter veros Philosophos & vulgares Alchymistas esse differentiam, quantum inter diem & noctem."

[25] *Works,* 6:79; quoted in Samuel Hartlib to Robert Boyle, 28 February 1653/54. See also Newman, *Gehennical Fire,* 62–75, esp. 70–71. Curiously, Boas Hall interprets this passage as witnessing Worsley's "disgust with alchemy" (*RBSCC,* 39), but it is quite the opposite.

desire to dignify a subject "judge[d] so much below a Philosopher, and so unserviceable to him" and to "beget a good understanding betwixt the Chymists and the Mechanical Philosophers."[26] In the *Sceptical Chymist* then, one of Boyle's arguments is that the manual practices, poorly constructed notions, and unproven assertions of practical chymists debilitate chymistry in the eyes of natural philosophers. Thus the successful elevation of chymistry and the exploitation of its usefulness to natural philosophy require the exploding of "vulgar" notions and their replacement with more philosophically acceptable thoughts.

Boyle explicitly names a third group of "vulgar chymists" (besides the cheats and laborants) in contrast to the adepti, namely, "those who are wont to write courses of *Chymistry* or other Bookes of that nature." Here Boyle clearly refers to the tradition of textbook writers—Jean Beguin, Estienne de Clave, E. R. Arnaud, Werner Rolfinck, Nicaise LeFebvre, and others—influential during the seventeenth century.[27] Although there is some diversity in format and selection of topics in these textbooks, they share a common emphasis on practical pharmaceutical preparations and reliance upon Paracelsian principles. Indeed, all these textbooks define chymistry primarily in terms of making medicines, and some explicitly restrict the scope of chymistry to pharmacy alone.[28] These very popular "courses of chymistry" broadly disseminated both chemical techniques and Paracelsian iatrochemical principles and consequently had great influence. Beguin's *Tyrocinium chymicum* went through no fewer than forty-one editions between 1610 and 1690. All of these authors lectured on chymical pharmacy (several at the Parisian Jardin Royale des Plantes), and their textbooks crystallized from these well-attended lectures. It is largely from such works or lectures that the "vulgar" laborants and apothecaries learned chymical operations and iatrochemical theory. These books also influenced the Oxford physiologists with whom Boyle associated during his composition of the *Sceptical Chymist.* Indeed, Antonio Clericuzio has argued that the too uncritical adoption of the chymical theory of the principles among these Oxford physiologists—as attested most clearly by Thomas Willis's 1659 *De fermentatione*—was a chief spur behind the publication of Boyle's manuscript.[29]

[26] Boyle, "Essay on Nitre," preface, [i] and [viii]. Note also how in this early work Boyle avers that the Elixir is "not at all in my aime," ([ii]) which, as we shall see, was probably not entirely true then and certainly not true in his later career.

[27] The classic study of these authors is Hélène Metzger, *Les doctrines chimiques en France au debut du XVIIe à la fin du XVIIIe siècle,* 2 vols. (Paris: Les Presses Universitaires de France, 1923).

[28] For example, Werner Rolfinck, *Chimia in artis formam redacta* (Jena, 1661), 28; "medicinae potius esse videtur pars, ad *Pharmakoitikēn* pertinens, & ab eâ nomine solùm differens." The attrition of chymistry down to mere pharmacy by iatrochemical textbook writers like Rolfinck may shed light on the origins of the definition of *chemist* preserved in Britain, namely, a pharmacist.

[29] Clericuzio, "Carneades and the Chemists," 80–81. On the Oxford physiologists, see

In the following section I shall endeavor to show conclusively, by a consideration of the content of Boyle's arguments, that the *Sceptical Chymist* is directed against the theories, practices, and beliefs of the laborants, apothecaries, textbook writers, and Paracelsian systematizers, and not against the proponents and adepti of chrysopoetic alchemy. Not only did these latter often hold views different from those Boyle attacks, but Boyle occasionally calls upon chrysopoeians to help in his attack on the "vulgar chymists."

Boyle's Arguments and Their Targets

The first proposition of the *Sceptical Chymist* is a denial of the opinion that fire is a universal and sufficient analyzer capable of dividing all bodies into their constituent ingredients.[30] Few traditional alchemists held this view. The analysis of metals and minerals for the purpose of isolating active substances from them is key to chrysopoetic theories, but few such authors assert that these substances could be separated by fire alone. Many of them, like Boyle, refer such beliefs to the vulgar and prescribe instead the use of various solvents, salts, and other compositions. The loudest champions of the use of fire alone are iatrochemical textbook writers like Jean Beguin, and Paracelsian systematizers like Quercetanus (Joseph DuChesne, ca. 1544–1609). The textbook writer E. R. Arnaud defines chymistry as separation "by means of fire," and LeFebvre insists at length that fire adequately separates true principles from mixed bodies.[31] When Boyle assails this belief, he turns for support to the Flemish Jan Baptista van Helmont, whose alkahest was reputed to be the universal analyzer of bodies, decomposing even coals, stones, and metals that common fire cannot alter into simple constituent bodies.[32] But Boyle also cites the strict chrysopoeian Gaston "Claveus"

Robert G. Frank, Jr., *Harvey and the Oxford Physiologists: A Study of Scientific Ideas and Social Interactions* (Berkeley and Los Angeles: University of California Press, 1980).

[30] Boyle, *Sceptical Chymist,* 48–102.

[31] Metzger, *Doctrines,* 60, 75–77; on Quercetanus and fire, see Allen G. Debus, *The French Paracelsians* (Cambridge: Cambridge University Press, 1991), on 52.

[32] Boyle, *Sceptical Chymist,* 74–79. On Boyle and van Helmont, see Debus, "Fire Analysis and the Elements," 128–47; Webster, "Water as the Ultimate Principle of Nature"; Clericuzio, "Robert Boyle and the English Helmontians"; "From van Helmont to Boyle."

The notion of the alkahest as a universal resolving agent was developed by van Helmont, who equated the alkahest with Paracelsus's *sal circulatum,* mentioned in his *De renovatione et restoratione,* which could reduce bodies to their primal substance. Van Helmont, consonant with his claim that water was the universal original substance, stated that the alkahest could reduce all substances first into their constituent principles, and then those into primal water by stripping them of the forms induced by their seminal principles. Van Helmont also terms the alkahest *ignis Gehennae,* alluding to its property of dividing and purging all substances like hellfire. On the alkahest, see Otto Tachenius, *Epistola de famoso liquore alkahest* (Venice, 1655); George Starkey, *Liquor Alchahest* (London, 1675); Hermann Boerhaave, *Elementa chemiae,* 2 vols. (Leiden, 1732), 1:848–68; Johann Kunckel, *Collegium Physico-Chymicum* (Hamburg and Leipzig, 1716), 500–506; Ladislao Reti, "Van Helmont, Boyle and the Al-

DuClo, from whose *Apologia chrysopoeiae* (1590) he borrows an experiment wherein DuClo kept gold and silver molten in a glass-furnace for two months in order to show that fire was powerless to divide them.[33]

Significantly for our argument about Boyle's real target, Boyle foresees that some chymists might object that they do not in fact claim fire to be a sufficient analyzer, and Boyle retorts that "it is not against those that . . . I have been disputing, but against those Vulgar Chymists, who themselves believe, and would fain make others do so, that the Fire is not only an universal, but an adaequate and sufficient Instrument to analyze mixt Bodies with."[34]

Alchemical Matter Theory—Mercury, Sulphur, and Sometimes Salt

Skepticism about the doctrine of the *tria prima*—Mercury, Sulphur, and Salt—makes up a significant part of the *Sceptical Chymist*. Boyle's refutation of the universality and elementary nature of these substances has generally been viewed as a direct attack on the foundations of traditional alchemy, and the *Sceptical Chymist* has been celebrated for "disposing of" this theory. The chief difficulty with this facile conclusion, however, is that it assumes that alchemy was universally built upon the doctrine of the *tria prima*. This assumption is, however, mistaken; it does not take into account the latitude of theories on the constitution of matter current among seventeenth-century chymists. An exploration of this diversity of opinion to display exactly who held the doctrine of the *tria prima as it was assailed by Boyle* reinforces the identification of Paracelsian iatrochemists—textbook writers, systematizers, and laborants—as the "vulgar chymists" Boyle attacks, and shows that the traditional alchemical adepti emerge mostly unscathed. Boyle himself later declares that it is specifically the "vulgar" view of the chymical principles the *Sceptical Chymist* attacks, and reveals how the attacks were viewed.

> I perceiv'd that no Author vouchsaf'd the *Scepticall Chymist* an answer; but a very Ingenious Man, from whom I chiefly expected it, told me, that he had indeed design'd to write one, but was hindered by considering, that I had so stated the case, that an answer could not confute that Book by any meer Justification of the Chymists Principles, since he would be obliged also to defend the Chymical Doc-

---

kahest" in *Some Aspects of Seventeenth Century Medicine and Science* (Berkeley and Los Angeles: University of California Press, 1969); Clericuzio, "From van Helmont to Boyle."

[33] Gaston DuClo, *Apologia chrysopoeiae et argyropoeiae*, in *TC*, 2:4–80, on 19; Boyle, *Sceptical Chymist*, 56–57. On DuClo, see Lawrence M. Principe, "Diversity in Alchemy: The Case of Gaston 'Claveus' DuClo, a Scholastic Mercurialist Chrysopoeian," in *Reading the Book of Nature: The Other Side of the Scientific Revolution*, ed. Allen G. Debus and Michael Walton (Kirksville, MO: Sixteenth Century Press, 1997), 169–85.

[34] Boyle, *Sceptical Chymist*, 80–81.

> trine *as 'tis generally taught by the vulgar Chymists;* and make good the Arguments by which they are wont to maintaine it. Since *'tis only that Doctrine and these Arguments, that I declare my self in that discourse to question;* and he himself did not think them sound and valid.[35] (Emphasis added)

The chymical doctrine Boyle assails is that all bodies are composed of a finite number of elemental substances—either three (Mercury, Sulphur, and Salt) in the usual formulation of the *tria prima,* or five (Mercury, Sulphur, Salt, Phlegm, and Earth) in an expanded seventeenth-century variant. The origins and tenets of these theories have been treated elsewhere, but it is worth reviewing them now.

These conceptions date back ultimately to Aristotle's attribution of the origin of minerals to two exhalations that arise from the center of the earth owing to the heat of the sun. One of these exhalations is wet, the other dry and smoky. Underground, these twin exhalations combine to produce stones, minerals, and metals, and when they ascend into the air, they cause lightning, thunder, hail, and other meteorological phenomena.[36] Medieval Islamic treatises, particularly those under the name of Jābir ibn-Ḥayyān, adopt the mineralogical parts of this theory and state that these dry and moist exhalations condense underground into two intermediate substances which they term Mercury and Sulphur by analogy to the common substances of those names. These subterranean substances then combine in different proportions and under differing conditions of warmth and purity to produce the metals.[37] Gold, for example, is composed of perfectly coequated amounts of pure Mercury and pure Sulphur combined in a warm, clean place. Ill-proportioned or impure Mercury and Sulphur, or the effects of subterranean cold or filth, produce less noble metals.

This theory of metallic composition provides a basis for the belief in metallic transmutation, for if all metals are composed of the same two substances and distinguished only by accidents, then the alchemist should be able to achieve transmutation by bringing the Mercury and Sulphur to the proper state of purity, maturity, or relative proportion. The Mercury-Sulphur theory was further developed in the Latin West by the early alchemical writers, particularly Geber (long confused with Jābir) in his highly influential thirteenth-century *Summa perfectionis.*[38]

This dyad theory—current in several slightly varied forms throughout the Middle Ages and continuing in some circles through the seventeenth century—was altered most substantially by Theophrastus von Hohenheim, called Paracelsus (1493–1541). Paracelsus's innovation was twofold: the addition of Salt as a third principle, and the expansion of the explanatory

[35] Boyle, *Producibleness,* preface, [iv].

[36] Aristotle, *Meteorologica* 341b, 6–12.

[37] Jābir ibn-Ḥayyān, *Book of Definitions* (*Kitāb al-iḍāḥ*), relevant passage cited in Karl Garbers and Jost Weyer, *Quellengeschichtliches Lesebuch zur Chemie und Alchemie der Araber im Mittelalter* (Hamburg: Helmut Buske Verlag, 1980), 14–15 and 34–35.

[38] See Newman, *The Summa Perfectionis of the Pseudo-Geber* (Leiden: Brill, 1991).

domain of the principles to include all bodies. These three principles, or *tria prima,* compose all bodies, but as with most things in Paracelsus, finding clear, consistent explanations is difficult.[39] Yet among his followers, notions about the Paracelsian *tria prima* became more defined (though not wholly consistent across the spectrum of Paracelsians). In general, Mercury represented a volatile, watery substance; Sulphur, an oily, inflammable one; and Salt, a dry, sapid one.

This triad theory was further extended by some chymists in the seventeenth century to include two additional principles—Phlegm and Earth—which represented both an insertion of the Aristotelian elemental water and earth, and a recognition of the experimental result that a watery distillate and water-insoluble ash are produced in the destructive distillation of most animal and vegetable substances. Several adherents to this pentad renamed Mercury, calling it spirit. This pentad, which became extremely popular among seventeenth-century iatrochemists, is generally attributed to Estienne de Clave, although it is clear that there had been intimations of such additional principles among Paracelsians since the late sixteenth century.[40]

It is important to note that although there is a temporal progression here—the dyad dating from the Middle Ages, the triad from the sixteenth century, and the pentad from the early seventeenth—the advancement of one system did not automatically displace its predecessor. Just as both Copernicans and Ptolemaists coexisted for an extended period of time, so too did adherents of all three chymical systems. These chymical doctrines had different attributes that attracted different groups. Having outlined these three theories, we must now contrast them and attach them, as far as is possible, to specific subsets of chymical practitioners.

The first of the two chief differences between the dyad on the one hand and the triad and pentad on the other is the breadth of their applicability. The traditional dyad has a very limited scope; it relates to only two general topics—the origin and composition of metals and some minerals, and the confection of the Philosophers' Stone. The triad, however, is claimed by Paracelsians to express the constitution of *all* substances. The pentad likewise extends to all substances.

The second chief difference is more complicated and involves claims made about the actual identity and status of the principles. In texts using the dyad, *Mercury* and *Sulphur* do not always refer absolutely to discrete materials; that is, they do not have fixed referents. Thus confusion (generally intentional) is ubiquitous owing to the almost innumerable ways in which the names *Mercury* and *Sulphur* are employed. While Mercury and Sulphur

[39] Paracelsus uses both the *tria prima* and the Aristotelian elements in confusing and inconsistent ways; see Reijer Hooykaas, "Die Elementenlehre des Paracelsus," *Janus* 39 (1935): 175–87. Boyle notes Paracelsus's inconsistencies in the *Sceptical Chymist,* 276–77.

[40] Metzger, *Doctrines,* 54–57; Estienne de Clave, *Les vrais principes et éléments de la Nature* (Paris, 1641), 159; Nicaise LeFebvre also adopted the pentad; see Metzger, *Doctrines,* 73–74; Reijer Hooykaas, "Die Elementenlehre der Iatrochemiker," *Janus* 41 (1937): 1–28.

were literally considered to be the constituents of metals, the terms *Mercury* and *Sulphur* could be extended analogically and attached to a vast number of other substances that bore some similitude to the metallic principles or their mutual activity. Thus they function also as *Decknamen* for any pair of substances (particularly those used in the preparation of the Stone) that can relate to one another in the way the subterranean Mercury and Sulphur do in the formation of metals—for example, in an active-passive, masculine-feminine, or coagulant-coagulated fashion.

As an illustration, consider the pair coagulant-coagulated, which is perhaps the most important pair, particularly in light of the alchemical axiom *solve et coagula* ("dissolve and coagulate") considered by some authors to be sufficient guidance for the preparation of the Stone. It had been well-known since at least the Islamic Middle Ages that the liquid metal mercury (quicksilver, Hg) when treated with sulphur (brimstone, S) is "coagulated" into a solid mass of cinnabar (mercuric sulphide, HgS). This simple reaction seems to me to be a major foundation for the analogical properties of the alchemical principles Mercury and Sulphur—Sulphur as the coagulant of Mercury. Under the earth, subterranean Sulphur coagulates the liquid subterranean Mercury into the metals. In the making of cinnabar one could say (at the risk of sounding tautological) that sulphur is Sulphur and mercury is Mercury. The making of cheese provides a classical alchemical trope used by the fourteenth-century Petrus Bonus and many subsequent authors. The addition of rennet coagulates milk into curds, and thus rennet could be called Sulphur and milk Mercury. Likewise, a small grain of the Philosophers' Stone cast upon hot mercury transmutes ("fixes") that liquid metal into solid gold; thus, in this transformation, the Philosophers' Stone is Sulphur, and mercury is Mercury. For this reason (among others) the completed Elixirs, red and white, were sometimes called fixed or perfect red and white Sulphurs.[41]

A biological example showcases other complementary dualities. According to traditional belief, male semen coagulates female menstrual blood into a fetus; thus in this pair, semen is Sulphur (as well as male and active), while menstrual blood is Mercury (as well as female and passive).[42] By continued extension of these sets of analogies, one of the origins of the vast network of *Decknamen* and interrelationships present in alchemical language and thought becomes considerably clearer. The alchemical Mercury and Sulphur in the dyad represent overlapping groups of substances categorized sometimes on the basis of physical properties but more frequently on the basis of relative activity toward each other.

In the Paracelsian triad the largely analogical, relative, and fluid categories of the Mercury and Sulphur of the dyad are made more rigid so that the *tria prima* become universal, constant, primary, elemental bodies. For ex-

[41] On coagulation of Mercury to form the metals, see Petrus Bonus, *Margarita preciosa novella*, in *BCC*, 2:1–80, on 1–2, 72–75.

[42] See ibid., 76–79.

ample, Quercetanus, one of the most important early expositors of the Paracelsian *tria prima,* insists in his 1603 *Liber de priscorum philosophorum verae medicinae materia* that Mercury, Sulphur, and Salt are present in *all* bodies, and that the scope of chymical philosophy includes not only minerals and metals but also plants and animals, weather, and even celestial bodies.[43] The principles are elemental substances or, as Quercetanus dubbed them, *hypostatical principles.* This trend toward universalization of the principles continues in the textbook tradition. These texts, which are designed to teach their readers in a didactically effective manner, are thereby constrained to make the concepts of Mercury, Sulphur, and Salt more concrete.[44]

Several seventeenth-century writers recognize the increased scope of the triad's principles and their more absolute and elementary nature. A chief chymist of the Académie Royale des Sciences, Samuel Cottereau DuClos (d. 1715), refers to this transformation of meaning in his "Dissertation on the Principles of Natural Mixed Bodies" presented to the academy in 1666.

> The Philosophers . . . gave the title and denomination of *principle* to Salt, Sulphur, and Mercury, not however absolutely and simply, for they did not recognize these materials as simple and primary substances, but only as parts materially and immediately constitutive of their Arcanum . . . they judged that the disciples of their Cabalistic School, accustomed to parables and allegories, would know to make the distinction between these immediate and particular principles of their grand metallic arcanum and the first and general principles of all mixed bodies. The vulgar chymists, to whom the grand arcanum was as little known as the true principles of mixed bodies, misled by the equivocation of the name of principle, took the Salt, Mercury, and Sulphur for general principles.[45]

Historians might question the specific details of DuClos's history of the principles, but his division of usage clearly illustrates his recognition of the expansion of the principles by later chymical schools. Note also that he, like Boyle, distinguishes the misguided "vulgar chymists" from the "Philosophers."

Van Helmont describes a similar expansion of meaning from the restricted use in metals to all bodies, attributing this development to Paracelsus. "Basilius Valentinus, a Benedictine monk, named the soul of a metal Sulphur or tincture, the body Salt, and finally he called the spirit Mercury. Theophrastus Paracelsus then, in a wonderous all-embracing fashion, carried over these things derived from Basilius as the principles in all bodies."[46]

[43] Debus, *French Paracelsians,* 51–58.

[44] See, for example, Beguin, *Tyrocinium chymicum* (London, 1669), 20–23.

[45] S. C. DuClos, "Dissertation sur les Principes des Mixtes naturals," in*Memoires de l'Academie Royale des Sciences, 1666–1699* (Paris, 1733), 4:1–30, on 13.

[46] Jan Baptista van Helmont, *Ortus medicinae* (Amsterdam, 1648; reprint, Brussels: Culture et Civilisation, 1966), 399. ". . . Basilius Valentinus, Monachus Benedictinus, Animam metalli, sulfur, sive tincturam nuncupavit, corpus verò sal, ac tandem spiritum dixit Mercurium. Quae sic à Basilio mutuata, deinceps in cuncta corporum principia, mirafica indagine

Again, regardless of the fact that van Helmont (like most of the seventeenth century) mistakenly believed that Basilius Valentinus predated Paracelsus by a century and was plagarized by him, he still indicates for us the recognition of expansion of the principles by Paracelsus to include *all* bodies.

Can the differing theories be attached to specific subgroups within chymistry? Although the lines are by no means rigid or distinct, we can in fact make some discriminations. The clearest assignment is that the triad, being a Paracelsian innovation, was supported by his followers. Further, since Paracelsus de-emphasized chrysopoeia in favor of chymical medicine, Paracelsians devoted more of their effort to medical preparations than to chrysopoeia. Thus the iatrochemists, chymical physicians, and apothecaries who followed these notions tended to adopt the related *tria prima,* or its expansion into a pentad, which correlated more closely with the kinds of operations with which they were involved. The textbook tradition, which is an expression of specifically iatrochemical interests, embraces without exception the triad or pentad as the true principles separable by fire from all bodies.

This division brings up the larger issue of what beliefs actually composed that which we term *alchemy.* It is crucial to understand that *the realm of alchemy is not coterminous with that of Paracelsus and his followers.* Paracelsus was by no means the chief of the alchemists but rather was as iconoclastic in regard to alchemy—especially chrysopoeia—as he was in regard to medicine or religion.[47] The often bizarre natural magical worldview of Paracelsus did not inform all of alchemy or even a major part of it. Much of the difficulty in understanding alchemy stems from a failure to differentiate its diverse schools of thought, for writers in the postalchemical period have been apt to group together any premodern chymical endeavors as alchemy. Yet neither chymistry nor its subset chrysopoeia is a monolith.

Accordingly, some seekers after transmutation were anti-Paracelsian or simply non-Paracelsian, and their works show little or no interest in chemiatria. One example is the aforementioned Gaston DuClo who, even though writing his defense of chrysopoeia in the midst of the vigorous Paracelsian revival in France, never once in all his writings makes reference to Paracelsus or a single Paracelsian notion, including the *tria prima.*[48] Since seekers after the Stone were naturally more interested in metals, and looked

---

transtulit Theophrastus Paracelsus . . ." The phrase "mirafica indagine" might also be rendered "by wondrous investigation," but the sense is clearly ironic.

[47] In his "Vom Terpentin" Paracelsus calls *alchemia* a pillar of medicine and distinguishes it from *alchemei,* which deals with transmutation (in German the ending *-ei* is used contemptuously with the sense of *useless repetition*); in *Sämtliche Werke,* ed. Karl Sudhoff, 14 vols. (Munich and Berlin, 1922–1923), 2:187. The distinction between Paracelsus and the alchemical tradition is nicely summarized in Massimo Luigi Bianchi, "The Visible and Invisible: From Alchemy to Paracelsus," in *Alchemy and Chemistry,* 17–50.

[48] See Principe, "Gaston DuClo,"181.

back to medieval and Renaissance authorities like Geber, the pseudo-Lull, and George Ripley who wrote before the conception of the *tria prima,* these later chrysopoeians often found the dyad to be sufficient, and its analogical nature suited to the writing of their allegorical, secretive works. These more traditional alchemists, while they might occasionally invoke a Paracelsian idea such as the *sal circulatum* (renamed as alkahest by van Helmont), sometimes did not adopt the doctrine of the *tria prima* at all or, if they did, did not much care to apply it outside of the metallic realm where their interests lay. Even the otherwise thoroughly Paracelsian textbook writer Jean Beguin seems to contradict himself in this area: for in his *Tyrocinium* he first claims that the *tria prima* constitute all bodies, but later when describing the composition of metals, he invokes only the two principles of the old dyad—Mercury and Sulphur.[49]

In spite of these distinctions, one must be careful not to err in the opposite direction by thinking of chrysopoeians and Paracelsian iatrochemists as wholly disjoined groups; in the seventeenth century there are quite a few notable overlaps—for example, the popular "Basilius Valentinus"—who adopted Paracelsian notions including the *tria prima* and sought both the Stone and chymical medicines.[50] The degree to which a given chymist adopted Paracelsian teachings varied widely; many saw their activities as part of a tradition that included Paracelsus only tangentially or not at all. This is not to say that Paracelsian principles were not widely employed in the seventeenth century; however, it does allow us to begin to segregate the Paracelsians as one subset in the larger sphere of chymistry.

[49] Although Beguin is best known for the *Tyrocinium chymicum,* whose emphasis is pharmaceutical, he also has much less publicized interests in chrysopoeian alchemy; for example, he edited the first edition of Sendivogius's *Novum lumen chymicum* in France (Paris, 1608). In fact, the first edition of the *Tyrocinium* (privately printed, 1610) begins with a definition of *alchymia* that mentions spagyrical medicines alongside transmutation and other operations; this introduction was removed from later editions. See T. S. Patterson, "Jean Beguin and His *Tyrocinium chymicum,*" *Annals of Science* 2 (1937): 243–98.

Boyle, of course, was not aware of Beguin's other interests in chymistry; his view of Beguin is provided solely by the *Tyrocinium.* Clearly, Beguin and other textbook writers are worthy of further study to integrate the totality of their activities with their publications. See also Gerald Schröder, "Neuere Ergebnisse der Beguin-Forschung," in *Die Vorträge der Hauptversammelung der Internationalen Gesellschaft für Geschichte der Pharmazie,* ed. George E. Dann (Stuttgart: Wissenschaftliche Verlag, 1966), 227–33.

[50] The wide interests of "Basil Valentine" may be partially due to the heterogeneity of the Valentinian corpus; it is unlikely that all the works which appeared under that pseudonym come from the same pen. See Claus Priesner, "Johann Thoelde und die Schriften des Basilius Valentinus," in *Die Alchemie in der europäischen Kultur- und Wissenschaftgeschichte,* ed. Christoph Meinel, Wolfenbütteler Forschungen 32 (Wiesbaden: Verlag Otto Harrassowitz, 1986), 107–118, and Lawrence M. Principe, "Apparatus and Reproducibility in Alchemy," in *Instruments and Experimentation in the History of Chemistry* (Chicago: University of Chicago Press, 1998).

Pruning Back the Chymical Principles

Returning to the *Sceptical Chymist* and revisiting its criticism of the *tria prima,* we can clearly identify Boyle's real targets. Boyle argues that some bodies provide more than three or five discrete substances while others provide fewer, and so the contention that *all* bodies are composed of exactly three or five substances is false.[51] But we have just seen that the understanding of the chymical principles as constituents of *all* bodies is a dogma of the Paracelsians emphasized in the textbook tradition, and not part of the dyad.

It might plausibly be objected that Boyle opposes the less comprehensive dyad implicitly along with the broader Paracelsian *tria prima* and pentad, but his comments concerning the mineral realm imply that he did not. For example, he is particularly critical of the notion of metalline Salt, and his doubts regarding it generally appear immediately following an affirmative mention of metalline Mercury and Sulphur. In one place, after admitting his own inadequate experience regarding the differences among the "Sulphurs of Metals," he asserts that "as for the salts of Metals," it is "much to be question'd whether they have any at all." Immediately thereafter, while he cites his own knowledge of the "Sulphur of Antimony" and the "strongly scented Anodyne Sulphur of Vitriol," he again expresses doubt "whether metals have any salt at all," noting that he has seen no such metalline Salt "not for want of curiosity."[52] In a further denial of metalline Salts Boyle touches upon a topic critical to chrysopoeians—the composition of gold itself:

> 'Tis not, that after what I have try'd myself I dare peremptorily deny, that there may out of Gold be extracted a certain substance, which I cannot hinder Chymists from calling its Tincture or Sulphur; and which leaves the remaining Body depriv'd of its wonted colour. Nor am I sure, that there cannot be drawn out of the same Metal a real quick and running Mercury. But for the Salt of Gold, I never could either see it, or be satisfied that there was ever such a thing separated, *in rerum natura,* by the relation of any credible eye witnesse.[53]

Here Boyle allows that substances denominable as Mercury and Sulphur can in fact be isolated from gold (he himself had prepared the Sulphur of gold and been shown its Mercury), thus acquiescing in the claims of non-Paracelsian chrysopoeians but at the same time denying the claims of the Paracelsians' *tria prima* by refusing to admit the Salt principle.[54]

[51] Boyle, *Sceptical Chymist,* 168–98.

[52] Ibid., 247–82, esp. 275–76.

[53] Ibid., 174–75; cf. 276. This claim to have separated a "Sulphur of gold" appears in several of Boyle's other works; the chrysopoetic import of these passages will be covered in the following chapter.

[54] The early version of the *Sceptical Chymist* indicates that Boyle had received testimony about the Mercury of gold from Kenelm Digby, although the reference to Digby is missing in the published text; Boas [Hall], "Early Version," 161–62.

A contemporaneous source recognizes not only the distinction between the Paracelsian triad and the pre- or non-Paracelsian dyad but also the *Sceptical Chymist*'s specific attack on the former. In 1672, Joel Langelott (1617–1680), physician to the duke of Holstein-Gottorp, published a letter addressed to the Academia naturae curiosorum concerning some alchemical operations that he had performed.[55] This brief tract provided the widely read polymath Daniel Georg Morhof (1639–1691) the opportunity to pen an *Epistola ad Langelottum de metallorum transmutatione,* which was published in 1673. In his letter, Morhof discusses the origin of metals as a necessary prerequisite to the study of their transmutation. After presenting various theories and citing Boyle's 1672 *Origine and Virtues of Gems,* Morhof presents the theory of the chymists.

> They impress upon us their Mercury and Sulphur. If you should examine what thing hides under these names (for they do not wish for them to be understood in the common signification) they would press upon you six hundred words, obscure descriptions, and enigmas, in which you would have to hunt down some sense to be collected here and there like the severed limbs of Hippolytus. Yet when you have done so you will find a consensus among them—except that the newest sect of the chymists (after the time of Paracelsus) has added a third principle, namely, Salt, to these two, Mercury and Sulphur; but I do not wish to inquire any further here regarding that number of principles, or whether they compose all substances, or how much they differ from the common, for that is most learnedly and copiously disputed by the illustrious Robert Boyle in his *Sceptical Chymist* . . . In truth, those old [chymists] before Paracelsus set up these two—Mercury and Sulphur—not as common principles as ours today wish to do, but rather only as metallic principles.[56]

The distinction between the dyad and the triad/pentad is repeated when Boyle argues at length that individual principles of the same species differ from one another (i.e., the Sulphur of one substance is not equivalent to the Sulphur of another), and therefore they cannot be true, universally elemental substances.[57] His discussion of Mercury, generally the most troublesome of the principles, is illuminating. Contrary to his usual custom of not citing authors by name, whether approvingly or disapprovingly, Boyle criticizes four writers sequentially by name in regard to their beliefs concerning Mercury, and then approves three others, two of them by name. Since Boyle so rarely names names, we should pay close attention to those occasions where he does. The harsher criticism falls upon Beguin and Quercetanus, but An-

[55] Joel Langelott, *Epistola ad praecellentissimos Naturae Curiosos de quibusdam in chymia praetermissis* (Hamburg, 1672). A German translation appeared as *Send-Schreibung an die hochberühmte Naturae Curiosos* (Nuremberg, 1672), reprinted in Friedrich Roth-Scholtz's *Deutsches Theatrum Chemicum,* 3 vols. (Nuremberg, 1728–1732), 2:381–406. Excerpts in English were published by Oldenburg in *Philosophical Transactions* 7 (1672): 5052–59.

[56] Daniel Georg Morhof, *Epistola ad Langelottum,* in *BCC,* 1:168–92, on 176.

[57] Boyle, *Sceptical Chymist,* 247–82.

dreas Libavius and Daniel Sennert fall close behind. Boyle first reiterates his dissatisfaction specifically with the extension of the principles: "What they can mean, with congruity to their own Principles, by the Mercury of Animals and Vegetables, 'twill not be so easie to find out."[58] He then notes what a "great difference is conspicuous" between the "Vegetable and Animal spirits" that are called Mercury by "many moderne Chymists" and the "running Mercury . . . which is separated from Metals." He sets these "moderne Chymists" off against "some Chymists that seem more Philosophers than the rest" who refer to metalline Mercury as "*Mercurius Corporum,*" or "Mercury of bodies," using *body* in the Geberian sense of *metals.* He names "Claveus"—that is, the non-Paracelsian chrysopoeian Gaston DuClo—as an example of the "more Philosophical" chymists.[59]

Boyle goes on to assert that even these metalline Mercuries differ from one another, and as evidence he cites DuClo regarding the properties of the Mercury of silver and the Mercury of tin, as well as the "Experienc'd *Alexander van Suchten*" regarding the Mercury of copper. Van Suchten, a late-sixteenth-century chrysopoeian, was highly influential; his writings constitute an important source for the alchemy of George Starkey, alias Eirenaeus Philalethes. Boyle adduces further testimony regarding the Mercury of lead from "an eminent person," who is identifiable as Sir Kenelm Digby. These chrysopoeians, like Boyle, had all asserted that the Mercuries of different metals were different substances with different properties. Strikingly, all three of the experiments that Boyle takes from these authors to use against Quercetanus, Beguin, Libavius, and Sennert are examples of the transmutation of specific metalline Mercuries into gold.[60]

Boyle also has much to say about the heterogeneity of the supposed principles, a fact that would rule out their being elemental.[61] This criticism directly attacks the *tria prima* as it is presented in textbooks but not the metallic dyad whose more fluid and analogical use did not require that metalline Mercuries and Sulphurs be elemental. Here again Boyle explicitly allows for the dyad in the mineral realm (but not in the vegetable or animal, and no Salt anywhere), while maintaining that the Mercury and Sulphur are not elemental: "there may sometimes either a running Mercury, or a Combustible Substance be obtain'd from a Mineral, or even a Metal; yet I need not Concede either of them to be an Element."[62] It is the overextension of the

[58] Ibid., 270.

[59] Ibid., 272–73; Boyle makes the same distinction between the "Quicksilver . . . whether it be common or drawn from Mineral Bodies" and "that immature and fugitive substance which in Vegetables and Animals Chymists have been pleas'd to call their Mercury," on 369.

[60] Ibid., 274–75; DuClo, *Apologia,* 37, 39, 61; Alexander van Suchten, *Of the Secrets of Antimony* (London, 1670), 86–88; on Suchten's influence on Starkey/Philalethes, see William R. Newman, "The Authorship of the *Introitus apertus ad occlusum regis palatium,*" in *Alchemy Revisited,* ed. Z.R.W.M. van Martels (Leiden: Brill, 1990), 139–44; the identification of Digby is made by Boyle himself in the early draft; see Boas [Hall], "Early Version," 163.

[61] Boyle, *Sceptical Chymist,* 209–41.

[62] Ibid., 351.

chymical principles in the triad and pentad that Boyle forcefully rejects. Boyle willingly admits that real Mercuries and Sulphurs may be isolated from the metals; however, he will not accept the extension to all substances or the assertion that the separated substances are elemental.

Criticism of the Chymical World-System

The *Sceptical Chymist* also criticizes the arrogant overextension of a few chymical experiments or notions into a grand system. Boyle states that if the *tria prima* were presented

> but as a Notion useful among Others, to increase Humane knowledge, they had deserv'd more of our thanks; and less of our Opposition; but . . . the Thing that they pretend is not so much to contribute a Notion toward the Improvement of Philosophy, as to make this Notion (attended by a few lesse considerable ones) pass for a New Philosophy it self.[63]

This criticism of the construction of a "New Philosophy" based upon an extended *tria prima* brings us to the Paracelsian systematizers. While textbook writers like Beguin, de Clave, Rolfinck, and Arnaud reduced chymistry to manual operations and pharmacy, thus lowering its status by making it the labor of "laborants," others went to the opposite extreme by inflating it into grand cosmological schemes. Such writers receive heavy treatment at Boyle's hands.

Boyle singles out Quercetanus, "the grand stickler for the *Tria Prima,*" by name for criticism on this score. He "scruples not to write, that if his most certain Doctrine of the three Principles were sufficiently Learned, Examin'd, and Cultivated, it would easily Dispel all the Darkness that benights our minds, and bring in a Clear Light, that would remove all Difficulties."[64] Quercetanus's writings are as broad as Paracelsus's in terms of their arrogance in subjecting the entire universe to a few chymical notions, and, from Boyle's point of view, probably worse than Paracelsus's as Quercetanus was more broadly comprehensive (and consistent). Indeed, Quercetanus's *Le Grand Miroir du Monde* presumes to encompass not only the elements but also the stars, angels, and God as well, and presents chymistry as the only true natural philosophy.[65]

Two further iatrochemical textbook writers fit Boyle's description as well. One is the expatriate Scot William Davisson (ca. 1593–1669), professor at the Jardin Royale des Plantes, who penned the equally lofty and obscure *Philosophia pyrotechnica seu curriculus chymiatricus* (1633–1635). While he exalts chemiatria in this "curriculus," practical techniques give way to a grand chymical cosmological system. Davisson's bizarre and rambling dis-

[63] Ibid., 305.

[64] Ibid., 179, 305.

[65] Debus, *French Paracelsians,* 51–52.

course combines Neoplatonism, cabala, numerology, and scriptural authorities into an esoteric Paracelsian chemical system.[66] The other is Annibal Barlet, whose 1653 *Vray et methodique cours de la physique* (reprinted 1657) is intended to allow the reader "to come to know the Ergocosmic Theotechne, that is, the Art of God in the work of the universe." Barlet's textbook shares Davisson's extravagant and incoherent nature (though employing Pythagorean symbolism rather more than Paracelsian cosmology) until in the last quarter it settles down to the recitation of recipes along the lines of Beguin.[67] Boyle's criticism of such pretentious "chymical philosophies" accords well with his general distaste for broad philosophical systems.[68] It is worth pointing out that Boyle's usual reticence to name his opponents may have caused him to suppress mention of recent authors like Davisson and Barlet (who were still living) and to use the names of only those who (like Beguin and Quercetanus) were dead by 1661.

Later in life Boyle declared that he wrote the *Sceptical Chymist* "to take those Artists off ther excessive Confidence in their principles and to make them a litle more Philosoph[ical] with their Art."[69] This brief testimony sums up Boyle's twin philosophical attacks on the "vulgar chymists." The "excessive Confidence" of some allowed the overextension of the chymical principles, now envisioned as elementary substances, to all bodies, and then into grand cosmological schemes. The rigid practicality of others—Beguin's *Tyrocinium chymicum,* for example, is little more than a pharmaceutical cookbook prefaced with broad declarative statements on the principles of all mixed bodies—kept chymistry in the ignoble status of a technical labor lacking sound philosophical principles.

One danger of "excessive Confidence" is the overlooking of facts that would lead to a truer natural philosophy, and here Boyle's well-known commitment to experimentalism and "diffidence" with respect to explanatory theories makes itself apparent.[70] But another danger is more rigidly practical, and Boyle cites it explicitly. "I fear," he writes, "that the too confident opinion of the Doctrine I question has made divers practitioners of *Physick,* make wrong estimates of Medicines."[71] This medical dimension, which is a significant but inadequately recognized part of the *Sceptical Chymist*'s message, directs us again to the iatrochemists rather than to the chrysopoeians and reinforces the link Clericuzio has made with the Oxford physiologists.

[66] On Davisson, see Metzger, *Doctrines,* 45–51; Debus, *French Paracelsians,* 124–25; Thorndike, *History of Magic,* 7:123–26. Davisson's ponderous treatise may have had some effect on the style of Boyle's *Sceptical Chymist,* for the fourth part begins with a lengthy dialogue between an Aristotelian and a chymist.

[67] Annibal Barlet, *Vray et methodique cours de la physique resolutive, vulgairement dite Chymie . . . pour connoistre La Theotechnie Ergocosmique, c'est a dire, L'Art de Dieu, en l'ouvrage de l'univers* (Paris, 1653); see Thorndike, *History of Magic,* 8:129.

[68] Sargent, *Diffident Naturalist,* 35–36, 39, and 56–57.

[69] In Hunter, *Robert Boyle by Himself,* 29.

[70] See Sargent, *Diffident Naturalist.*

[71] Boyle, *Producibleness,* preface, [ii and iv].

Tria Prima *versus Corpuscularianism?*

Before leaving the topic of matter theory, we must broach the important subject of Boyle's corpuscularianism. Some readers might object that the unifying thread of Boyle's chymical thought—a corpuscularian explanatory system—is itself sufficiently divergent from traditional alchemical thought to render any concessions to alchemy small by comparison. Indeed, Hélène Metzger's classic studies divide alchemy from chemistry on the basis of the introduction of mechanical corpuscularianism. This notion, however, rests upon a false dichotomy. Corpuscularian matter theories were in fact widely accepted in alchemical circles, as William Newman has clearly shown. Newman has traced the tradition of alchemical corpuscularianism from the influential writings of the late medieval pseudo-Geber down to the seventeenth century. Both van Helmont himself and the extremely popular Eirenaeus Philalethes (alias George Starkey and sometime collaborator with Boyle on chymical projects) employ corpuscularian explanatory systems. Many chrysopoeians from pseudo-Geber to Philalethes use an explicitly corpuscularian system to explain metallic transmutation. Further, Newman has provided good evidence that Boyle's own early adoption of corpuscularianism comes not solely from the renowned revivers of ancient atomism, such as Gassendi and Descartes, but at least partially from the tradition of chymical corpuscularianism dating back to Geber.[72] Regardless of how much of his own corpuscularianism Boyle derived from alchemical sources, the presence of a corpuscularian tradition in alchemy suffices to show that Boyle's particulate matter theory per se cannot be seen as a break from alchemical matter theory.

Obscurity and Secrecy

There is one more topic of criticism in the *Sceptical Chymist*—Boyle's attack on the obscurity of chymical writers.[73] Indeed, the section of the *Sceptical Chymist* that deals with the topic of ambiguous and obscure terms is the sharpest of all Boyle's criticisms. The civility and roundness of expression that characterize Boyle's style momentarily falter as barbed expressions of his "Indignation" surface. Certainly the adepti are highly susceptible to attacks upon their obscurity and doublespeak. Allegorical, metaphorical, and emblematic alchemical texts are common and well known. Basil Valentine's *Zwölff Schlüssel* (*Duodecim clavibus*), Michael Maier's *Atalanta fu-*

[72] Newman, "Boyle's Debt to Corpuscular Alchemy"; "The Corpuscular Transmutational Theory of Eirenaeus Philalethes"; "The Corpuscular Theory of J. B. van Helmont and Its Medieval Sources," *Vivarium* 31 (1993): 161–91; "The Alchemical Sources of Robert Boyle's Corpuscular Philosophy." Boyle's positive use of Sennert here may seem to conflict with his criticisms of Sennert in the *Sceptical Chymist,* but Boyle's rejection of Sennert's notion of the elements does not require a rejection of his arguments for corpuscularianism.

[73] Boyle, *Sceptical Chymist,* 199–209.

giens, the *Rosarium philosophorum,* and numerous other texts present the preparation of the Philosophers' Stone in terms of allegorical illustrations accompanied by cryptic texts. The traditional English alchemical works published by Elias Ashmole in 1652 as the *Theatrum chemicum britannicum* are essentially riddles in verse. Likewise, all the alchemists, from classical authorities like the pseudo-Lull to Eirenaeus Philalethes, employed complex sets of *Decknamen* sometimes woven into allegorical texts to guard the secrecy of their processes from unworthy readers. Certainly, Boyle had an overwhelming amount of material at his disposal to criticize.

Close examination shows, however, that Boyle's criticisms—while real enough—are not directed primarily at traditional alchemical authors and their mysteries. Rather, he exculpates the adepti and shows his chief target when he declares that

> much may be said to Excuse the Chymists when they write Darkly and Aenigmatically, about the Preparation of their *Elixir,* and Some few other grand *Arcana,* the divulging of which they may upon Grounds Plausible enough esteem unfit; yet when they pretend to teach the General Principles of Natural Philosophers, this Equivocall Way of Writing is not to be endur'd.[74]

This passage shows that the issue is twofold—unnecessary ambiguity of expression on the one hand and intentional secrecy on the other. The latter, chiefly the fault of chrysopoeians, Boyle tended to excuse in part, although there is no question that he still flings sarcastic statements in their direction—"As for the Mystical Writers scrupling to Communicate their Knowledge, they might less to their own Disparagement, and to the trouble of their Readers, have conceal'd it by writing no Books, then by Writing bad ones."[75] Most of Boyle's criticism, however, is directed at obscurity of expression—the failure of chymists to teach *principles* clearly and coherently—not at secrecy. The explanations of these chymists are in effect no explanation at all; the "Chymists have written so obscurely of their three Principles" because they do not have "Clear and Distinct Notions of them themselves."[76] Interestingly, while the modern reader would naturally associate the allegorical books regarding the Philosophers' Stone with what Boyle terms "Metaphoricall Descriptions," this is not at all what he means. When Boyle explicitly links his dissatisfaction to the unclear exposition of philosophical principles, "which for their being Principles ought to be defin'd the more accurately and plainely," he again names the iatrochemists Quercetanus and Beguin, making reference to the "very many . . . faults that may be found with their Metaphoricall Descriptions."[77]

[74] Ibid., 203. This indulgence granted specifically to the adepti is mentioned also by Clericuzio, "Carneades and the Chemists," 82–83.

[75] Boyle, *Sceptical Chymist,* 202; cf. 204. Note that here Boyle uses the word *mystical* (cf. 269) in its archaic sense of "secret" and not in the modern sense; see chapter 1.

[76] Boyle, *Sceptical Chymist,* 202–3.

[77] Ibid., 267–68. Beguin also peppered his *Tyrocinium* with secret phrases, as when in medi-

Again we can turn for support to a contemporaneous account of how the *Sceptical Chymist* was read. George Castle (1635 or 1636–1673) was a physician, a fellow of the Royal Society, and an associate of the Oxford physiologists along with Boyle in the 1650s. In 1667 he published *The Chymical Galenist*, an attack particularly on Marchamont Needham's 1665 *Medela medicinae*, but extending to the whole of the then ongoing battle between the "Chymical Physicians" and the Royal College of Physicians.[78] While Castle charts a middle course between the opposing sides, he, like Boyle, criticizes the "conjuring, unintelligible words of the *Chymists*" and provides a list of them: all are terms of Paracelsian or Helmontian medicine. Castle also quotes the *Sceptical Chymist* at length and clearly interprets Boyle's attack on obscurity as relating specifically to chymical medicine.[79]

Methodology and "The Requisites of a Good Hypothesis"

Boyle's "Indignation" at the unclear exposition of "General Principles" brings up a methodological issue that undergirds the *Sceptical Chymist*—the framing of good hypotheses from experiments. The chymists' principles are not sufficiently evinced by their experiments, and a few chymical experiments cannot be legitimately extended into grand world-systems though they "cry up their own Sect for the Invention of a New Philosophy."[80] This fact lies behind Boyle's often-quoted sentiment that he valued the chymists' contributions "upon the score of their experiments, not upon that of Their Speculations," and that there is "great Difference betwixt the being able to make Experiments, and the being able to give a Philosophical Account of them."[81] He assails the chymical philosophy as narrow and notes that the best that could be said of the doctrine of the *tria prima* is that it may be "useful to Apothecaries," but not to philosophers who desire the "Knowledge of Causes."[82] This point, tied up closely with Boyle's promotion of experimentalism and central to his contribution to early modern science, has been well treated by other authors, most recently and notably by Rose-Mary Sargent.[83] Presently, however, I think there is reason to tie this preoccupation of Boyle's even more closely to his early chymistry and his dissatisfaction with "vulgar chymists."

I think it likely that the *Sceptical Chymist* is linked with a programmatic work on which Boyle labored for thirty years but never completed—"The

---

cal receipts he prescribes the use of the "venenate scum of two dragons" and "spirit of balsamic salt" (89).

[78] P. M. Rattansi, "The Helmontian-Galenist Controversy in Restoration England," *Ambix* 12 (1964): 1–23.

[79] George Castle, *The Chymical Galenist* (London, 1667), 6 and 16–17.

[80] Boyle, *Sceptical Chymist*, 299–340, on 299.

[81] Ibid., 307.

[82] Ibid., 334.

[83] Sargent, *Diffident Naturalist*, 70–75.

Requisites of a Good Hypothesis." Scattered fragments of the work dating from the 1650s to the 1680s survive among the Royal Society Boyle Papers, and Boyle catalogs this work repeatedly in his inventory lists of unpublished papers from 1667 to 1691.[84] He notes also that his *Excellency and Grounds of the Mechanical Hypothesis* (1674) is written in relation to "what the Author had written (by way of *Dialogue*) about the *Requisites of a good Hypothesis*" although he is "not willing to let *that,* at least quickly come abroad."[85] Again in the *Imperfection of the Chymist's Doctrine of Qualities* (1675) Boyle borrows "an illustration from our unpublished *Dialogue of the Requisites of a good Hypothesis.*"[86] A manuscript summarizing the "Requisites" was published by R. S. Westfall in 1956, and two other fragments by Boas Hall in 1965.[87]

Several pieces of evidence suggest a link between the *Sceptical Chymist* and the "Requisites." First, many of the programmatic statements in the *Sceptical Chymist* revolve around this issue of drawing good hypotheses. Boyle speaks of the inadequacy of the "*Chymical Hypothesis*" as well as the "*Aristotelian Hypothesis,*" and in one place begins, "one of the chief Requisites of a good hypothesis . . . ," as if a section of the work had been spliced in (a typically Boylean method of revision).[88] Second, Boyle explicitly states that other dialogues (and we know the "Requisites" to have been a dialogue) were related to the *Sceptical Chymist.* These associated dialogues were "betwixt the same speakers (though they treat not immediately of the Elements)."[89] Third, the early draft of the *Sceptical Chymist* contains none of the material on hypotheses; thus these programmatic parts were added later, perhaps as the early essay was recast into dialogue format.

These three observations are consolidated by the most important piece of evidence—a hitherto uncited manuscript among the Boyle Papers entitled "The REQUISITES of a Good HYPOTHESIS." The manuscript is written in Boyle's 1650s autograph and in secretarial hands datable to a similar period and associated with Boyle's residence at Oxford.[90] Thus it is possible to date this fragment to a period before the publication of the *Sceptical Chymist.* Boyle's inscription presents the work as "a Dialogue between Carneades,

[84] BP, 36:177 (19 November 1667; only here listed as "a short Epistle," elsewhere as a dialogue); 91; 60 (7 July 1684); 72 (3 July 1691). There are several drafts of each of these catalogs; their existence was first noted in Hunter, *Guide,* xxiii–iv.

[85] Boyle, *Excellency and Grounds of the Mechanical Hypothesis* (London, 1674), "Publisher's Advertisement," [i]; see also 1.

[86] Boyle, *Imperfection of the Chymist's Doctrine of Qualities* (London, 1675), 40–41.

[87] Richard S. Westfall, "Unpublished Boyle Papers Relating to Scientific Method," *Annals of Science* 12 (1956): 63–73, 103–17, on 69–70 and 116–17; Boas Hall, *Robert Boyle on Natural Philosophy,* 134–35.

[88] Boyle, *Sceptical Chymist,* 306, 326. On Boyle's method of splicing works, see Lawrence M. Principe, "Style and Thought of Early Boyle: Discovery of the 1648 Manuscript of *Seraphic Love,*" *Isis* 85 (1994): 247–60.

[89] Boyle, *Sceptical Chymist,* preface, [ii] and [xii]. About 1742, Henry Miles recorded a then-surviving portion of "Requisites" as being "freely Considered by Carneades," BP, 36:158.

[90] BP, 38:37–39. Hands G and F; see Hunter, *Guide,* xxxvii–xxxviii.

Eleutherius, Themistius, and Zosimus." These characters are the same as those in the *Sceptical Chymist,* except for the use of Zosimus (a fourth-century Hellenic Egyptian alchemist) as a spokesman for the chymical view instead of Philoponus. The inclusion of the alchemist Zosimus in the dialogue implies that Boyle saw the drawing of good hypotheses as closely related to chymistry, and the close similarity of the cast of characters to that in the *Sceptical Chymist* argues for at least a conceptual, if not a direct, link between the two works. This interrelationship both explains the origin of methodological matter in the *Sceptical Chymist* and highlights one of Boyle's criticisms of the vulgar chymists—their failure to make good, sound hypotheses and theories. Furthermore, it showcases the primary importance of Boyle's thought about the status and deployment of chymistry in developing the methodology of his natural philosophy.

If indeed, as seems likely, the "Requisites" and the early *Sceptical Chymist* were part of a connected set of dialogues, it is intriguing to speculate on what lies behind the change of name from Zosimus to Philoponus. It is tempting to see this alteration as another manifestation of Boyle's recognition of different classes of chymists coupled with his concern not to give offense. Zosimus is a highly respected classical figure of traditional alchemy; perhaps Boyle realized that Zosimus would be seen as the representative of traditional alchemy and of the adepti rather than of the comparatively new Paracelsian system he was actually attacking. He may also have felt it disrespectful or insulting to use Zosimus's name in this connection, potentially drawing the displeasure of the adepti (who felt themselves to be part of the lineage back to Zosimus), by whom he still wished to be "both willingly and thankfully instructed." We shall see that when Boyle *did* use Zosimus as a character in a dialogue, that character's party expresses Boyle's own chrysopoetic convictions and wins the day.

Evolution of Boyle's Chymical Thought 1661–1680

An analysis of the *Sceptical Chymist* is incomplete without an examination of the appendix to the 1680 second edition. A deficiency of much of the earlier literature on Boyle is its concentration on the Boyle of the 1660s, and the implicit assumption that Boyle never developed or changed his mind outside of that period—it is not coincidental that the Boyle of the 1660s fits best with the portrait of a "modern." Boyle's post-1660s books present a troublesome miscellany, consisting as they generally do of "promiscuous" brief tracts on diverse topics, often haphazardly bound together. As such they are difficult to subordinate under a comprehensive historical model or description—indeed, this was a problem with Boyle's writings from the very beginning, as witnessed by the generally unsuccessful attempt of Peter Shaw to "methodize" them. At any rate, the 1680 appendix—entitled *The Producibleness of Chymical Principles*—reflects several important developments. Most strikingly, whereas the *Sceptical Chymist* showed a polite but rather noncommittal respect for the adepti, parts of *Producibleness* are explicitly positive on their behalf. Indeed, one of the points I will argue throughout this study is that Boyle became *more* devoted to traditional alchemy as he grew older, not less—quite the opposite of Boas Hall's claim that he "rapidly eschewed" alchemy.[91]

The publication of *Producibleness* was (as is usual of Boylean texts) not straightforward. It received the imprimatur on 30 May 1677 but bears the publication date of 1680, which, as a rare advertisement declares, is an erroneous date, as the work was actually printed in late 1679 and offered for sale in early 1680.[92] In light of the imprimatur, we may conclude that the contents are a product of the mid-1670s; a time at which, as we shall see, Boyle became particularly preoccupied (and public) with alchemical apologetics.

Producibleness is more orderly and comprehensible than the chaotic *Sceptical Chymist.* The appendix is divided into six sections, each one devoted to the producibleness (i.e., nonelementary nature) of a chymical Principle—Salt, Spirit, Sulphur, Mercury, Phlegm, and Earth. It may not be immediately apparent to the modern reader that Mercury is out of place in this list. The remaining principles constitute the pentad of Spirit, Sulphur, Salt, Phlegm, and Earth that the *Sceptical Chymist* considers preferable to the triad in the case of bodies of vegetable and animal origin.[93] In this pentad, *Spirit* had assumed the role of *Mercury;* therefore, to write of both Spirit and Mercury is strictly redundant. What Boyle does here is segregate the broad-based Mercury of the Paracelsian *tria prima* or pentad under the title of *Spirit,* away from the earlier, strictly metallic and mineral notion of Mercury of the dyad, for which he retains the title *Mercury.* This section on Mercury also differs from the others in terms of its structure, for it contains three independent tracts, which Boyle states "were written for differing *Vertuosi,* at severall times, and on distinct occasions."[94] Besides its heterogeneity in terms of structure, this section is the most interesting from the point of view of alchemical studies, for it shows Boyle coming strongly to the defense of the adepts' metalline Mercury.

[91] Boas, *RBSCC,* 26.

[92] John F. Fulton, *A Bibliography of the Honourable Robert Boyle* (Oxford: Clarendon Press, 1961), 29.

[93] On the unimportance of water and earth, see Beguin, *Tyrocinium,* 23–24; Eleutherius proposes the inclusion of Phlegm and Earth as principles and the renaming of Mercury as Spirit and possibly Sulphur as Oil (thus constructing the pentad); Carneades declares this view "in some respects more defensible then that of the Vulgar Chymists," although it is still susceptible to the same objections regarding the operation of fire in dividing such putative principles from mixed bodies and must be limited to animal and vegetable bodies; Boyle, *Sceptical Chymist,* 286–98, on 287.

[94] Boyle, *Producibleness,* 141. It should have contained *four* tracts, but "by an oversight, some leaves were left behind," a typically Boylean problem; "Publisher's advertisement," [xxi–xxii].

Defending Metalline Mercuries and Secrecy

The first of these mercurial tracts—"Whether Mercury may be obtained from the Metals and Minerals?"—is built around four propositions. In the first three propositions Boyle notes that some receipts for metalline Mercuries cannot be followed or are false, and that some supposed Mercuries actually come from a mercurial ingredient added in the preparation. In the fourth statement, however, Boyle writes clearly that "It is possible to obtain, att least from some metals and Minerals, true running Mercury, that cannot be justly thought to come meerly from the additament."[95] While Boyle only blandly or concessively accepts metalline Mercuries in 1661, here he more forcefully asserts their real existence and describes his change of heart. He writes that "the indisposition I had," and "my longe backwardnes to beleeve" in such Mercuries, have been overcome by both his own experience and that of other trustworthy persons.[96] In the following tract he asserts that he himself possessed "portions of the *Mercury's* of more than one or two metalls."[97] While he claims such Mercuries are Magisteries (a Magistery being a substance prepared from another without any separation of parts) rather than elementary principles, he nonetheless confirms their existence. It should be noted that Boyle's unequivocal affirmation of metalline Mercuries not only links him closely with a key feature of chrysopoetic alchemy but also flatly contradicts the Halls' assertion that Boyle "rejected the whole concept" of metalline Mercuries.[98] Boyle makes a similar affirmation of belief in metalline and mineral Sulphurs in the corresponding section on Sulphur, where he claims them also to be real and preparable, as well as being Magisteries rather than elemental principles.[99]

In *Producibleness* Boyle actually criticizes those who deny the existence of such Mercuries. He censures the "severe opinion" of writers such as Angelus Sala who brand accounts of metalline Mercuries as the empty boasts of cheats.[100] Interestingly, he extends his censure even to "most of the mechanicall Philosophers," who approve of this negative opinion. Here, as in his criticisms of the vulgar chymists, Boyle is consistent in his opposition to

[95] Boyle, *Producibleness,* 151.

[96] Ibid., 153, 161.

[97] Ibid., 188.

[98] Boas and A. Rupert Hall, "Newton's Chemical Experiments," 148.

[99] Boyle, *Producibleness,* on Mercuries as Magisteries, 166 and 171; on Sulphurs, 130 and 137. Clericuzio notes this identification as Magisteries ("Carneades and the Chemists," 88) but slightly misinterprets Boyle's definition (borrowed, as he rightly notes, from Paracelsus) of a Magistery as "substances produced in an intermediate stage during the process of transmutation." An alchemical Magistery is actually any new body produced from some other body without any separation of parts. This is clear from Boyle's examples of Magisteries: "As when Iron or Copper by an acid *Menstruum,* that corrodes and associates it selfe with it, is turn'd into Vitriol of *Mars* or of *Venus* . . . or . . . when *Quicksilver* . . . is by the lasting operation of the fire . . . turn'd into a red Powder, that Chymists call *Precipitate per se*" (172–73).

[100] Boyle, *Producibleness,* 148.

dogmatic pronouncements based upon insufficient evidence. He begins the tract by noting with dissatisfaction how

> divers of the more learned of the Spagyrists themselves, have look'd upon the pretension of other Chymists to the art of making these *Mercury's* as but a Chymical brag: and some judicious modern Writers, applauded therein by most of the mechanicall Philosophers, have proceeded so far, as to explode all these Mercury's of body's as meer *non entia Chymica,* nay some of them have not scrupl'd to censure all those who pretend to have seen or made any of them, as credulous or Impostors.[101]

The context for Boyle's tract and the identity of at least two of the "modern Writers" he criticizes is suggested by Boyle's word choice. The phrase *non entia Chymica* (chymical nonentities), which Boyle uses several times in *Producibleness* in connection with his defense of alchemical arcana, strongly suggests a link to two dogmatic, antialchemical tracts published in the early 1670s. The first is by Werner Rolfinck (1599–1673), a physician and anatomist at Jena. Rolfinck edited and revised the textbook of his teacher Zacharias Brendel (1592–1638) *Chimia in artis formam redacta* (originally published in 1630) and then (quite confusingly) wrote one himself under the same title in 1661.[102] The sixth part of Rolfinck's *Chimia in artis formam redacta* is entitled "De effectis seu operibus imaginariis, & non entibus chimicis," which received greater attention when it was published separately at Jena in 1670 as *Non ens chimicum, mercurius metallorum, et mineralium* (A chemical nonentity, the Mercury of metals and minerals), and reprinted at Berlin in 1674.

The second of the "modern Writers" criticized is an anonymous author, hidden behind the pseudonym of Utis Udenis (presumably from the Greek "nobody at all"). This writer authored the similarly titled *Non-Entia Chymica sive catalogus eorum operum operationemque Chymicorum, quae, non sint in rerum natura nec esse possint* in 1645. Like Rolfinck's work, this tract became more widely known after it was reprinted at Frankfurt in 1670 (in response to the Rolfinck reprint edited by his pupil Georg Wolfgang Wedel) and reprinted again in 1674 together with Rolfinck's work.[103]

Rolfinck's flat denial of the existence of metalline Mercuries elicited a lengthy refutation from Johann Joachim Becher (1635–1682) in the form of a supplement to his *Physica subterranea.* This supplement, published in 1671, was entitled *Experimentum chymicum novum . . . loco . . . responsi*

[101] Ibid., 143–44, 148.

[102] On Rolfinck and Brendel, see *Chymia Jenensis: Chymisten, Chemisten, und Chemiker in Jena* (Jena: Friedrich-Schiller-Universität Jena Verlag, 1989), 21–25 and 17–19, respectively.

[103] Utis Udenis is identified as Michael Kirsten by Ferguson, *BC,* 2:489, and as George Kirsten (no relation) by Thorndike, *History of Magic,* 7:196–97. The 1674 Berlin edition of both *Non-entia* was as a companion to Johann Sigismund Elsholtz, *Destillatoria curiosa* (Berlin, 1674); a further edition appeared in 1683 at Nuremberg.

ad D. Rolfincii schedas de non entitate mercurii corporum and took the form of a rather lengthy defense of metalline Mercuries supported by both traditional authorities and Becher's own experiments.[104] Johann Kunckel's *Chymische Brille contra non-entia chymica* in his 1677 *Chymische Anmerckungen* likewise asserts the reality of metalline Mercuries against Rolfinck; his defense reappeared at London in 1678 in a Latin translation.[105] Thus it is clear that the 1670s witnessed considerable debate regarding the existence of metalline Mercuries, and that these discussions employed Rolfinck's and Utis's phrase *non-entia chemica.* Boyle's tract in *Producibleness* (as well as his *Philosophical Transactions* paper on "incalescent Mercury" to be detailed in chapter 5) should be seen as part of this discussion. It is noteworthy that Boyle sides with the *defenders* of the alchemical Mercuries of the metals.

Producibleness also shows development of Boyle's notions regarding chymical nomenclature and the related issue of secrecy (on which latter topic I will have much to say in chapter 5). In 1661, Boyle reserved some of his harshest criticism for obscure terminology, and although much of his attack was directed toward the textbook writers and Paracelsian systematizers, even chrysopoeian alchemists were not exempt from his "Indignation" at their riddles and *Decknamen.* But by the late 1670s Boyle had undergone at least a partial change of heart. In several places he speaks respectfully of "those who understand the mystical writings of some of the best Chymicall Philosophers of former times," contrasting them with "other Men" and "Vulgar Chymists."[106] Now, in the late 1670s, he does not criticize the obscure and confusing speech of alchemical writers. Indeed, even while making mention of the difficulty of understanding their mysterious writings, Boyle praises traditional chrysopoetic writers, calling Raymond Lull "one of the greatest Chymicall Philosophers" and Basil Valentine the "Intelligentest Spagyrist."[107] He more forcefully distinguishes the warranted secrecy of the adepti from the ambiguity, incoherence, or error of the "vulgar chymists," castigating the latter while casually and nonjudgmentally recognizing the former. For example, in the first Mercury tract, Boyle tries to explain the cause of the widespread disbelief in metalline Mercuries. Some processes "are so darkly deliver'd, that the generality of Chymists cannot sufficiently understand them," for, being "set down in termes of Art," they "are not to be understood but by the authors themselves, or those who are vers'd in the more mysterious parts of *Hermetick* Philosophy." Boyle contrasts these processes with others that "are either false, or . . . unfitt to be trust'd." As an example of the latter, Boyle cites Beguin's *Ty-*

[104] I have used the collected edition of Becher's *Physica subterranea,* ed. Georg Ernst Stahl (Leipzig, 1738), in which this first supplement occupies 281–346.

[105] Johann Kunckel, *Chymische Anmerckungen* (Wittenberg, 1677); *Utiles observationes* (London, 1678).

[106] *Producibleness,* 57.

[107] Ibid., 171 and 113.

rocinium chymicum and censures it.[108] The first kind of processes, those of the adepti, may be obscure, but those of the vulgar chymists are simply *false.* Boyle not only leaves the adepti's processes uncriticized but tacitly commends them when he appends entire processes verbatim out of Paracelsus and Lull as examples. Boyle's new understanding of the "mystie speech" of the chrysopoeian authors becomes strikingly clear in his preface to the process from Lull.

> I must give you this advertisement, that besides the obscurities, and imperfections, that a moderate degree of attention may enable you to discover in these processes understood in the literall sense, there are, if I much mistake not, some affected Equivocations in terms that seem very plain, and free from suspicion of ambiguity. As for instance, though the word *Sal Armoniacum* seem to be of this sort, yet amongst *Hermetick* Philosophers it often signifies not common *Sal Armoniack,* which is far from being able to perform the effects they ascribe to theirs, but is a very differing and much more noble and operative thing, which because it may be sublim'd like common *Sal Armoniack,* they are pleased to call by that name: and though sometimes they give it the title of *Sal-Armoniacum Philosophorum,* yet oftentimes they omitt the discriminating Epithite, especially in Philosophical processes, (that is, such as those wherein they deliver their higher *Arcana,*) of which sort are many of *Paracelsus's* processes, and more (not to say most,) of *Lully's.* What is meant by *Sal Armoniacum Philosophorum,* I think it needless to tell you here, (but may perchance do it on another occasion,) since that composition requires an Ingredient that neither of us is furnish'd with, and that you cannot procure. There may be other Ambiguities in the following processes, that will not be easily discover'd, but by such as are vers'd in the mysterious language, which some would call *canting,* of the *Hermetick* Philosophers.[109]

The implications of this passage are really quite remarkable. Boyle matter-of-factly states why the receipts of the adepti are difficult to follow—without negative comment—and then claims to know the hidden meaning of this philosophical salt, but rather slyly demurs from naming it, thinking it "needless" to do so. Boyle knows the secret meaning and can obtain the hidden ingredient, unlike his addressee, but refuses to name it. The passage is not a condemnation of intentional alchemical obscurity but rather a warning to naive readers. But Boyle is clearly no longer one of those naive readers, for the effect of this circumspect revelation that Boyle is "vers'd in the more mysterious parts of *Hermetick* Philosophy" is that Boyle sets himself up, not yet as an adept, but at least as an advanced Son of the Art. Clearly, Boyle has come a long way in the previous two decades; his frustration at the obscurity and "affected Equivocations" of the adepti evident in the *Sceptical Chymist* is absent here, being replaced with not only understanding but, so it seems, a willingness to continue the mystery. This passage

[108] Ibid., 145–48.
[109] Ibid., 194–95.

opens up two major dimensions of Boyle's alchemy to be pursued in subsequent chapters—his secrecy, and his adoption of the techniques and rules of the "community" of adepti.

Summary and Ramifications

In this chapter I have endeavored to reassess the *Sceptical Chymist*—a work that has too casually been presumed to be antialchemical. I have consciously avoided using other Boylean works and papers (save when strictly necessary) in order to examine the *Sceptical Chymist* by itself, in an attempt to "level the playing field" for the subsequent treatment of Boyle's chrysopoeia. I have shown that the vulgarly received notion of the *Sceptical Chymist* is faulty and cannot be immediately and uncomplicatedly adduced against the thesis that Boyle was deeply involved in traditional transmutatory alchemy.

It should now be clear that Boyle distinguishes classes among chymists, differentiating the lower, "vulgar chymists"—identified here as laborants, iatrochemical pharmacists and textbook writers, and Paracelsian systematizers—from the "Chymical Philosophers" or alchemical adepti. The prime target of the *Sceptical Chymist* is not those we would today tend to call alchemists, but rather the Paracelsian chymists (although Boyle accords some respect to Paracelsus himself). Additionally, Boyle's respect for, belief in, and knowledge of the arcana of the chrysopoetic adepti increased with time; the 1680 appendix to the *Sceptical Chymist* shows an increased commitment to important alchemical ideas and new criticism of those who reject them.

The somewhat surprising result that the *Sceptical Chymist* is aimed primarily at the textbook writers and technicians has several ramifications. Boyle's approval of the "high esoterics" (chrysopoeians) rather than the "low technicians" (Paracelsian chymists and apothecaries) requires greater appreciation and scrutiny in the historiography of the origins of modern science, which may have focused too heavily on the rhetoric (from the Royal Society, for example) regarding learning from tradesmen, artisans, and laborants. Additionally, this same division exists throughout traditional alchemical literature. The books of professed adepti show constant criticism of those sometimes called "Geber's cooks"—the cloudy-minded lower order of aspirants, who labor without understanding the principles of the Art. Thus Boyle's *Sceptical Chymist* might be viewed in one sense as continuous with an alchemical literary tradition—the purification of the art from the errors and confusions of the "vulgar" who do not understand aright.[110]

[110] The distinction between *cognoscenti* and vulgar is of course a common trope applicable by anyone to anything; the same distinction was made within the Paracelsians; see, for example, Allen G. Debus, *The English Paracelsians* (London: Oldbourne Press, 1965), 140–45.

Further, by directing his criticism at the iatrochemical systematizers and textbook writers, Boyle opposes figures who have generally been viewed as having made contributions to the history of chemistry that were not only significant but substantially more positive than those of traditional chrysopoeians. Specifically, the identification of textbook writers as a chief target of the *Sceptical Chymist* seems to conflict with the model of the "didactic origins" of modern chemistry advanced by Owen Hannaway's *Chemists and the Word*. Hannaway's thesis hinges on the contrast between the books of two early-seventeenth-century writers—Andreas Libavius and Ostwald Croll. Croll's *Basilica chymica* (1609) presents a broad, visionary Paracelsian cosmology along with its iatrochemical receipts. Libavius's *Alchemia* (1597) presents chymistry as a set of laboratory operations and preparations cast in a didactic form, partly along Ramist lines. Hannaway argues that chemistry became defined as a discipline when it became a body of information that could be taught. Libavius's "methodized didactic" represents a chief source of the textbook tradition, for Beguin's *Tyrocinium*, the earliest of the textbooks according to Metzger, is in fact extensively cribbed from the *Alchemia*. According to Hannaway, the crucial step for chemistry was the turning away from the "enthusiastic alchemical ideology" of Croll and his "all-embracing chemical world-view" toward the invention of a rigorously verbal, teachable, methodical chemistry.[111]

Hannaway's thesis has received much attention from historians of early modern science, and rightly so. Findings from this chapter, however, bear upon it in several ways. The cosmological Paracelsian scheme visible in Croll (and dependent upon both Paracelsus himself and Quercetanus) did not end with him. I have noted that the midcentury textbooks of Davisson and Barlet share many aspects of Croll's chemical worldview. The textbook tradition is, then, not an unproblematic whole, for it contains lines of descent from both the technical, didactic books of Libavius and Beguin and the grander cosmological world-systems of Quercetanus and Croll. Boyle recognizes both aspects of the textbook tradition and criticizes both, rejecting those who debilitate the status of chymistry by insisting upon its practical operations and technical applications (the Libavius-Beguin dynasty), as well as those who overextend chymistry into extravagant world-systems (the Croll-Quercetanus dynasty). Further, it seems slightly artificial to grant the origin of a rigorous, teachable chemistry to Libavius, for the late medieval scholastic alchemical treatises of the pseudo-Geber, Albert the Great, and others share the clarity, freedom from "mysticism," and didactic tone of the *Alchemia*.[112]

Now if in fact both Boyle and the textbook tradition are important in the development of early modern chymistry—and there is every reason to be-

[111] Owen Hannaway, *The Chemists and the Word: The Didactic Origins of Chemistry* (Baltimore: Johns Hopkins University Press, 1975); Andrew Kent and Owen Hannaway, "Some New Considerations on Beguin and Libavius," *Annals of Science* 16 (1960): 241–50.

[112] See Newman, *Summa Perfectionis*.

lieve that they are—then there is a tension between Hannaway's thesis and Boyle's rejection of the textbook tradition. Initially this apparent conflict must remind us of the complexity of seventeenth-century chymistry and ultimately return us to the question of what set of beliefs constitute chymistry or its subdivisions. Need we choose one path in the maze of seventeenth-century chymistry—whether passing through textbooks, Boyle, or something else—as a primary route to modern chemistry? Does the very attempt to identify a unique trail of descent involve too large a measure of essentialism—the assumption that it must be possible to recognize or denote something as chemistry in any time period—for the contemporary historian of science? To feel obliged to choose either Boyle or the textbook tradition he disliked is a symptom of excessive reductionism. It seems to me that the development of chemistry is neither straightforward nor linear, and that there is sufficient latitude within it to allow for the coexistence of the textbooks' significance and Boyle's as well.

Hannaway's notion of chemical textbooks' inventing chemistry as a discipline by creating a methodized language for teaching is taken up (with some reservations) by Jan Golinski, who applies it along with some postmodernist notions on language and rhetoric to the issue of Boyle's relationship to the textbooks.[113] Golinski claims that the skepticism of the *Sceptical Chymist* is primarily "directed against the entire structure of seventeenth-century chemical discourse" and that Boyle's design was to replace it with "a new order of chemical discourse." Boyle's dissatisfaction with the *tria prima* as elementary constitutive substances is subsidiary; Golinski implies that Boyle purposefully cast the principles as "the axiomatic basis of chemical discourse" in order to attack the "systematic *form* of that discourse." Golinski then identifies "traditional chemical discourse" with that of the textbooks, and Boyle's "new order" with that of his experimental essay.[114]

While Golinski is correct to note that Boyle favored the style of the experimental essay (as exemplified by "Essay on Nitre") for some purposes and disliked the systematic approach to knowledge, his thesis is not without problems. Most notable is the uncomplicated way in which he identifies "traditional chemical discourse" with the textbook tradition as if there were no other kinds of chymical writings. We have seen in fact that Boyle was not anxious "to disown" *all* of "his chemical forebears" but was quite precise in his identification and targeting of specific groups within chymistry for praise or blame. Further, Antonio Clericuzio has rightly pointed out that Boyle's

[113] Jan V. Golinski, "Chemistry in the Scientific Revolution: Problems of Language and Communication," in *Reappraisals of the Scientific Revolution,* ed. David C. Lindberg and Robert S. Westman (Cambridge: Cambridge University Press, 1990), 367–96, esp. 372–77; "Robert Boyle: Scepticism and Authority in Seventeenth-Century Chemical Discourse," in *The Figural and the Literal: Problems of Language and Communication,* ed. Andrew E. Benjamin, Geoffroy N. Cantor, and John R. R. Christie (Manchester: Manchester University Press, 1987), 58–82.

[114] Golinski, "Robert Boyle," on 61–62, 68, 72, 78.

criticism of books like Beguin's is related not to their rhetorical form but rather to their presentation of "chemistry as a purely practical discipline" contrary to Boyle's arguments in favor of a philosophical status for chymistry. Clericuzio also rejects Golinski's contention that the *Sceptical Chymist*'s purpose was to replace "traditional discourse" with experimental essays, and favors instead Boyle's own explicitly enunciated natural philosophical motives regarding the chymical principles.[115]

Indeed, the reduction of natural philosophical enterprises to the rhetorical motivations of "discourse" is profoundly unsatisfying, for it raises and then evades the question of whether the "persuasive power" from "particular narrative and representational techniques" is something writers wish to gain in its own right, or because there is a preexistent, nonrhetorical conviction that the author wishes to narrate, represent, and be persuasive about. In this specific instance, one might question why, if Boyle were really so rhetorically driven, he failed to provide a polished and coherent text for this allegedly rhetorical work, leaving it instead so obviously "maim'd and imperfect." Moreover, one might inquire why, if rhetoric, skepticism, and methodology were primary in Boyle's mind (rather than the status of the *tria prima*), the dialogue format and rhetorical/methodological issues are absent from the originally circulated manuscript version of the *Sceptical Chymist,* and why he bothered to entitle it "Reflexions on the Experiments Vulgarly Alledged to Evince the 4 Peripatetique Elements, or the 3 Chymicall Principles of Mixt Bodies."

There is one more thesis regarding seventeenth-century chymistry with which the present study of the *Sceptical Chymist* may seem to conflict. Allen Debus, in numerous learned books and articles, has convincingly argued for the importance of the Paracelsians in the early modern period. The Paracelsians not only influenced medicine, pharmacy, and chymistry but also constructed a view of the universe and man's place therein that competed with other contemporaneous views to replace faltering classical systems. Indeed, as Debus has argued, the fervent chymical activities of the Paracelsians alone refute the notion of a stagnant sixteenth- and seventeenth-century chymistry whose "revolution" was "delayed" (relative to that of the other physical sciences) until the eighteenth century.[116] Yet these very Paracelsians, so active as iatrochemists, apothecaries, textbook writers, and systematizers, are specifically the "vulgar chymists" whom Boyle attacks while elevating the chrysopoeians above them.

Debus's thesis of the historical importance of the Paracelsians need not conflict with Boyle's rejection of them unless we adopt a positivist and reductionist view of the development of chemistry wherein the only figures of

[115] Clericuzio, "Carneades and the Chemists," 82; this motive correlates perfectly with Boyle's own private explanation for writing the *Sceptical Chymist* mentioned in the "Burnet Memorandum."

[116] Debus, "Iatrochemistry and the Scientific Revolution," in Martels, *Alchemy Revisited,* 51–66.

positive importance are those whom an important historical figure of assumed (or even proven) importance himself viewed as important. We have a signal here again that chymistry contains crosscurrents more conflicting than has been widely recognized. But in another, possibly more important sense, Boyle's discrimination between the Paracelsians and the chrysopoeians can help correct what is an unexpected by-product of Debus's notable success in rightly arguing for the importance of the Paracelsians. The scholarly attention brought to bear upon the Paracelsians has allowed the chrysopoeians to fall yet further out of the spotlight so that among nonhistorians of chemistry Paracelsianism sometimes becomes equated with alchemy as a whole. Boyle's more positive relations with chrysopoeians serve to return attention to that branch of chymistry and its historical importance.

Conclusion

The identification of the real targets of the *Sceptical Chymist* as the Paracelsians primarily interested in chemical medicine makes Boyle's avid interest in transmutational alchemy less jarring. Boyle routinely makes concessions to the chrysopoetic adepti and expresses belief in some of their important tenets. It is now no longer necessary, for example, to strive to show a "rational," nonalchemical origin to Boyle's belief in transmutation lest it conflict with the *Sceptical Chymist.*[117] An enemy to alchemy (as Boyle has generally been portrayed) would not have been so heavily involved in alchemical correspondence and experiments, as we shall see he was. Furthermore, I have shown that although it has been claimed that Boyle appreciated chymical experiments but rejected chymical theories—even Boyle seems to suggest so at times—that view is too simplistic. Boyle's corpuscularian hypothesis is itself derived partly from the alchemical corpuscularian tradition stemming from the Geberian *minima.* His theoretical notions on the problem of mixture seem to derive as well from a thoroughly chrysopoetic source.[118] More dramatically, while he rejects the Paracelsian *tria prima* understood as elementary constituents of all bodies, he never rejects the reality of metalline Mercury and Sulphur (although he denied their elemental nature) and positively asserts them in 1680. Boyle even uses the Philosophers' Stone's transmutation of lead into gold as a defense of his own views. With the *Sceptical Chymist* now more accurately understood and contextualized, I can proceed to the documentation and analysis of Boyle's overtly chrysopoetic pursuits.

[117] Boas, *RBSSC,* 102–9; see also Lawrence M. Principe, "Boyle's Alchemical Pursuits," in *RBR,* 91–105, on 92.

[118] Principe, "Gaston DuClo," 176–78.

CHAPTER III

The *Dialogue on Transmutation,* Kinds of Transmutations, and Boyle's Beliefs

IN THE foregoing chapter I endeavored to show that we cannot view Boyle's *Sceptical Chymist* (especially in view of the 1680 appendix) as a break from traditional alchemical ideas. Indeed, several points of continuity with the alchemical tradition were identified, and the real targets of the *Sceptical Chymist* more accurately defined. My arguments hitherto, however, have been largely negative—showing that Boyle *did not reject alchemy*—rather than overtly positive, that is, describing Boyle's actual views on traditional chrysopoetic alchemy. The assessment of Boyle's views on this topic has been hampered by the fact that Boyle's corpus lacks a sustained and explicit treatment of his views on this subject. Boyle occasionally promises such discussions, but these promises remained unfulfilled.[1] While Boyle did not in fact publish any works devoted primarily to stating his views on traditional alchemy, he did work on several in manuscript. A list datable to ca. 1680 inventories seven works on alchemical subjects:

> A List of my Tracts Relating to the Hermetical Philosophy
>
> Of the Transmutation & Generation of Mettals to which belongs the Degradation of Gold by an Anti-Elixir
> Of the obscure & enigmaticall stile of Chymical Philosophers
> Of the difficulty of understanding the books [of] Hermeticke Philosophers
> Of chemicall arcana Medicinall or of Chimico-Medica arcana
> Of Chymical arcana not Medicinal or of Chymico-Physical arcana
> Of the supernatural arcana pretended by some Chymists
> Of the liquor Alchahest and other Analizing Menstruums[2]

Clearly, Boyle's pen was not idle with regard to "the Hermetical Philosophy"; this list testifies to the effort and continuing interest he directed toward it. I shall return eventually to several items mentioned in this list but now will concentrate on the first tract.

The first item on Boyle's list—*Of the Transmutation & Generation of Mettals*—provides further information about the context of the *Anti-Elixir* tract. As I mentioned in chapter 1, *Anti-Elixir* has been one of those perennial reminders of Boyle's alchemical interests. The list indicates that Boyle did not intend for *Anti-Elixir* to stand alone but wrote it as part of a longer

[1] For example, Boyle, *Origine of Formes and Qualities,* 365.
[2] [3]RSMS, 198:143*v.* On the dating of this document, see Hunter, *Guide,* xxix.

Dialogue on Transmutation.[3] This full work on transmutation appears frequently in catalogs of Boyle's manuscripts drawn up in the 1680s, and there it is specifically cited as a dialogue or as "conferences." The first mention of this *Dialogue on Transmutation* outside of Boyle's private papers was in 1744, with the publication of Thomas Birch's edition of the *Works.* Using Boyle's own catalogs, Birch's assistant, the Reverend Henry Miles, compiled a list of titles that he then checked against Boyle's surviving papers. Miles's list was published in the *Works;* fifteenth on that list stands the *Origin of Minerals and Metals—Conferences about the Transmutation and Melioration of Metals,* and it is cited as not extant.[4]

In our own century, Marie Boas Hall mentions this work briefly in her study of Boyle's chemistry and cites a surviving manuscript of the "Heads of the Dialogue" (my fragment 2a), but she dismisses the work contemptuously, claiming that it "was probably better left unfinished." It is interesting to note that Boas Hall dates the fragment as "probably before 1662" but gives no indication whatsoever of how she arrives at that conclusion, for there is nothing about the document to suggest an early date.[5] Rather, it is likely that Boas Hall's negative view of alchemy and her need to portray Boyle as a modern dictated that the document be assigned to a date sufficiently early to allow it to be dismissed as mere juvenilia. She might well have made the seemingly arbitrary choice of "before 1662" in order to suggest that the *Dialogue* predates the *Sceptical Chymist* (1661) and is thus implicitly repudiated by it. Actually, however, the *Dialogue* is a mature work—more mature in fact (both temporally and compositionally) than the *Sceptical Chymist.* More recently, in the course of cataloging the Royal Society Boyle Papers, Michael Hunter recognized several fragments from the work, noted that the *Anti-Elixir* tract was actually a part of this dialogue, and provided a brief indication of the *Dialogue's* content.[6]

Unfortunately, no complete manuscript of the dialogue has survived. Nonetheless, I have culled twenty-three extant fragments—generally first or second drafts—from Boyle's surviving papers and reconstructed much of the *Dialogue* from them. The previous lack of works by Boyle explicitly on the subject of chrysopoeia is now filled by his *Dialogue on the Transmutation and Melioration of Metals,* presented in print for the first time as an appendix to this volume. It represents the most important primary source for Boyle's views on transmutational alchemy.

[3] The "publisher's note" (which was actually written by Boyle himself; see appendix 1, 226–27, 288–89) states clearly that *Anti-Elixir* was "but a Continuation of a larger discourse." This fact is also noted in Aaron Ihde, "Alchemy in Reverse: Robert Boyle on the Degradation of Gold," *Chymia* 9 (1964): 47–57.

[4] For Boyle's citations of the *Dialogue,* see appendix 1, introduction; on Miles's list, *Works,* 1:ccxxxvi–ccxxxviii; on ccxxxvii.

[5] Boas, *RBSCC,* 102.

[6] Hunter, "Alchemy, Magic and Moralism," 400.

I have deferred all discussion of textual issues—dating, development, coherence, loss, and reconstruction—to appendix 1, where that study serves as an introduction to the *Dialogue* itself. I will now summarize briefly that the dialogue was written over an extended period from ca. 1675 to the mid-1680s, and that the fragments assembled in the appendix are from various stages of the project and should not be imagined to fit together seamlessly like pieces of a single puzzle. Nonetheless, the document as we now have it offers us much useful and otherwise unobtainable information about Boyle. The plot, the setting, the characters, and the method and sources of the argument all provide intriguing insights not only on Boyle's views regarding chrysopoeia but also on his epistomology, his literary style, his method of composition, and his view of the Royal Society.

Synopsis of the *Dialogue*

The idea of a plot and a setting in a Boylean dialogue may strike the knowing reader oddly, for in Boyle's published dialogues there is generally little of either to speak of; however, the storytelling aspect of the *Dialogue on Transmutation* is surprisingly strong. As the text opens, a narrator describes the events that took place at that day's meeting of a "Noble Society." Philetas, a society member, exhibits a piece of gold that had been presented to him as alchemical. Another member, Erastus, objects that unless some test could demonstrate that the metal is really alchemical gold and not merely natural gold, Philetas's offering ought not to be considered anything special. Besides, continues Erastus, the whole story of the Philosophers' Stone is a fiction, and transmutation by projection is an impossibility. This exchange sparks a lively debate. Heliodorus, the society president, questions Erastus's flat rejection of alchemical claims and decides that the society should spend the meeting discussing the matter "since this controversy over the Transmutation of the Baser metals into the more noble ones, especially by Projection, has not been (as far as I know) aired by Authors versed in the principles of the modern Philosophy."[7]

Heliodorus regulates the debate by allowing the assembly to divide itself into three groups—those denying the existence of the Philosophers' Stone (the anti-Lapidists) headed by Erastus, those affirming the Stone (the Lapidists) headed by Zosimus, and a neutral party headed by Eleutherius. The members move about the room into the three groups, and Erastus commences. He cites the unlikelihood of transmutation and how many authors refute it. He rejects accounts of transmutations, because "I am far from believing that these accounts are either true or verified by sufficient testimony."[8] The adepti did nothing useful to commend themselves, and their

[7] Boyle, *Dialogue on the Transmutation and Melioration of Metals,* appendix 1, 243.
[8] Ibid., 247.

identities are often unknown. Further, their theories of the metals and of the operation of the Philosophers' Stone are unclear and undemonstrated.

Zosimus then begins his defense of chrysopoeia by agreeing with Erastus that traditional theories on metals and the Stone are unclear and undemonstrated, but adds that

> just because no chymist steeped in the narrow and lean principles of the Peripatetic Schoolmen has yet given an explanation, it does not follow that one could not be given in the future when a deeper investigation into chymistry is carried out by rational philosophers instructed in both fertile principles and hermetic experiments.[9]

Zosimus illustrates his argument by asserting that people never require that brewers be able to give a philosophical account of fermentation and malting before they will believe that brewers can make beer. Zosimus then praises the value of experimental narratives in proving rare phenomena. "I have never said," explains Zosimus, "that I accept the Philosophical Stone because I can clearly explain how its properties and operations can be performed, but rather because I have been convinced by sufficient testimony that such operations have actually been carried out."[10]

Heliodorus congratulates Erastus and Zosimus on their opening arguments, and Erastus continues by summarizing his four arguments against the Stone. The existence of the Stone is rendered unbelievable, he says, because (1) the changes involved in transmuting lead into gold are too great, (2) the nature of the metals is too stubborn to be thus changed, (3) the proportion of the Stone to the metal is too small for such massive changes, and (4) such changes occur with too great a speed. Zosimus counters that these objections render the Stone merely *improbable* rather than *impossible,* and the reality of these changes, though improbable, is manifested by actual experience.

A lacuna causes us to miss Zosimus's answer to Erastus's first two objections, but the next fragment (fragment 4) begins with his answer to his third and fourth objections. Zosimus employs a corpuscularian hypothesis, arguing that some "parts [of the Philosophers' Stone] may be so commodiously shap'd that their effects are greater than one might easily imagine," and he adduces the rapid curdling of milk by a small quantity of rennet as an analogous example of such a process.[11] This explanation is reminiscent of the theory of the action of the Philosophers' Stone in transmuting mercury into gold as advanced in *Origine of Fixtness* (1675).[12] The next six fragments, constituting a significant fraction of the text, consist of experimental accounts given by various Lapidists in answer to Erastus's objections. Some tell of unusual changes wrought upon metals; others give accounts of trans-

[9] Ibid., 255.

[10] Ibid.

[11] Ibid., 257.

[12] Boyle, *Of the Mechanical Origine or Production of Fixtness* (London, 1675), 15–18.

mutations of base metals into gold. The most striking of these latter is a highly detailed account of the transmutation of lead into gold as witnessed by Boyle himself. I shall discuss all these experiential accounts in the following chapter.

A lacuna of uncertain length intervenes, after which we find the discussion suspended while the president entertains an "illustrious stranger" who has just been admitted to the meeting. The members wander about, and our narrator falls in with Philetas, Cleanthes, and Eleutherius, just as Eugenius, from the neutral party, asks Philetas why the alchemists are so secretive. Here, however, Boyle is specific: he does not question secrecy in regard to the Stone but rather in regard to "Medicinal *Arcana*, . . . which would be highly beneficial to Mankind."[13] Philetas begins to answer, but Erastus breaks in, retorting that Philetas could not possibly give satisfactory excuses for an inexcusable practice. Philetas, "a little surpriz'd at this briskness," asks Erastus to explain himself. Unfortunately, Erastus's explanation is not extant. The next set of fragments, which form a coherent unit, do not follow clearly from the question about medical secrecy but instead assail the authority of alchemical writers by saying that their writings (especially in the *Theatrum chemicum*) have no value, and that the adepti are frauds and hypocrites.

These last fragments are problematic, for the vigor of their attack on the adepti is somewhat at odds with conclusions drawn throughout this study from the rest of the *Dialogue* and Boyle's other writings. It might be argued that they form part of one of the other dialogues listed in Boyle's catalog—but the use of Erastus and the Lapidists links them to the *Dialogue on Transmutation*. Even if they belong to another work, they seem to show a dismissive attitude on Boyle's part toward the adepts. To resolve the tension between these fragments and Boyle's frequent expressions of respect for adepts found elsewhere, we must bear two factors in mind. First, these fragments may have been deleted from the final version, for a toned-down and shortened version of their contents appears in Erastus's opening statement, which is a draft of later date. Second, even if they express some of Boyle's own exasperation in trying to understand chrysopoetic texts, we lack the Lapidists' full rejoinder to them. Consider in comparison how distorted an image of Boyle's attitude toward transmutation we would get if only Erastus's opening statement had survived. Just as the serried ranks of experimental testimonies summoned by Zosimus and his colleagues refute Erastus's a priori arguments, so these last fragments (if they were even retained by Boyle) may have been balanced or dismissed by a Lapidist response.

Finally, we must turn to the *Anti-Elixir* tract, which was the conclusion of the *Dialogue* in its original format, even though it might well have been altered or entirely replaced in the final version. There Pyrophilus tells the assembly how he has himself transmuted gold into a base metal by projec-

[13] Boyle, *Dialogue on Transmutation*, appendix 1, 270.

tion. Erastus, "with a disdainful smile," condemns this as a "useless discovery and a prejudicial practice." Aristander rebukes Erastus "with a Countenance and Tone that argued some displeasure," telling him that his comment is suitable only for "Goldsmiths and Merchants," and unworthy of an "Assembly of *Philosophers* and *Virtuosi,* who are wont to estimate Experiments, not as they inrich Mens Purses, but their Brains."[14] Pyrophilus continues by recapitulating the main objections that Erastus had set down at the outset of the conference, and he adds another of his own, namely, the change in specific gravity that must occur when base metals are transmuted into gold, which is of a much greater density than any other metal. Pyrophilus then delivers an account of how he received from a foreign traveler a minute quantity of powder, which when cast upon molten gold transformed it in a few minutes into a "lump of Metal of a dirty color."[15] This lump was brittle, appeared like old bell-metal, and was endowed with a specific gravity considerably less than that of gold. Thus this "Anti-Elixir" has made a *great change* in a *very stable metal* that *exceeded it in weight by 1,200 times* in a period of only *a few minutes;* all of Erastus's objections are soundly defeated.

Pyrophilus remarks that "if . . . Art has produc'd an *Anti-Elixir,* that is able in a very short time, to work a very notable, though deteriorating, change upon a Metal; I see not why it should be thought impossible that Art may also make a *true Elixir.*" The neutral Aristander then concludes

> I hope, that though it be very allowable to call Fables, Fables, and to detect and expose the Impostures or Deceits of ignorant or vain-glorious Pretenders to Chymical Mysteries, yet we shall not by too hasty and general censures of the sober and diligent Indigators of the *Arcana* of Chymistry, blemish (as much as in us lies) that excellent Art it self, and thereby disoblige the genuine Sons of it.[16]

Heliodorus rises from his seat, approves this conclusion in favor of the Lapidists, and the dialogue ends.

Besides its clear statement of support for traditional transmutatory alchemy, the *Dialogue* is a rich source of information and insight about Boyle, his beliefs and commitments, his milieu and social relations, and the intricacies of his literary style. These rewards amply repay the effort of making a fine-grained analysis of the text.

The Setting

Boyle's choice of the dialogue style for his work on transmutation has several points of significance. The dialogue format allows for distance between

[14] Ibid., 278.

[15] Anonymous "travelers" figure prominently in accounts of alchemical transmutations; in chapter 4 I will suggest the identity of this traveler.

[16] Boyle, *Dialogue on Transmutation,* appendix 1, 280, 288.

the author and the opinions expressed by interlocutors. For a man as reticent to engage in controversy as Boyle, especially in a case as controversial as alchemical transmutation, the dialogue format is particularly appropriate. This dialogue format harkens back to both the classical dialogues of Plato and Cicero and the more recent natural philosophical dialogues of Galileo. Yet another contributor to Boyle's choice of the dialogue, however, is his youthful addiction to romances and his attempts to achieve a smoother prose style by grafting fictional forms onto otherwise drier (natural philosophical or moral) subjects.[17] But further, the *Dialogue on Transmutation* recalls famous alchemical dialogues. The most important of these is the *Turba philosophorum,* a text cast as a meeting of adepti to discuss the Stone. The *Turba* was very highly regarded for its supposed antiquity and authority. Later examples of alchemical dialogues, like Michael Sendivogius's *Dialogus mercurii, alchemistae et naturae* or Egidius de Vadis's *Dialogus inter naturam et filium philosophiae,* are not uncommon; the collection *Theatrum chemicum* alone contains seven.[18]

The location Boyle chooses for the *Dialogue* has special significance. Whereas Boyle's other dialogues have little or no action outside of the intellectual discourse (often little more than a monologue) and take place in a vaguely characterized or undefined space, this dialogue has story elements, movement, and a well-defined setting. Rather than a chance meeting of friends (as in Carneades' garden in the *Sceptical Chymist*) or the more amorphous settings in Boyle's other dialogues, the *Dialogue on Transmutation* presents a discussion carried on at a meeting of a "Most Noble Society" constituted for the purpose of investigating and discussing topics of natural philosophy. The rules and characteristics of this society as cited in the text identify it as either the Royal Society itself or at least a very closely analogous society invented for the sake of the *Dialogue.*[19] For example, the presi-

[17] See Lawrence M. Principe, "Virtuous Romance and Romantic Virtuoso: The Shaping of Robert Boyle's Literary Style," *Journal of the History of Ideas* 56 (1995): 377–97. On the dialogue form of *Things Above Reason,* see Wojcik, *Boyle and the Limits of Reason,* 101; on the *Sceptical Chymist,* see Clericuzio, "Carneades and the Chemists," 81.

[18] Michael Sendivogius, *Dialogus mercurii, alchemistae et naturae,* in *TC,* 4:448–56, and also *MH,* 590–600; Egidius de Vadis, *Dialogus* in *TC,* 2:81–109. Other dialogues in *TC* include Micreris, *Tractatus suo discipulo mirnefindo,* 5:90–101; (pseudo-)Plato, *Quartorum cum commento Hebuhabes Hamed,* 5:101–9 (a dialogue interspersed with Plato and glosses thereupon); Christopher Hornius, *De auro medico philosophorum,* 5:869–912; Bernard Penotus, *Chrysorrhoas, sive de arte chemica dialogus,* 2:139–50 (a dialogue between Theophrastus and Chrysophilus); Thomas Moufet, *De iure et praestantia chemicorum medicamentorum dialogus,* 1:64–89 (a dialogue between Philerastus and Chemista).

[19] The Royal Society is mentioned in the third person in fragment 7, which might militate against the supposition that the *Dialogue*'s society is the Royal Society. That fragment, however, is not compatible with the prologue where the society is introduced, for the majority of fragment 7 has been transferred to Heliodorus's opening comments and therefore is likely to have been deleted from the final version; see introduction to appendix 1.

Secondary literature on the Royal Society is enormous; see, for example, Michael Hunter, *The Royal Society and Its Fellows 1660–1700: The Morphology of an Early Scientific Institu-*

dent, Heliodorus, enumerates rules and regulations for the discussion and polices adherence to them. He first reminds the company that

> this Society is not one of Scholastics but of Physicists; thus it will be unfitting for the disputators to spend time and hours in vulgar arguments supported by the authority of Aristotle or others little versed in metalline experiments, or of those who suppose the existence of substantial forms, or finally with other vulgar doctrines of the Schools that this most noble Society has very often refuted as obscure, begging, or false.[20]

The reader will be amused that at one point Heliodorus actually interrupts the proceedings to correct Zosimus who, by slipping into apologetic and complimentary speech, becomes a "transgressor of the accustomed regulations." Heliodorus reminds him that empty speech is a violation of the society's rules, and tells him start over. Later, when the "illustrious stranger" arrives, he too is made acquainted with the "Laws & Customs practis'd by their Society for the regulation of their Meetings." His presence in itself is not unlike that of the numerous foreign visitors who were admitted to meetings of the Royal Society. Additionally, just as the Royal Society maintained a repository for curiosities (into which Boyle occasionally deposited unique items), Heliodorus advises that Philetas's lump of alchemical gold should be "placed in our Repository," until it can be better examined.[21] In the same way as Philetas is commanded to place the alchemical gold in the repository, so Boyle was requested to deposit there an aeolipile crushed by atmospheric pressure.[22]

Had the *Dialogue* been published, it is doubtful that any reader could have missed its setting as a portrayal of the Royal Society. It might possibly have been intended to serve as an illustration of the functioning and regulation of their meetings. It is also possible that the suppression of the *Dialogue* may have stemmed partly from uncertainty about the suitability of portraying the society discussing alchemical transmutation and, especially, vindicating it at the conclusion. While the society had a number of members deeply involved in alchemical endeavors (Elias Ashmole, for example), many natural philosophers and the general public were becoming increasingly suspicious of alchemical claims. Boyle may have felt that he did not have the

---

tion, 2d ed. (Oxford: British Society for the History of Science, 1994); *Science and Society in Restoration England* (Cambridge, Cambridge University Press, 1981); *The Early Royal Society and the Shape of Knowledge* (Dordrecht: Kluwer Academic, 1991); Marie Boas Hall, *Promoting Experimental Learning: Experiment and the Royal Society 1660–1727* (Cambridge: Cambridge University Press, 1991); K. T. Hoppen, "The Nature of the Early Royal Society," *British Journal for the History of Science* 9 (1976): 1–24, 243–73.

[20] Boyle, *Dialogue on Transmutation*, appendix 1, 243.

[21] Michael Hunter, "Betweeen Cabinet of Curiosities and Research Collection: The History of the Royal Society's Repository," in *Establishing the New Science* (Rochester: Boydell Press, 1989), 123–55. The collection was cataloged in Nehemiah Grew, *Musaeum Regalis Societatis* (London, 1681).

[22] Boyle, *Animadversions upon Mr. Hobbes's Problemata de vacuo* (London, 1674), 94.

right to seem to speak on behalf of the Royal Society in a matter of such controversy.

This setting is rhetorically unique among scientific dialogues—a natural philosophical discussion is carried on in an institutional arena designed specifically for that purpose. The choice of venue invests the discussion with a greater air of authority—a "Society of Physicists," not a group of friends of unstated qualifications, argues an issue. The choice also allows for the enumeration of specific regulations and codes of conduct, many of them dear to Boyle himself as well as to the Royal Society. Some of these codes of discourse were briefly enumerated in the opening dialogue of the *Sceptical Chymist,* though they exist there in a less digested form, and their introduction seems quite artificial for a group of friends talking in a garden; in the dialogue's society, such regulations seem more natural coming from the lips of the president.[23]

The Royal Society and Manners in the Dialogue

While the dialogue's society resonates with the Royal Society, there is one aspect that is at variance with some current notions in the literature regarding disputes at the Royal Society and Boyle's thoughts on managing them. Namely, the *Dialogue*'s society is a much more contentious and voluble arena than recent scholarship would allow for Boyle and the Royal Society. For example, at Erastus's flat denial of chrysopoeia, the society is thrown into confusion, as those "better inclined toward chymistry . . . began to *inveigh* against [Erastus] *vehemently.*" The narrator describes the members' "raging ardor of the spirit that the biting speech of Erastus had been able to stir up." Later, one member answers another "with fierceness in his looks," and Philetas is surprised by Erastus's "briskness." Perhaps most dramatically, the membership divides itself up into factions and actually moves about the room into three separate groups to illustrate their differing allegiances after the manner of a scholastic debate or even of a parliamentary "division."[24]

In his recent *Social History of Truth,* Steven Shapin asserts that scientific discourse was constructed by the Royal Society and its fellows (particularly Boyle) according to the model of civil conversation—disputes and vigorous assertions were to be avoided. The importance of "truth-claims" was thus, according to Shapin, subjugated to the necessity of maintaining gentlemanly decorum; "no conception of truth could be legitimate if pursuing and maintaining it put civil conversation at risk." In order to prevent indecorous dispute and factionalization, members freely and casually accepted "knowledge-claims," rather than subjecting them to the scrutiny,

[23] On the rules of discourse, see Shapin and Schaffer, *Leviathan,* 74–75; Golinski, "Robert Boyle," 66–67; Peter Dear, "Totius in verba: Rhetoric and Authority in the Early Royal Society," *Isis* 76 (1985): 145–61.

[24] Boyle, *Dialogue on Transmutation,* appendix 1, 241, 243–45, 270, 275; emphasis added in the first quotation.

and perhaps offensiveness, of rigorous examination. Shapin asserts that "the Royal Society was a place whose inhabitants had learnt to accomplish the assessment and modification of the great majority of knowledge-claims without doing anything visible as negation."[25]

The *Dialogue* presents quite another picture—here we have "vehement inveighing," "raging ardor," and "fierce looks." Disputants speak sharply to one another, ill temper is displayed, the society splinters into competing sects, and the claims of the anti-Lapidists are firmly and conclusively negated. Erastus vanishes silently after Aristander rebukes him "with a Countenance and Tone that argued some Displeasure," and it is hard to imagine this character not skulking off grumbling after this personal rebuke for "unphilosophical" behavior. Erastus's "knowledge-claims" are rejected. Negation occurs. One might argue that such vigor is a liberty of Boyle's fictitious form rather than a portrayal of a typical Royal Society meeting—but then why would Boyle have bothered to cast his text into what is so clearly an archetype of a Royal Society meeting? Even if we concede *arguendo* that the *Dialogue*'s society does not accurately reflect the situation at the Royal Society, the dialogue is still Boyle's creation and must display a dispute as designed by Boyle. As such it demonstrates Boyle's acceptance of vigorous dispute and negation regarding "knowledge-claims" in the pursuit of truth.

In terms of civility, the *Dialogue* does state that "there are certain rules . . . of justis equity & decency that are to be observd thô it were but for our own sakes, when we speak of the authors of books"; still, disputes are by no means forbidden or discouraged. Boyle objected to neither dispute nor refutations—both were essential for uncovering and demonstrating the truth about the natural world. What Boyle *does* find objectionable is the name-calling and broad condemnations that filled earlier writings. Alchemical disputes were fertile ground for such acrimony, and an anti-Lapidist does not hesitate to take even adepti to task for unfair condemnations:

> 'tis certainly a great want not onely of civility but of common equity to insult over studious & industrious men for not haveing bin succesfull in very difficult atempts, especialy since many of those who were misled by the dark & envious wrightings of these that dispose them, and these insolent adepti are perhaps the onely sort of wrighters in the world that think a man cannot be mistaken or make a paralogisme about an abstruse subject without being a sophister or a Duns.[26]

While the *Dialogue* witnesses Boyle's willingness to admit vigorous dispute and negation in the investigation of natural philosophy, his esteem of

[25] Steven Shapin, *Social History of Truth* (Chicago: University of Chicago Press, 1994), esp. 121–24; see also Mordechai Feingold's essay review, "When Facts Matter," *Isis* 87 (1996): 131–39, and the expanded version in *History of Science,* forthcoming 1998.

[26] Boyle, *Dialogue on Transmutation,* 276. Compare Boyle, *Physiological Essays,* 26–27: "I love to speak of Persons with Civility, though of Things with Freedome." In the *Dialogue on Transmutation,* the civility sometimes wears a bit thin.

appropriate behavior in so doing also appears. In the case of Erastus, variance from Boyle's ideal is used to advantage. Erastus's behavior becomes increasingly sarcastic and unmannerly, he dismisses alchemy out of hand, using harsh language against alchemical believers, he rudely interrupts another speaker, and he responds to Pyrophilus with disdain. This behavior is requited at last when he is rebuked and silenced for being unphilosophical and mercenary—the same charge he had earlier leveled at chrysopoeians. Erastus's comportment underscores his eventual defeat; heated debate and dissension are acceptable in the pursuit of truth, but unnecessary rudeness is not.

Boyle's view of chrysopoeia is made clear by the defeat of Erastus and his anti-Lapidists, which is far more resounding than the lukewarm (and incomplete) refutations of the *Sceptical Chymist.*[27] Boyle leaves no doubt about who is in the wrong—this dialogue, particularly in its conclusion, takes on perhaps the broadest epideictic tone of all Boyle's writings. Thus Boyle's position is clearer here than in most of his other works on natural philosophy. The *Dialogue on Transmutation* unambiguously expresses Boyle's firm belief in the reality of the Philosophers' Stone and chrysopoetic transmutation, as well as his dissatisfaction with those who are ready to dismiss it out of hand as a nonentity.

The Characters

The names that Boyle chooses for his interlocutors are not arbitrary—some further insights are to be gleaned by a consideration of them.[28] A few characters are old friends. Eleutherius occurs in most of Boyle's dialogues; his name, deriving from the Greek *eleutheria* (liberty, freedom), sufficiently defines his role as the character to be convinced. Eleutherius is the one willing to be won over by cogent arguments; he is thus in a sense Boyle's analogue to Galileo's Sagredo. Eugenius ("good spirit") joins Eleutherius on the neutral party, and Themistius, the Aristotelian who makes a brief appearance in the *Sceptical Chemist,* shows up in the list of anti-Lapidists, but any fragments that mentioned him are now lost.[29] Philoponus ("lover of work"), like Themistius, made a brief appearance in the *Sceptical Chymist,* where, although he makes only a short speech in the fragmentary first dialogue as published, he defends the chymists. Here he continues his alchemical affections by siding with the Lapidists, and it is in his mouth that Boyle places the

[27] On the incompleteness of these refutations, see Boyle, *Sceptical Chymist,* 432–36.

[28] See fragment 1, *Dialogue on Transmutation,* appendix 1, for a list of the characters.

[29] Themistius continues his Aristotelian cause in Boyle's *An Examen of Antiperistasis* (printed with *History of Cold* [London, 1665]), and somewhat less dogmatically in *A Sceptical Dialogue about Cold* (published with *Saltness of the Sea* [London, 1674]).

most significant experiential account.[30] Pyrocles and Pyrophilus are also reused characters.[31]

The Lapidists' ranks are filled with several new characters. Zosimus, who heads the party, carries the alchemically respected name of a famous Hellenic Egyptian alchemist of the fourth century.[32] It will be remembered that Zosimus made an appearance in the 1650s fragment of "The Requisites of a Good Hypothesis," and if we accept the link between that dialogue and the *Sceptical Chymist,* he was later replaced by Philoponus. The characters of Arnoldus, Guido, and Bernardus are all important chrysopoetic authorities: Arnold of Villanova (thirteenth century), Guido de Montanor (fourteenth or fifteenth century), and Bernard Trevisan (fifteenth century).[33] Philetas, who begins the whole dialogue by proffering a piece of alchemical gold, has special importance. The classical Philetas was not an alchemist but a poet and grammarian, a native of Cos, born about 320 B.C., and a tutor to Ptolemy Philadelphus at the court of his father, Ptolemy I. Why would Boyle choose so obscure and unlikely a character to initiate this alchemical discussion? Philetas's chief claim to fame rests not upon his poetry but rather upon his physical condition—severe emaciation caused by excessive study. Athenaeus's *Deipnosophistae* portrays Democritus advising a friend:

> Like Philitas of Cos, therefore, who pondered what he called "the deceitful word," you run the risk some day of being quite dried up, as he was, by these worries. For he became very much emaciated in body through these studies, and died, as the epitaph on his monument shows: "Stranger, I am Philetas. The deceiving word caused my death, and studies of riddles late at eve."[34]

[30] Philoponus appears in other locations where he is not a chymist, e.g., *A Sceptical Dialogue about Cold,* although he has very little to say on that occasion.

[31] Pyrophilus was the addressee of early works such as *Some Considerations Touching the Usefulnesse of Experimental Natural Philosophy* (1663) and *Origine of Formes and Qualities* (1666), where he is identifiable as Boyle's nephew Richard Jones. In the *Dialogue* it is doubtful that he is anything more than a reused Boylean interlocutor. Pyrocles appears in *Things Above Reason* (1681).

[32] Several manuscripts attributed to Zosimos occur in Marcellin Berthelot, *Collection des anciens alchemistes grecs,* 3 vols. (Paris, 1887–1888; reprint, London: The Holland Press, 1963), 117–242. For an improved edition of the works of Zosimos, see Michele Mertens, *Les alchimistes grecs* (Paris: Les belles lettres, 1995); "Project for a New Edition of Zosimus of Panopolis," in *Alchemy Revisited,* ed. Z.R.W.M. van Martels (Leiden: Brill, 1990), 121–26. See also Zosimos, *On the Letter Omega* (Missoula, MT: Scholar's Press, 1978).

[33] For the classical details regarding these figures, see their entries in *BC* and the references therein. For more recent work, see (on Arnold) Antoine Calvet, "Les *alchimica* d'Arnaud de Villeneuve à travers la tradition imprimée," in *Alchimie: Art, histoire, et mythes,* ed. Didier Kahn and Sylvain Matton, Textes et Traveaux de Chrysopoeia I (Paris: S.E.M.A., 1995), 157–90; "Alchimie et joachimisme dans les *alchimica* pseudo-Arnaldiens," in *Alchimie et Philosophie à la Renaissance,* ed. Jean-Claude Margolin and Sylvain Matton (Paris: Vrin, 1993), 93–107; and "Le *De vita philosophorum* du pseudo-Arnaud de Villeneuve," *Chrysopoeia* 4 (1990–1991): 34–79; (on Bernard), Newman, *Gehennical Fire,* 103–6 and 158–61.

[34] Athenaeus, *Deipnosophistae* 9.401; quoted from *Athenaeus,* trans. C. B. Gulick, Loeb Classical Library (London: William Heinemann, 1930), 4:319.

The most common anecdote about Philetas describes how he became so thin from study that he had to wear leaden soles to prevent being overturned by the wind.[35] Philetas's reputed physique makes it likely that he is a humorous reference to Boyle himself. John Evelyn describes Boyle as "rather tall and slender of stature, for the most part valetudinary, pale and much emaciated," and Caspar Lindenburg writes that he "looked as thin and miserable as a skeleton."[36] Boyle corroborates his similarity to the unfortunate Philetas of Cos in referring to the apparently popular explanation of his poor health "that I brought myself to so much Sickliness by over-much Study."[37] Olaus Borrichius wrote to Boyle that "your body is worn away and exhausted by sleepless nights."[38] It is tempting to link the "deceitful word" and "studies of riddles" that so afflicted Philetas to alchemy. This use of "Philetas" is a reminder of Boyle's sense of humor, one of those "too human" characteristics that tend not to appear in portrayals of important historical characters.

Among the anti-Lapidists are the classical figures Crates (a fourth-century Cynic) and his pupil Cleanthes (a Stoic), as well as the unidentifiable Julius; of these, only Cleanthes appears in the extant fragments. Two additional characters represent modern authors. Guibertus is certainly Nicholas Guibert, a physician from Lorraine (ca. 1547–ca. 1620), who was a vehement opponent of alchemy. He had studied alchemy for many years in his youth but became convinced of its falsity and then wrote several antialchemical works whose strong sentiments are apparent from their titles alone—for example, *Alchemy, Together with Its Fallacies and Delusions Which Make Fools of Men, Impugned and Assailed So Strongly by Reason and Experience, That It Shall Never Again Be Able to Raise Itself Up* and *On the Burial of Transmutatory Alchemy.*[39] His antialchemical tirades provoked a bitter battle with Andreas Libavius.

The name of the leader of the anti-Lapidists, Erastus, refers to Thomas Erastus (1523–1583), an anti-Paracelsian who wrote a lengthy condemnation of Paracelsian medicine and principles in which he referred to Paracelsus as a "grunting swine."[40] More germane to Erastus's role in Boyle's

[35] Ibid. 12.552; on 5:507. The story is repeated in Claudius Aelianus, *Varia historia* 9.14; a seventeenth-century popular edition exists: *Claudius Aelianus his Various History,* trans. Thomas Stanley (London, 1665), on 178.

[36] John Evelyn, *Diary and Correspondence,* 4 vols., ed. William Bray (London, 1854), 3:351; Z. C. von Uffenbach, *Merkwürdige Reisen durch Neidersachsen, Holland und Engelland,* 3 vols. (Ulm and Memmingen, 1753–1754), 2:68: "Herr Boyle seye auch sehr schwächlicher Constitution gewesen, und habe so dürr und elend, wie ein Sceleton, ausgesehen."

[37] Boyle, *Medicinal Experiments* (London, 1692), preface. Boyle claims, however, that this explanation of his sickliness is not true.

[38] Olaus Borrichius to Robert Boyle, 30 March 1664; BL, 1:88–89*v.*

[39] *Alchymia ratione et experientia ita demum viriliter impugnata & expugnata, una cum suis fallaciis et deliramentis, quibus homines imbobinarat, ut numquam in posterum se erigere valeat* (Strasbourg, 1603) and *De interitu alchimiae metallorum transmutatoriae* (Toul, 1614).

[40] On Erastus, see Charles D. Gunnoe, Jr., "Thomas Erastus and His Circle of Anti-

dialogue is his 1572 essay denying transmutation, which eminently qualifies him for the leadership of the "Adversaries of Chymistry."[41] Gaston DuClo's 1590 *Apologia chrysopoeiae,* which Boyle cites approvingly in the *Sceptical Chymist,* was written as a response to Erastus's denial of chrysopoeia.

An ambiguity regarding the name of this chief anti-Lapidist highlights Boyle's dissatisfaction with his opinion. In part of the prologue and in the *Anti-Elixir* tract, this character is called not Erastus but Simplicius. Boyle's translator, confused at this exchange of names midway through the document, scribbled in the margin: "note well, for just a moment ago he was called Erastus." Since this chief anti-Lapidist is no promoter of Aristotle, the reference is not to the sixth-century commentator but rather to that hapless character so often made to play the fool in Galileo's dialogues. In choosing such a loaded name, and portraying his engaging in inappropriate behavior, Boyle underscores the character's error and presages his eventual defeat.

Only in the *Dialogue on Transmutation* does Boyle use modern authors, as opposed to little known ancients or fictitious persons, as characters. This distinguishing feature alone makes it most regettable that no fragments exist where any of the modern chrysopoeians speak; it would have been fascinating to see whether Boyle would have used their names in only an evocative, generic manner or have summoned up the shades of these authors to present their own views. In the case of Erastus, Boyle does preserve the sneering, declamatory style of the historical figure, and the opinions of the interlocutor Erastus do conform to those of the historical Thomas Erastus. Though it is rather otiose to conjecture on this point since we lack a complete text where all these characters speak, the inclusion of authentic modern authors implies that the *Dialogue on Transmutation* was conceived of as a meeting of minds or battle of the books where rival, noncontemporaneous authors set their ideas against one another. Such an arrangement is highly uncharacteristic of what we know of Boyle's writing, for he generally avoided citing authors (living or dead) by name in either praise or criticism and often protested (primarily rhetorically) that he did not bother to read other authors closely. Additionally, the use of such evocative figures makes Boyle's *Dialogue* resemble even more closely the traditionally respected alchemical *Turba philosophorum* where classical authorities are the interlocutors.

Boyle and Varieties of Transmutation

It is clear from the *Dialogue* that Boyle sides strongly with the Lapidists in favor of the reality of alchemical transmutation. One crucial point about

---

Paracelsians," in *Analecta paracelsica,* ed. Joachim Telle (Stuttgart: Franz Steiner Verlag, 1994), 127–48, and Johannes Karcher, "Thomas Erastus (1524–1583), der unversöhnliche Gegner des Theophrastus Paracelsus," *Gesnerus* 14 (1957): 1–13.

[41] Thomas Erastus, *Explicatio quaestionis famosae illius, utrum ex metallis ignobilibus aurum verum et naturale arte conflari possit,* in vol. 2 of *Disputationum de medicina nova Philippi Paracelsi,* 3 vols. (Basel, 1572).

which there should be no doubt is the *type of transmutation* in which Boyle expresses belief. Although historians have long recognized Boyle's general belief in the possibility of metallic transmutation, they have in most cases striven to distinguish such belief from the transmutations spoken of by traditional alchemists by depicting Boyle's belief as no more than a logical consequence of his corpuscularian hypothesis—and as such, disconnected from the alchemists' transmutation.[42] Thus historians have been able to have their cake and eat it too by concluding that "Boyle was always interested in this possibility [of transmutation], but not in the sense of alchemical projections."[43]

Boyle's acceptance of "nonalchemical" transmutation has been instanced with an oft-quoted passage from the *Origine of Formes and Qualities.*

> since Bodies, having but one common Matter, can be differenc'd but by Accidents, which seem all of them to be the Effects and Consequents of Local Motion, I see not, why it should be absurd to think, that . . . almost of any thing, may at length be made Any thing: as, though out of a *wedge* of Gold one cannot immediately make a *Ring,* yet by either Wyre-drawing that Wedge by degrees, or by melting it, and casting a little of it into a Mould, That thing may easily be effected.[44]

This passage relies upon Boyle's well-known commitment to the notion of a "catholick matter" that gives rise to diverse substances by being formed into corpuscles of differing sizes and shapes, and endowed with differing motions, and Boyle explicitly ties this notion to the explication of metallic transmutation elsewhere.[45] The *Dialogue on Transmutation,* however, does show Boyle's belief in transmutations explicitly "in the sense of alchemical projections." It is worthwhile now to examine the different understandings of "transmutation" in Boyle's mind and in chrysopoetic alchemy; we shall see that Boyle adopts exactly the distinctions maintained by chrysopoeians and affirms the reality of each.

Universal and Particular Transmutations

The means of effecting transmutation is the standard way of classifying transmutations. Chrysopoeians thus commonly divide transmutations into *universal* and *particular.* Universal transmutation occurs only by means of the Philosophers' Stone—the "universal metallic medicine" and the adepts' greatest secret. This is the most powerful kind of transmutation, and it is

[42] Boas, *RBSCC,* 102–6. A similar labeling of Boyle's belief in transmutation as "a consequence of [his] being a corpuscularian" occurs in More, *Life,* 221; similarly, see Sarton, "Boyle and Bayle," 160–63, and Kuhn, "Robert Boyle and Structural Chemistry," 21–23. Clearly, the fact that many transmutatory alchemists themselves conceived of matter in corpuscularian terms further undercuts the notion that Boyle's corpuscularianism somehow distances him from the alchemical tradition (see chapter 2).

[43] Maddison, *Life,* 104.

[44] Boyle, *Origine of Formes and Qualities,* 95–96.

[45] Ibid., 350.

called "universal" for two reasons. First, the completed Stone is capable of transmuting *any* metal into gold. Second, the Stone is liable to infinite "multiplication"; that is, once a single sample of the Stone has been prepared, it can be augmented in both quantity and quality. For example, if a successful alchemist prepared half an ounce of the Philosophers' Stone, he could then proceed to "multiply" the Stone in quantity by adding the appropriate matter and redigesting his mixture for a specified amount of time; the Stone would then transform the added material into its own substance, and the alchemist could thus obtain (say) a full ounce of the Stone, and so on. Alternatively, he could augment the Stone in quality—that is, transmuting power. In this case, if the alchemist's Stone was at first capable of transmuting ten times its weight of metal into gold, the alchemist could join that Stone to a somewhat different matter, digest it appropriately, and thus prepare a Stone that could then transmute a hundred times its weight of base metal. These processes could be carried on ad infinitum.[46]

Particular transmutations are a somewhat easier matter to effect since they do not require the Philosophers' Stone. The means of transmutation in this case are a variety of substances—some liquids, some solids—called *particularia,* whose power was far inferior to that of the Stone. While the Stone could transmute all metals, these *particularia* could transmute only one or a few metals; many could advance only silver to gold and remained too feeble to ennoble the base metals at all. The universal Philosophers' Stone might be augmented to a stupendous degree of virtue; the portion of the Stone supposedly found by John Dee and Edward Kelley was able to transmute 272,330 times its weight of base metal into gold.[47] *Particularia,* on the other hand, were incapable of augmentation either in quality or quantity and could transmute at best only five or six times their weight of metal. Some *particularia* were of such low potency that their authors warned they were not even profitable to prepare.[48]

Clear illustrations of this difference between universal and particular transmutations are given in the heterogeneous corpus that gained great popularity in the seventeenth century under the name of Basil Valentine. Valentine's "Stone of Fire" mentioned at the conclusion of his *Triumpf-Wagen antimonii* (1604) is an example of a *particularium.* After he provides ob-

[46] On multiplications, see, for example, Eirenaeus Philalethes (George Starkey), *Introitus apertus ad occlusum regis palatium,* in *MH,* 647–99, on 697; George Ripley, *Compound of Alchymie,* in *TCB,* 107–93, on 181–83, *The Bosom Book,* in *Collectanea chemica* (London, 1684; reprint, London, 1898), 121–47, on 141–42; and Gaston DuClo, *De recta et vera ratione progignendi lapidis philosophorum,* in *TC,* 4:388–413 on 408–10.

[47] Elias Ashmole, Annotations to *TCB,* 478–84, on 481.

[48] On the profitability of particular transmutations, see, for example, the remarkable *Coelum philosophorum* (Dresden and Leipzig, 1739), esp. 60, 125–26, where the anonymous author actually provides calculations of the profit margins for various processes based on the cost of materials, glassware, and charcoal and the value of the noble metals produced! See also DuClo, *De triplici praeparatione argenti et auri,* in *TC,* 4:371–88, on 374–75, where he dissuades readers from bothering with particulars because they are not worth the cost.

scure directions for preparing the Stone of Fire from antimony, Valentine notes that

> the Stone of Fire does not tinge [i.e., transmute] universally like the Philosophers' Stone . . . but rather only particularly, namely, Luna [silver] into Sol [gold], and likewise with tin and lead; Mars [iron] and Venus [copper] it does not touch; . . . further, one part of this Tincture can transmute no more than five parts of metal . . . unlike the true, ancient, famous Stone of the Philosophers that can do infinitely much; likewise, it cannot be brought so high in its augmentation.[49]

In annotating this passage, Valentine's commentator Theodore Kerckring illustrates the difference between universal and particular transmuting agents, or Tinctures, by exclaiming that "as far as the heavens are raised above the earth, so far is the true Philosophers' Stone from this Stone of Fire."[50] The third book of the Valentinian *Letztes Testament* deals with the "Universal of this whole world," the Philosophers' Stone, while the fourth book covers the preparations of the "*particularia* from all seven metals."[51]

The pseudo-Lull also distinguishes the Philosophers' Stone as an agent of universal perfection (in all things) from the *particularia* whose action was much more limited.[52] Philip Muller, a professor at Freiburg, includes a chapter specifically on *particularia* ("to be used by those who can spare the cost, time, and labor") in his *Miracula chymica,* immediately following his advice on making the Philosophers' Stone.[53] Even Werner Rolfinck, a fierce opponent of transmutation, recognizes the same distinction into universal and particular in his condemnation of alchemy, where he divides chrysopoeia into these two categories and then denies the reality of each in turn.[54]

The means of applying these two classes of transmuting agents also differed. Transmutation using the Philosophers' Stone was carried out by *projection*—a tiny fragment of the Stone is projected (i.e., cast upon, from *proiecere*) into a crucible full of molten lead (or some other metal) or hot mercury, and the base metal is thereby converted in a few minutes into pure gold. The *particularia* are very rarely projected; rather, they are generally ground, digested, or fused with the metal to be transmuted. One subset of

[49] Basilius Valentinus, *Chymische Schrifften,* 2 vols. (Hamburg, 1677; reprint, Hildesheim: Gerstenburg Verlag, 1976), 1:442. "Es *tingirt* aber der Stein *Ignis* nicht *universaliter,* wie der *Lapis Philosophorum,* . . . sondern er *tingirt particulariter,* nemlich *Lunam in Solem,* neben dem Zinn und Bley / *Martem* und *Venerem* läst er wol bleiben / . . . / auch kan dieser Tinctur ein Theil über fünff Theil nicht *transmutiren* / . . . da dargegen der rechte uhralte grosse Stein der Weisen / unzehlich viel thun kan / dergleichen kan er in seiner Augmentation so hoch nicht bracht werden . . ."

[50] Theodore Kerckring, *Commentarius in currum triumphalem Basilii Valentini* (Amsterdam, 1671), 315.

[51] Basilius Valentinus, *Schrifften,* 2:227–318.

[52] (Pseudo-)Lull, *Testamentum,* in *BCC,* 1:763, 776–77.

[53] Philip Muller, *Miracula chymica* ([Wittenburg], 1611), 39–48, on 39.

[54] Rolfinck, *Chimia,* 22–26.

particulars, the *gradators,* are special solvents supposedly able to transmute some portion of a metal in the course of dissolving it. In summary, chrysopoeians recognized two general kinds of transmutations: the universal, accomplished by projection of the Stone, and the particular, accomplished by the application of a variety of low-potency transmuting mixtures.

Boyle's Transmutations

Returning to Boyle, we find that he too explicitly differentiates *two separate types of transmutation.* A key experiment for Boyle's chrysopoetic pursuits occurs in *Origine of Formes and Qualities* (1666). There Boyle describes the preparation and use of a liquid he calls his *menstruum peracutum,* a highly corrosive solvent he prepared by distilling together a mixture of aqua fortis (nitric acid) and butter of antimony (antimony trichloride). Boyle records that when he poured this solvent over pure gold, the gold dissolved quietly, leaving behind a white powder. This white powder, when fused with borax (as flux) provided tiny metallic globules that Boyle found to be silver—thus Boyle concluded that his *menstruum* was capable of converting gold into silver, a metallic transmutation, "though Deteriorating." Boyle asserts that "there may be a real Transmutation of one Metal into another," but quickly points out that "I speak not here of Projection, whereby one part of an Aurifick Powder is said to turn I know not how many 100 or 1000 parts of an ignobler Metal into Silver or Gold, . . . because, though Projection includes Transmutation, Transmutation is not all one with Projection, but far easier then it."[55]

This initially rather confusing statement becomes clearer when we recognize that Boyle is recapitulating the traditional distinction between particular and universal transmutations. Any conversion of one metal into another classifies as transmutation, but projection refers only to the use of the "Aurifick Powder" or Philosophers' Stone, which by its universality and potency can transmute "many 100 or 1000 parts of an ignobler Metal." While Boyle claims success in this simple transmutation, he also acknowledges that such transmutation is "far easier" than projection, which requires the preparation of the Philosophers' Stone—the most difficult and most hidden of all alchemical processes. We shall return to Boyle's *menstruum peracutum* and its transmuting properties shortly.

Boyle's use of the traditional alchemical distinction into universal and particular is further evidenced by his recognition that a large array of *particularia* can be prepared. In fact, he assembled a collection of "som things that relate to those Particulars as Chymists are wont to call them"; only the preface to these processes survives (see appendix 2, text D). This collection was composed of "Experiments communicated to me that were affirm'd by the Imparters to be capable of exalting, ripening, or at least of separating

[55] Boyle, *Origine of Formes and Qualities,* 365.

from inferiour metalls & minerals some portions of one or both of the two noblest, nor did I take them all upon trust having had tryals successfully enough made of more than one of them."[56]

Boyle assembled this collection for several reasons. The first he mentions is to "convince diverse virtuosi" that they are wrong "to think that there is really no gold or silver to be obtain'd by any Chymical Art from the baser mettals, or from Minerals." But another reason is to allow "poor Artists" to make themselves a living by them, and here Boyle's understanding of the lesser potency of particulars is made clear when he writes that "the greatest number of Particulars are not considerably Lucrative unles made in great quantitys." This manuscript may be related to the "Hermetic legacy" that Boyle mentions in a similar preface published by Thomas Birch in 1744 and included in appendix 2. There Boyle again refers to "those gainful [experiments], that chemists call *particulars*."[57] Some comments in this published text parallel those of the above manuscript, although the manuscript more explicitly recounts Boyle's reasons for compiling the collection. Birch dates the text to 1689 (Boyle says it was written after he had "grown old"), noting that the collection of receipts to which it was originally annexed was not to be found among the Boyle papers. Since these texts from the latter part of Boyle's life retain the usage and distinctions made in *Origine of Formes,* it is clear that Boyle's use of the distinction between the *particularia* and the universal Stone persisted throughout his life.

The subset of particulars known as *gradators* are also mentioned specifically by Boyle. These are corrosive solvents that are capable of transmuting, or "graduating," a small portion of the metals they dissolve into a more noble one. In a manuscript that is probably an unpublished part of *Usefulnesse* dating from the 1660s, Boyle notes that "various liquors can be made . . . which do not only separate the silver which they dissolve from the gold which they leave behind undissolved, but are also (as the chemists speak) *gradators,* that is, they are designed to advance some parts of the silver into *luna fixa* or gold."[58] While all these examples speak of Boyle's recognition of and belief in the transmuting powers of particulars, the *Dialogue on Transmutation* primarily concerns universal transmutation, that is, projection of the Philosophers' Stone. Throughout the dialogue the characters repeatedly and explicitly remind the reader that they are discussing transmutation by *projection.* And as the *Dialogue* amply dem-

[56] Appendix 2, text D.

[57] *Works,* 1:cxxx; appendix 2, text C.

[58] BP, 24:399–419, on 415. *Luna fixa* is a metal that has the weight and chemical properties of gold (i.e., resistance to aqua fortis) but lacks its color. In "Un-succeeding Experiments," in *Physiological Essays,* Boyle again mentions "*Luna fixa* (as Artists call that Silver, which wanting but the tincture of Gold abides the trial of *Aqua fortis,* &c)," 70. It is defined also by Daniel Georg Morhof, *Epistola ad Joelum Langelottum de transmutatione metallorum,* in *BCC,* 1:168–92, on 178. Boyle mentions gradators as well in "Of Un-succeeding Experiments," in *Physiological Essays,* 68.

onstrates, Boyle clearly and forcefully affirms the reality of this universal transmutation.

Thus, to summarize this account of the varieties of transmutation and Boyle's response to them, we have seen that Boyle recognized, maintained, and utilized the traditional alchemical distinctions between particular and universal transmutations. In *Origine of Formes* he claims his own success with a particular transmutation ("though Deteriorating"), in an unpublished part of *Usefulnesse* he asserts the reality of gradators, and late in life he compiled collections of *particularia.* In the *Dialogue,* Boyle specifies universal transmutation ("by projection"), which he also affirms. Thus we find Boyle affirming both sorts of transmutations distinguished by chrysopoeians. Further, it is not tenable to see Boyle's belief in transmutation as merely an unavoidable consequence of his corpuscularianism, somehow unrelated to the alchemical transmutations. If this were so, there would be no reason for Boyle to adopt the divisions and even the very terminology used by chrysopoeians. Moreover, the distinction between a corpuscular basis for transmutation and some supposedly "alchemical" one rests upon a false dichotomy, for alchemical accounts of transmutations often employ explicitly corpuscularian terms. Thus the transmutation treated in the *Dialogue* is not *merely* a corollary of Boyle's corpuscularianism; it is identical with the central tenet of traditional chrysopoeian alchemy.

Boyle's Menstruum peracutum, *the Soul of Gold, and Transmutation by Transplantation*

In spite of the *Dialogue*'s concentration on projective alchemy using the Philosophers' Stone, the first extended discussion of transmutation (in Erastus's opening speech) actually concerns a *particular* means of chrysopoeia. In examining this section of the *Dialogue,* we should first return to the account of Boyle's *menstruum peracutum,* which is, in effect, an "antigradator." The significance of this experimental account, however, extends beyond its exemplification of a particular transmutation attested by Boyle. If instead of focusing on the *residue* of white powder that Boyle was able to fuse into a white metal and identify as silver, we direct our attention to the *extract,* we shall see that this process is tied to a special *particularium* mentioned not only in the *Dialogue* but widely in alchemical literature and elsewhere in Boyle's corpus.

When Boyle explains the ability of his *menstruum peracutum* to transmute yellow gold into white silver, he suggests that

> however the Chymists are wont to talke irrationally enough of what they call *Tinctura Auri,* and *Anima Auri;* yet, in a sober sense, *some such thing* may be admitted . . . there may be some more noble and subtle Corpuscles, being duely conjoyn'd with the rest of the Matter, whereof Gold consists, may qualifie that Matter to look Yellow . . . yet these Noble parts may either have their Texture destroy'd by a very piercing Menstruum, or by a greater congruity with its Cor-

puscles, then with those of the remaining part of the Gold, may stick more closer [*sic*] to the former, and by their means be extricated and drawn away from the latter.[59]

As an illustration of his thought, Boyle then compares gold to vermilion (mercuric sulphide), which although it acts (in sublimation, for example) as "one Physical Body," yet one can reveal its compounded nature by heating it with salt of tartar (potassium carbonate) to separate its component mercury and sulphur; clearly, Boyle's meaning is that gold is compounded of distinct and separable substances, and his *menstruum* can separate them.[60] It is possibly significant that out of all possible illustrations, Boyle chooses vermilion as his example; for that substance, being composed of mercury and sulphur, bears an analogical identity with gold and the other metals that are composed of Mercury and Sulphur.

Boyle's usage of the phrase *anima auri,* or soul of gold, in reference to his extract of gold links it to the alchemical Sulphur of gold, for Sulphur and soul are considered synonymous in alchemical parlance. Boyle in fact uses the term *Sulphur* when referring to this extract in a brief intimation of the same process made in the *Sceptical Chymist.*[61] Boyle also refers to these "noble and subtle Corpuscles" as "Tinging parts"; and chrysopoeians generally attribute the color of gold to its mature Sulphur.[62]

Boyle's process in *Origine of Formes* has many precedents in alchemical literature. An example of such an extraction of the yellow "soul of gold" occurs as the first process in the Valentinian treatise on *particularia.* There the pseudonymous author describes a deeply colored extract made from gold that leaves behind a white residue; this colored extract (there termed the *sulphur, tincture,* and *soul of gold*) is to be used to "tinge" silver into gold.[63] I have already shown elsewhere how Boyle's process of extracting a tincture of gold with the *menstruum peracutum* draws its origins from Valentine's popular *Von dem grossen Stein* (1599), an enigmatic and emblematic treatise on preparing the Philosophers' Stone. Since the *menstruum peracutum* process dates from the 1650s and originates in Valentine's book on making the Stone, it argues for Boyle's early attempts to prepare the Stone, in opposition to his contrary protestations in his contemporaneous "Essay on Nitre."[64]

Gaston DuClo recounts an identical experiment in which he prepared a solvent able to extract a citrine tincture from gold, leaving behind "a white mass of gold, which when reduced to a body by the fire appears to be sil-

[59] Boyle, *Origine of Formes and Qualities,* 358–59.

[60] Ibid., 359–60.

[61] Boyle, *Sceptical Chymist,* 174, 179; note that Eleutherius expresses some concern about the use of the word *Sulphur* as a name but does not at all deny the reality of this extraction of "highly colour'd parts" of the gold.

[62] Boyle, *Origine of Formes and Qualities,* 360.

[63] Basilius Valentinus, *Schrifften,* 2:279–92.

[64] Principe, "The Gold Process"; "Robert Boyle's Alchemical Secrecy: Codes, Ciphers, and Concealments," *Ambix* 39 (1992): 63–74, on 68–69; and the introduction to this volume.

ver."[65] Daniel Georg Morhof, in his epistolary essay on transmutation, also cites such experiments on the extractability of the tincture of gold and quotes Boyle in particular.[66]

These processes in chrysopoetic texts (and in Boyle's *Origine of Formes*) and the beliefs behind them are exactly what Erastus assails in his opening statement. There he notes that some chymists

> suppose that there is in vulgar gold a certain part or portion that is far more noble than the rest, which being separated (although it is very little), the residual mass is deprived not only of its color but of its very nature, and ceases to be gold, by being degenerated into another substance, which is allowed to be silver, or some other white metal, or perhaps a mineral to which no appropriate name has yet been assigned. Furthermore, these chymists . . . believe [this separated substance] is the true tincture, or rather, the soul of the gold [*anima auri*].

Note that Erastus uses the same terminology (*tinctura* and *anima auri*) that occurs in the *Origine of Formes*. These chymists, he tells us, also believe that this tincture or soul of gold can be joined with a base metal to provide pure gold. Erastus distinguishes this process explicitly from the projective transmutation using the Stone and even provides a specific name for the process. This particular transmutation is "only a certain (if one might so call it) transplantation of an aurific tincture out of one metallic body into another, while our chrysopoeians allege that their Quintessence [Philosophers' Stone] can convert I know not how many hundred times its weight of lead into perfect gold." Erastus here argues that the extracted soul of gold could produce a quantity of gold only equal to the quantity of gold destroyed in the making of the extract. This was in fact one argument against profitable transmutation held by some of Boyle's contemporaries, most notably, Prince Rupert.[67]

But this denial of the profitability of chrysopoeia is refuted not only by several alchemical works but by Boyle himself. First, several chrysopoeians assert that this transplantation is very different from their much more powerful projection. Van Helmont carefully distinguishes the *transmutation by projection* from *transplantation*. He writes that the transmuting powder once given him "was not an extract from gold, which would have converted only so much mercury as there had been of the gold from which it was extracted."[68] Similarly, George Starkey's *Marrow of Alchemy* points out the error of those critics who understand only the particular transmutation using the *anima auri* and not the awesome power of the Stone.

[65] DuClo, *Apologia*, 60. "exhausta omni tinctura subsidet corpus auri album, quod in corpus igne agente reductum conspicitur argentum."

[66] Morhof, *Epistola ad Langelottum*, 178.

[67] The prince's belief is cited in the introduction to Johann Joachim Becher, *Magnalia naturae* (London, 1680), on [i–ii]; this book was printed "at the request of . . . Mr. Boyl," and I will return to it in the following chapter.

[68] Van Helmont, *Ortus medicinae*, in "Arbor vitae," 793. "Non erat autem extractum ex auro, quod totidem argenti vivi pondera commutaret, quot erant auri, unde extractum fuisset."

For thus they think that we of gold the soul
Extract, which from a masse a substance small
Is had, though it tinge without controul,
Yet scarce so much t'abide the trials all
Of fire and test, of gold there will proceed,
As first was us'd to yeeld that tinging seed.[69]

Starkey here admits that this transplantation can produce only as much gold "as first was us'd" to provide the transmuting substance; an unprofitable labor far different from the highly lucrative action of the universal Philosophers' Stone.

While Boyle does not mention any attempts of his own to transplant the *anima auri* he extracted with his *menstruum peracutum,* he elsewhere affirms an instance of that very process, noting that it was even done with profit. In "Of Un-succeeding Experiments" (1661) Boyle relates an account of his friend "Dr. K." who treated gold with a certain aqua fortis that separated

> the Tincture or yellow Sulphur from [gold], and made it volatile, (the remaining body growing white) and that with this golden Tincture he had, not without gain, turn'd Silver into very perfect Gold . . . when I demanded whether or no the Tincture was capable to transmute or graduate as much Silver as equall'd in weight that Gold from whence the Tincture was drawn, he assur'd me, that out of an ounce of Gold he drew as much Sulphur or Tincture as sufficed to turn an ounce and a half of Silver into that noblest Metall.[70]

Note how Boyle was careful to ask if the labor actually produced more gold than it destroyed, and so learned that a 50 percent increase was obtained. Unfortunately for Dr. K., when that parcel of aqua fortis was depleted, he could not find another sample that could do the same. The identity of "Dr. K." is revealed in Johann Friedrich Helvetius's popular *Vitulus aureus* (The golden calf, 1667), a work to which we will return shortly owing to its detailed account of a projection wrought by a mysterious traveling adept. Helvetius prefaces his own story with an account that is surely the *very same story* as Boyle's, excerpted from a letter by one Dr. Küffler. This person is presumably the Hague chymist Johannes Küffler (1598–1657). Küffler's testimonial letter reprinted by Helvetius relates that

> I found an aqua fortis in my laboratory . . . I poured this over Solar calx, namely, gold prepared in the common way, and after the third cohobation, it sublimed the Tincture of gold with itself into the neck of the retort, which tincture I mixed with silver precipitated in the common way, and saw that one ounce of the sublimed Tincture of Gold transmuted . . . one and a half ounces of silver into the best gold.[71]

[69] Eirenaeus Philoponus Philalethes (George Starkey), *The Marrow of Alchemy* (London, 1654–55), pt. 1, 48.

[70] Boyle, *Physiological Essays,* 68.

[71] Johann Friedrich Helvetius, *Vitulus aureus,* in *MH,* 815–63, on 831.

The similar details of the story and the identical proportions of transmuted metal attest to the identity of Helvetius's account with that related earlier by Boyle. It is interesting to note that the same story of transmutation occurs in Boyle and in a classic work of transmutational alchemy.[72]

Clearly, the process of the transplantation of the "soul of gold" mentioned by DuClo, Valentine, van Helmont, and Starkey is the same as that recounted by Boyle in *Origine of Formes* and in "Un-succeeding Experiments." Although Erastus claims that there is "no real possibility" of such an extraction of an *anima auri* and its transplantation, Boyle's citation of his own and Küffler's experiences indicates that he believed in such a process. But we must be careful to note exactly what it is that Boyle believes. He seems certain of the "matter of fact," that is, that a menstruum has separated a yellow material from gold, leaving a residual white metal, and that yellow material has been successfully employed in transmuting silver into gold. But Boyle, as is typical, is less certain of the *explanation* of process. I must now digress briefly to comment upon Boyle's style of offering explanations of experimental "matters of fact," before going on to ascertain whether Boyle adopts only the *experiment* from chrysopoetic sources or takes up *theoretical considerations* from them as well.

Boyle and the Diffident Explanation: The Dangers of Picking and Choosing

Boyle's attempts to explain the action of his *menstruum peracutum* bring up the much larger issue of how historians treat Boyle and his texts. Boyle's first explanation, which is based upon the extractability of "Tinging parts," comes directly from traditional alchemical sources that posited the color of gold in its Sulphur. But Boyle subjoins a second explanation that

> it is not impossible, but that the Yellowishness of that rich Metal may proceed not from any particular Corpuscles of that Colour, but from the Texture of the Metal . . . consequently, the Whiteness, and other Changes, produc'd in the new Metal we obtain'd, may be attributed not to the extraction of any tinging Particles, but to a Change of Texture, whereon the Colour, as well as other Properties of the Gold did depend.[73]

Now this explanation seems more "modern" and is in accord with the attribution of color to the texture of bodies that Boyle makes elsewhere. But it is important to note that both explanations—one based upon the al-

[72] The identification of Dr. K. is corroborated by Boyle's mention of Dr. K.'s "famous Scarlet Dye" (for which the aqua fortis was required); Johannes Küffler's scarlet dye (made from cochineal and aqua fortis) was widely admired in Boyle's day; see Thorndike, *History of Magic,* 7:497–98. Küffler was also the son-in-law of Cornelius Drebbel, the rather eccentric inventor who gained notoriety for supposed navigation of the Thames in a submarine.

[73] Boyle, *Origine of Formes and Qualities,* 360–61.

chemically informed concept of the extractability of "Tinging parts" and the other upon the effects of texture—exist side by side in Boyle's text. It is Boyle's intentional ambiguity, or rather his hesitancy to commit to a single, exclusive explanation of observable phenomena, that has contributed to selective readings of his writings. In a corpus as prolix and as diverse as Boyle's it is quite possible to pick and choose scraps to support almost any model, but this practice requires the simultaneous dismissal or neglect of points and opinions as valid to their author as those we choose to take up. Historians wishing to emphasize Boyle's commitment to the "mechanical philosophy" can readily take up the texture explanation and neglect the "Tinging parts" explanation.[74] Such selectivity would then show Boyle as more "modern" and more assertive about explanatory models than he himself was willing to be.[75] There are many instances of parallel but mutually exclusive explanations throughout Boyle's corpus. For example, in a manuscript on changes in the declination of the compass Boyle suggests in parallel both a more modern-sounding explanation of moving masses of subterranean iron (as in the vicinity of Mount Etna) and an alchemically based explanation of the decreasing magnetic virtue of subterranean iron as it matures into more noble though less magnetic metals (e.g., silver and gold).[76] If Boyle declined to decide between rival explanations, we ought not do it for him on the basis of subsequent scientific knowledge or our own notions and historiographic models, but rather recognize his own diversity of opinion. This is the reality of the historical record, and it is according to this record that Boyle must be presented in our historical studies.

If we now return to the question of the origin of color in gold, we find that in the *Dialogue* Erastus denies the extractability of any "Tinging parts" from gold and adopts instead the "texture" explanation. He asserts that

> the color in gold, as well as that in many other metals, does not proceed from any particular or separable portion, but actually from the texture of the whole body, which consists of particles whose concourse composes the body in such a way that it is fit to alter the light which it reflects to our eyes so that in the requisite way it can produce yellowness in them.

Does Erastus's rejection of extractable "Tinging parts" in favor of a mechanical alteration of the incident light reflect the revised attitudes of Boyle in the 1680s (perhaps altered by the intervening studies of color by Hooke and Newton), or is this insistence on texture merely the character Erastus playing his role as an anti-Lapidist? Indeed, it is fair to extend this question to other parts of Erastus's arguments—for example, to his explicit denial of the transplantation theory that Boyle clearly entertained in the 1650s and

[74] It must be stressed that the "Tinging parts" explanation is equally mechanical, but not as "modern."

[75] Boyle's "diffidence" in terms of explanatory models is a unifying theme in Sargent, *Diffident Naturalist.*

[76] BP, 21:155–89, on 179.

1660s. While such questions are not fully answerable from the text alone (one of the very reasons an author might choose to write in a dialogue format!), there is ancillary evidence to support the view that much of Erastus's certainty is his own rather than Boyle's. We are able to get a handle on this issue by examining a very closely related experiment investigated by Boyle—the analogous extraction of a tincture from copper.

In *Origine of Formes,* Boyle explicitly states that his maintenance of the extraction explanation (alongside the change of texture model) for the whitening of his gold is supported by accounts given to him of the success of "a known Person in . . . the Netherlands" in extracting a blue tincture from copper, which left behind a white residue fusible into a white metal. Boyle adds that he "elsewhere relate[s]" his trials with such a "White and Malleable Copper."[77] This "elsewhere" to which Boyle refers is possibly an unpublished essay presumably intended to be part of the extended editions of *Usefulnesse of Experimental Natural Philosophy.* The composition of this essay is datable to the 1660s, but the original English version is now lost and only a Latin translation prepared in the 1680s survives. Boyle relates the story told him by a reliable Belgian or Dutch refiner (*Purificator Belga*) wherein a blue tincture was extracted from copper leaving behind a white powder that was then fused into a malleable white metal "exactly comparable to silver."[78] Boyle had the story confirmed by a "remarkable experiment" which he designed to satisfy himself that "it was a genuine tincture of copper." Boyle continues that on another occasion another traveler brought him a piece of white metal which he found was actually copper, and that yet another friend of his, following directions Boyle himself designed, was able to produce such a white metal by extracting the color from copper. Interestingly, van Helmont describes a similar production of white copper through the extraction of its color with a solvent.[79] These citations demonstrate that during the 1660s Boyle believed in the extractability of a tincture (or Sulphur) from copper, as he did contemporaneously in regard to gold.

Several later papers demonstrate that Boyle sustained this belief throughout his life, during the writing of the *Dialogue* and well after. There exist among the Boyle Papers several unfinished compositions on the natural histories of various metals. Chapter headings for a text on copper survive on a sheet dated 10 May 1687, and these headings include

[77] Boyle, *Origine of Formes and Qualities,* 361–63.

[78] BP, 24:399–417, on 401–3. *Belga* may refer to an inhabitant of either of the lands we now call Holland and Belgium, both of which were "the Netherlands" in the seventeenth century; thus this story is probably the same as that given Boyle by "the known person in . . . the Netherlands." Boyle's insistence on the malleability of the white copper arises from the knowledge that copper is easily rendered white when it is alloyed with a small quantity of arsenic, but that white alloy is brittle, and so Boyle wants to emphasize that this white copper is not a trivial product.

[79] Jan Baptista van Helmont, "De lithiasi," in *Opuscula medica inaudita* (Amsterdam, 1648; reprint, Brussels: Culture et Civilisation, 1966), 9–110, on 69.

Whether from Copper may be obtained a T[inctu]R[e], especially so as to leave the Body white? And if so, Of what Colour that T[inctu]R[e] will be?
Of a surprising Effect of this last nam'd Tincture.
Of the white Body of Copper that would remain after the extraction of the tincture.[80]

The "remarkable experiment" of the extracted tincture that Boyle mentioned in the unpublished 1660s *Usefulnesse* fragment recurs here twenty years later in these headings as a "surprising Effect."[81] Since Boyle goes on to describe this effect in the second heading, the answer to his first question about the extractability of a tincture from copper must be "yes." Thus this outline of a lost or unwritten composition argues that as late as 1687 Boyle continued to believe in the extractability of highly colored tinctures from the metals leaving residual white metals. This continuity not only evidences Boyle's persistent belief in the heterogeneity of the metals and the maintenance of alchemical theories of their composition (i.e., that they contain an extractable, color-donating Sulphur) but also answers our original question by indicating that Erastus's dismissal of the extraction/transplantation theory is not representative of Boyle's own views. At most, Erastus is giving voice to Boyle's typical reticence to commit to a single hypothesis regarding the true explanation of phenomena. Certainly, it is clear that Erastus's negative views on transmutation by projection are thoroughly repudiated in the main thrust of the *Dialogue*.

Conclusion

This chapter has introduced a highly significant addition to the corpus of Boyle's writings—the *Dialogue on the Transmutation and Melioration of Metals*. While it survives only in fragmentary form, the extant text provides a hitherto unavailable resource for assessing Boyle's real attitudes toward chrysopoeia. I have indicated Boyle's authorship of alchemical tracts and underscored the significance and rhetorical uniqueness of the *Dialogue on Transmutation*. Boyle's knowledge and maintenance of traditional alchemical principles, particularly as regards types of transmutations and the analysis of heterogeneous metals, have also been showcased. The foregoing arguments and evidence demonstrate beyond doubt Boyle's abiding belief and activity in both chrysopoeia and metallic spagyria. Furthermore, this chap-

[80] BP, 21:198; a copy exists at 36:80. See also a note (among others bearing dates in the 1660s) that reads "The distill'd liquor being digested with our p[re]pared Copper will, from the calxe draw a fine blew tincture easily made volatile leaving the rest a metalline substance of more difficult fusion then before and deprived of its native colour" (BP, 27:86).

[81] Presumably, if Boyle was continuing along the lines of alchemical theory, this "surprising Effect" was the recombination of the tincture and the whitened copper to recompose red copper.

ter has highlighted the necessity of recognizing in our historical studies Boyle's reticence to commit to a single explanation of phenomena. The following chapter continues the scrutiny of the *Dialogue*, focusing on its continuity of form with the alchemical tradition, and Boyle's contacts with adepti.

CHAPTER IV

Adepti, Aspirants, and Cheats

IN THE previous chapter, I introduced a crucial new source for the study of Robert Boyle and alchemy, the *Dialogue on the Transmutation of Metals.* This work strongly asserts Boyle's belief in the claims of traditional alchemy regarding the existence of the Philosophers' Stone and its ability to effect chrysopoetic transmutations by projection. It is now worthwhile to focus upon the methodology that Boyle employs to carry his cause. As alchemical apologetics have a long history before the late seventeenth century, it will be enlightening to compare Boyle's probative methods with the foregoing tradition. It is important to answer the question of how Boyle was himself convinced of the truth of chrysopoetic claims and how he endeavored in turn to convince the *Dialogue*'s readers of their truth. This inquiry will open up an important dimension to Boyle's alchemy, namely, his contacts with practicing alchemists. We shall see how the alchemists whom Boyle encountered were both many and diverse. The title of chrysopoeian was attached to a wide variety of characters in the seventeenth century, many of whom Boyle was eager to meet, and others of whom he would have been equally eager to avoid.

Debates over the validity of chrysopoeia are not a seventeenth-century development. While the intensity of such disputes waxed and waned periodically over the course of the generations, these arguments were carried on in every century that alchemy was practiced. The first high point of these debates in the Latin West occurred in the late thirteenth century, and the results are fairly conveniently summarized in the *Margarita preciosa novella* of Petrus Bonus of Ferrara, a ponderous scholastic work composed around 1330.[1] From that time until the final disappearance of transmutational alchemy in the late eighteenth century, it was common for alchemical writers to begin their works with a defense of their art against those who denied its reality or licitness. In many cases, the same arguments (pro and con) recur like a refrain over the course of the centuries, and several classical ones occur in the *Dialogue.* Arguments from nature, from classical authorities, from experiment, from analogy, and from testimonial accounts were used on both sides of the controversy.

The strategy of the *Dialogue*'s argument is quite straightforward. Erastus's arguments against the Philosophers' Stone are negative and a priori; the Stone's action seems incredible, the adepti who assure us of it are un-

[1] Newman, *The Summa Perfectionis,* 1–47, and Chiara Crisciani, "The Concept of Alchemy as Expressed in *Preciosa margarita novella* of Petrus Bonus of Ferrara," *Ambix* 20 (1973): 165–81.

known or untrustworthy, and there is no adequate theory to explain its effects. Zosimus is not satisfied with such reasoning, and he undertakes first merely to demonstrate the *nonimpossibility* of the Philosophers' Stone and subsequently (with help from fellow Lapidists) uses experiential accounts to prove its *reality*. This strategy is consonant with Boyle's universal unwillingness to make general programmatic statements. Boyle, as a "diffident naturalist," was well aware of the philosophical impossibility of disproving the existence of a thing. In this particular case, his reticence to rule out the possibility of a thing was supported by what he himself saw as irrefutable evidence.

Boyle (under the guise of Erastus) first summarizes the arguments against projective transmutation—the incredibility of the change, the stubborn nature of the metals, the small proportion of Stone employed, and the celerity of the action. None of these are original arguments: some occur in Petrus Bonus, and the rest in later antichrysopoetic writings.[2] Boyle then (under the guise of Zosimus) divides his defense into two parts, stating that transmutation by the Philosophers' Stone can be first "countenanced by things Analogicall in nature or Art" and then "prov'd, by particular Histories and other testimonies."

Fragment 4 contains the analogies with which Boyle chose to refute the latter two objections. Two of these, the small quantity of rennet used to coagulate a great deal of milk, and the ability of a small flame to propagate itself through a vast amount of wood, are canonical alchemical images. Alchemists often compared the Philosophers' Stone to rennet, for it can coagulate quicksilver or molten lead into gold. Likewise, it is also called a fire, for it can diffuse itself through imperfect metals like fire through kindling.[3] Boyle adds to these classical images his own observations regarding the tiny amount of venom that produces the "sad effects of the Teeth of angry vipers" in killing both "Brutes and men." He mentions also the (chemically accurate) observation that only a trifling amount of tin can significantly alter the physical properties of metals, as well as the "unanimous tradition of the Chymists" that the mere vapor of lead can cause mercury to solidify. The alchemical writers themselves had used this latter belief to assert that quicksilver can be coagulated into a metal solid like the others, thus supporting the possibility of analogously fixing it into gold; thus Boyle's usage is quite analogous to theirs. (In fact, Boyle expended some effort in attempting to verify this "unanimous tradition.")[4] In general, Boyle had a particular

[2] Petrus Bonus, *Margarita preciosa novella,* in *BCC* 2:1–80, on 9–16.

[3] The rennet analogy is remarked upon by Bonus, *Margarita,* 50; see also, for example, Starkey (Philalethes), *Marrow of Alchemy,* pt. 1, 46: "[The Philosopher's Stone] On metal it like wax is guided / To enter to the center, just as milk, / Is penetrated by the Rennet sour, / And curdled in the minute of an hour." Likewise on fire, 15: "Much like to fire which kindled doth not cease / It self to multiply, nor ever shall / An end be found of its encreasing might, / If fed with fuell new, . . ."

[4] See Robert Boyle, "A History of Fluidity and Firmnesse," in *Physiological Essays,* on 218–19.

interest in disproportionately large effects produced by apparently trifling or "unheeded" agents; his tracts on *Cosmicall Suspitions* (1671) are full of such inquiry, and he mentions the topic again in *Christian Virtuoso* (1690).[5]

Transmutation Histories

It is clear both from Zosimus's opening statement and from the small amount of space devoted to "things Analogicall" that Boyle gives probatory preeminence to testimonial accounts. The next six fragments, as well as the clinching argument of the conclusion (*Anti-Elixir*), present experimental accounts of processes that refute one or all of Erastus's objections. While theoretical considerations played a major role in alchemical apologetics throughout the history of alchemy in the Latin West, by the seventeenth century accounts of transmutations had risen to an equal or even greater importance. Indeed, we can point to an entire genre of alchemical writings that may be termed "transmutation histories."[6] Transmutation histories were published either singly or as collections, and the apologists' arsenal was increasingly stockpiled with them throughout the century. Many involve stories of traveling adepti who performed projection privately before one or more aspiring alchemists or alchemical skeptics. Some such accounts are so romantic or outlandish that they rarely fail to provoke a wry smirk from the modern reader. Many are painstakingly precise, noting exact times, places, persons (often of rank and station) in attendance, the quantity of gold or silver produced, and so forth. Other examples are public, having taken place at some court or assembly, and were sometimes commemorated by the striking of coins or medallions. By the end of the century, a sufficient number of such coins had been minted that an entire dissertation was written about them by Samuel Reyher.[7] An early example of the transmutation history is Ewald de Hoghelande's *Historiae aliquot transmutationis metallicae,* which recounted the projections made by the elusive Alexander Seton in the course of his tour through central Europe during the opening years of the seventeenth century.[8]

During Boyle's life, two detailed stories of projections garnered especially widespread attention. The first was an account published in 1667 by Johann

[5] John Henry, "Boyle and Cosmical Qualities," in *RBR,* 119–38; Boyle, "Fluidity and Firmnesse," on 243; *Of the Reconcileableness of Specifick Medicines to the Corpuscular Philosophy* (London, 1685), 11–14, 24–31; *Christian Virtuoso* (London, 1690), 62–64.

[6] For an introduction to the genre with examples, see Newman, *Gehennical Fire,* 3–11.

[7] Samuel Reyher, *Dissertatio de nummis quibusdam ex chymico metallo factis* (Kiel, 1692); see also Wolf-Dieter Müller-Jahncke and Joachim Telle, "Numismatik und Alchemie: Mitteilungen zu Münzen und Medaillen des 17. und 18. Jahrhunderts," in *Die Alchemie in der europäischen Kultur- und Wissenschaftgeschichte,* ed. Christoph Meinel (Wiesbaden: Harrassowitz, 1986), 229–75, and V. Karpenko, "Coins and Medals Made of Alchemical Metal," *Ambix* 35 (1988): 65–76.

[8] Ewald de Hoghelande, *Historiae aliquot transmutationis metallicae . . . pro defensione alchymiae contra hostium rabiem* (Cologne, 1604).

Friedrich Helvetius (1625–1709) and entitled *Vitulus aureus* (The golden calf).[9] According to the story, on the afternoon of 27 December 1666, a stranger visited Helvetius, physician to the prince of Orange, at his house in The Hague. Helvetius had written skeptically about the validity of alchemy, and the visitor engaged him in conversation on this topic. After some discussion, the visitor took out a small ivory box and showed Helvetius its contents—three heavy lumps of a glassy yellow substance—which he claimed to be the Philosophers' Stone in sufficient quantity to produce twenty tons of gold. The mysterious adept then left but promised to return in three weeks. Helvetius, who had been allowed to hold the precious Elixir in his hand for some minutes as they spoke, surreptiously dug his fingernails into the lump, and after the stranger had left, the enterprising physician dislodged a minute particle of the Stone from under his fingernail. Helvetius projected this particle upon molten lead but, since he knew neither the proportions to employ nor the method of projection, obtained only a glassy mass.

Upon the adept's return, Helvetius admitted what he had done, provoking laughter from the adept, who jibed that had he been as proficient as a projector as he had been as a thief, his results would have been better. The adept then told Helvetius the proper method of projection and presented him with a tiny particle of the Stone, "smaller than a rapeseed," promising to return the next day to direct the projection. The adept never returned, and so Helvetius himself performed the projection on six drachms of lead, transmuting it successfully into gold. To test the authenticity of the gold, Helvetius brought it to the mint-master of his province, who assayed it by quartation. In quartation—a standard assaying technique—gold is melted into an alloy with thrice its weight of silver, the alloy beaten into plates and treated with aqua fortis (nitric acid). The silver and any base metals dissolve leaving the gold pure as a black powder. This black powder is then dried and fused to metallic gold, and the weight of the resultant metal can be compared with that of the initial sample. In the test of Helvetius's metal, however, the assayer obtained *more* gold at the end than he had used as a sample—a two-drachm sample provided two drachms and two scruples of pure gold (an increase of 33 percent). Apparently, the melting of the alchemical gold with the silver resulted in a transmutation of some of the silver "on account of the abundance of Tincture" present in the alchemical gold.[10] But suspecting that the excess weight was owing to some silver that had not been properly separated in the quartation, the assayer retested the augmented gold by melting it with antimony (antimony trisulphide)—a common method for the purification of gold that functions by oxidizing all metals except gold to sulphides, which are skimmed off as a scum leaving the pure gold beneath. But this test showed that all two drachms and two scruples were in fact true gold.

[9] Helvetius, *Vitulus aureus*.

[10] Ibid., 844; "videlicet drachmae Auri duae transmutaverant duos Argenti scrupulos prae Tincturae abundantiâ in Aurum sibi homogeneum seu similare."

This account soon achieved great notoriety; for example, Benedict Spinoza went to both the assayer and to Helvetius to confirm its veracity.[11] A letter to Boyle from one Michael Behm (dated 2 October 1668) discusses a range of alchemical matters and refers at one point to Helvetius's very recently published *Golden Calf* expressing some uncertainty as to the reliability of the account, "particularly when he speaks of the short time for completing the Stone, and its color."[12] Helvetius's visitor claimed the Stone could be made in three days, and the Stone he showed Helvetius was of a yellow color; traditionally, alchemical writers claimed that the confection of the Stone required at least nine months and resulted in a red material. Helvetius himself questioned the traveler in regard to the color of his material and was assured that the color was not crucial. One sign of the influence of Helvetius's story on Boyle is clear from the *Dialogue* when the president of the society enumerates the possible means of distinguishing natural from alchemical gold. He states that one of these might be "by an excess of tincture whereby [alchemical gold] may fix some small part of the lead added to it in cupellation," which is strictly analogous to what happened to the silver in Helvetius's quartation.[13]

A second account well-known in Boyle's time was that of the shady Wenzel Seyler, a projector at the imperial court of Leopold I during the 1670s. Very little is known of Seyler's life with certainty; whatever we can now reconstruct is invariably intermingled with the numerous contemporaneous rumors that circulated about him. It is fairly certain that he arrived at court in 1675, was ennobled as Ritter von Reinberg on 16 September 1676 and appointed a mint-master. Of particular note was his spectacular transmutation of a huge medallion before the emperor in 1677.[14] This medallion, which now rests in the Kunsthistorisches Museum in Vienna, is composed of a silver-gold-copper alloy, with its surface enriched in gold, probably by selective dissolution of the silver and copper with nitric acid.[15] Clearly this is not a projective transmutation using the Stone; however, Seyler had previously performed actual projective transmutations. In 1675, gold ducats were coined from metal purportedly made

[11] *Spinoza Opera im Auftrag der Heidelberger Akademie der Wissenschaften* (Heidelberg, n.d.), *Epistolae*, 196–97.

[12] BL, 1:59–61, on 60.

[13] Cupellation is a somewhat different assaying technique. A precious metal (either gold or silver) is melted with a large excess of lead in a shallow dish, and a blast of air directed over the molten metal. The air oxidizes the lead and any base metals and blows away their oxides, leaving the noble metal pure.

[14] Müller-Jahncke and Telle, "Numismatik und Alchemie," 252, 269–70; Pamela Smith, "Alchemy as a Language of Mediation in the Habsburg Court," *Isis* 85 (1994): 1–25; *The Business of Alchemy: Science and Culture in the Holy Roman Empire* (Princeton: Princeton University Press, 1994), 180–82.

[15] R. Streibinger and W. Reif, "Das alchemistische Medaillion Kaisers Leopold I," *Mitteilungen der Numismatischen Gesellschaft in Wien* 16 (1932): 209–13. An alternative means of its apparent conversion to gold is supplied by the testimony to the Royal Society of St. George Ashe; see 97.

by Seyler by projection. These coins bore the bust of Leopold on one side, and on the other the verse "Aus Wenzel Seylers Pulvers Macht, Bin ich von Zinn zu Gold gemacht" (By the power of Wenzel Seyler's powder, I was made from tin into gold).[16]

Boyle sought out the imperial ambassador, Count Waldstein, while he was in London (from 5 June 1677 to 9 March 1679) to inquire about Seyler's transmutations, which the count had witnessed. Notes from this interview survive among the Boyle papers and are presented in appendix 2, text A.[17] Further, Boyle's conversations with Waldstein provided details for supporting evidence in fragment 7 and an account that constitutes all of fragment 9 of the *Dialogue*. In 1680, Johann Joachim Becher (1635–1682) published *Magnalia naturae*—a rather lurid account of Seyler's rise to fame—"at the Request, and for the Satisfaction of several Curious and Ingenious, especially Mr Boyl."[18] Becher had been one of the inspectors appointed by the emperor to investigate the veracity of Seyler's transmutations. Becher verified the reality of the transmutations but denied that Seyler had actually produced the powder, claiming instead that he was an ignorant thief. According to Becher's account, Seyler was an Austrian who lived as an Augustinian monk at Brünn in Moravia (now Brno in the Czech Republic). While scheming to find some way to flee the monastery, he came into the possession of a magical wax ball that would roll toward any hidden treasure. Seyler, however, could not understand the use of the ball. He showed the object to an old and wise monk who understood its employment, and together they discovered that it rolled toward an old pillar in the abbey church. Later, during a violent storm, the foundations of the church were shaken, and the abbot ordered this old pillar removed lest it collapse on the heads of the assembled monks. Seyler and the old monk then watched its demolition and saw the workers uncover an old chest walled up in it. They took the box, and upon opening it, Seyler was bitterly disappointed to find only a cryptic manuscript and four parcels of red powder; he had been hoping for gold. But the old monk soon deciphered the manuscript and thus learned both the identity and the use of the red powder—which was, of course, the Philosophers' Stone—and performed several projections with Seyler. Soon thereafter the old monk had a seizure, and as he lay dying, Seyler stole the Stone from him and fled the monastery. He attempted to get protection from several noblemen (all of whom tried to kill him for the powder) and eventually arrived at the emperor's court. There he made numerous projections, squandering the proceeds in dissolute living, and finally, having used up most of the powder, fell into cheating practices and incurred the displeasure of the court. Again according to Becher, the emperor, unwilling to let it be known that he had been duped by a thief and a swindler,

[16] Kopp, *Die Alchemie*, 1:158–59.

[17] Hunter, "Alchemy, Magic and Moralism," 401–3.

[18] Johann Joachim Becher, *Magnalia naturae, or, the Philosophers' Stone lately expos'd to Publick Sight and Sale* (London, 1680).

removed Seyler from court, elevated him to the status of a baron, and sent him away.

Boyle refers to Seyler in his important letter to Joseph Glanvill regarding accounts of witchcraft.[19] Boyle remarks to Glanvill (who was collecting testimonial accounts of the activity of spirits and witches) on the absolute necessity of having thorough circumstantial details and verifiable testimony of instances of spirit activity, and draws an analogy to accounts of transmutation. Referring to Seyler, Boyle wrote that

> supposing the truth of this, I say, this one positive instance will better prove the reality of what they call the philosophers' stone, than all the cheats and fictions, wherewith pretending chemists have deluded the unskilful and the credulous, will prove that there can be no such thing as the elixir, nor no such operation as projection.[20]

Accordingly, Boyle strove to learn more of Seyler's activities and compiled his own collections of such events. In addition to the accounts used in the *Dialogue,* transcripts of Boyle's interviews with Count Waldstein regarding Seyler bear (as Hunter has noted) contemporary pagination which indicates that the fragments we have are only part of a larger collection.[21] This collection might conceivably be identified with the "Severall Papers concerning the Transmutation of Metalls, Quarto red cover strings, sad red" listed separately from the *Dialogue* in Boyle's catalog dated 10 July 1684.[22] Still further information about Seyler is contained in Boyle's copy of a letter from St. George Ashe to the Royal Society, dated 9/19 July 1691.[23] There the author states that Seyler used a red powder to transmute the huge medallion before the emperor. This red powder (for Seyler did not know how to prepare it) was obtained from the prior of an Augustinian convent, who obtained it from one "Count Sellich" who was privy councillor to Rudolf II, who in turn is purported to have gotten it from Edward Kelley, the English scryer and friend of John Dee.[24] (According to other contemporaneous texts, Kelley is supposed to have obtained the powder of projection from the tomb of a bishop in the ruins of Glastonbury Abbey!)

[19] Boyle to Joseph Glanvill, 18 September 1677; *Works,* 6:57–59.

[20] Ibid., 58.

[21] Hunter, "Alchemy, Magic and Moralism," 402 n. 45.

[22] BP, 36:120.

[23] BP, 25:281.

[24] On Kelley and Dee, see Nicholas H. Clulee, *John Dee's Natural Philosophy: Between Science and Religion* (London: Routledge, 1988); C. L. Whitby, "John Dee and Renaissance Scrying," *Bulletin of the Society for Renaissance Studies* 3 (1985): 25–36; Peter J. French, *John Dee: The World of an Elizabethan Magus* (London: Routledge & Kegan Paul, 1972), esp. 113–22; Deborah E. Harkness, *Talking with Angels: John Dee and the End of Nature* (Cambridge: Cambridge University Press, 1998); Francis A. Yates, *Theatre of the World* (Chicago: University of Chicago Press, 1969). The story of Kelley's discovery of the powder at Glastonbury is told by Elias Ashmole, Annotations to *TCB,* 481. Morhof, *Epistola ad Langelottum,* on 189–90, tells a different story; see chapter 6.

In the *Dialogue* Boyle also uses an account of transmutation gathered from Olaus Borrichius, who saw a silver coin transmuted into gold by an Anabaptist in Amsterdam (fragment 5). Borrichius had visited Boyle at Oxford on 4 August 1663. Their conversation during that meeting, as recorded in Borrichius's travel diary, was full of chymical topics, including the making of metallic Mercuries and the transmutation of metals.[25] The surviving Boyle correspondence preserves one letter from Borrichius, which is likewise full of alchemical discussion.[26] Whether the account used in the *Dialogue* was told to Boyle by Borrichius when they met or in some later (now lost) correspondence is unclear.

Boyle's Witness of Projection

The secondhand nature of all these accounts, regardless of the identity of the witnesses, renders them inferior to direct experience. The account offered by Pyrophilus (published in *Anti-Elixir*) is clearly written as Boyle's firsthand account of an experiment, but that is not an example of projection of the Philosophers' Stone. The most powerful evidence for Boyle (or for anyone else, for that matter) of the reality of projective transmutation would naturally be the sight of such an event before his own eyes. Indeed, Boyle was not at all lacking in firsthand experience of chrysopoeia. Fragment 10 of the *Dialogue*, a lengthy narrative spoken by the Lapidist Philoponus, becomes the most significant of the transmutation accounts when we recognize that it actually represents Boyle's first person eyewitness description of a transmutation of lead into gold. While other fragments also represent Boyle's testimony of experiments, these pale in importance next to an account of Boyle's own meeting with a traveling adept who showed him projective transmutation.

According to this document, Philoponus (while visiting a "forraigne doctor of Phisick") is introduced to a certain foreigner who attempts to show him a method whereby lead can be turned into running mercury. This experiment, however, miscarries when the crucible is upset, spilling its contents into the fire. In recompense the foreigner agrees to demonstrate another experiment, which Philoponus mistakenly assumes to be a mere repetition of the miscarried mercurification experiment. Philoponus's servant is sent to obtain more lead and crucibles; when the lead is melted, the foreigner, after offering to allow Philoponus to perform the operation, casts a small quantity of red powder onto the molten metal and covers the crucible. After heating the crucible strongly for fifteen minutes, they allow it to cool, and, upon knocking out the cooled mass, Philoponus is amazed to find a ponderous yellow metal, which his later tests show to be pure gold.

[25] *Olai Borrichii Itinerarium 1660–1664*, ed. H. D. Schepelern, 4 vols. (Copenhagen: Danish Society of Language and Literature, 1983), 3:65–66.

[26] Olaus Borrichius to Robert Boyle, 30 March 1664; BL, 1:88–89*v*.

Philoponus's fluency in French, his knowledge of assaying, and his liveried "footman at the house dore" all point to his identity with Boyle. But the mention in the fragment of an "accidentall indisposition" in his eyes that made them "Dazelld & offended by the glowing fire" confirms this identity by citing an infirmity Boyle often describes of himself. This meeting is probably the same as that only briefly and allusively described by Boyle in *Producibleness of Chymical Principles.* There, in the context of a discourse on the mercurification of metals (which was the first experiment the traveler tried to demonstrate for Philoponus), Boyle writes that he was

> once in a place, Where a forreiner, that was a stranger to me, was showing a freind of his, with whom I had some little acquaintance, a Metalline experiment, that I confesse, I could not but admire (for this Forreiner was so civil because I came so luckyly in, as to let me be present att the experiment though not to discover any thing of the drug he imployed about it).[27]

While this published text is notably sparse in detail, its juxtaposition to a discourse on the mercurification of metals and its mention of a chance meeting with a foreigner and his friend link it to the Philoponus account. But this brief and ambiguous allusion is not the only other mention of this momentous event in Boyle's life. Dramatically, the story told here by Philoponus is summarized in the surviving notes taken by Boyle's confidant Bishop Gilbert Burnet during an interview with Boyle in which he dictated important autobiographical details. Burnet's memorandum recalls an incident that occurred when Boyle was

> going to Visit a forreigner that was known to him [and] found one with him of whom he had no knowledge so the forreigner after he had whispered the other [(]it is like to ask his leave to let him see the Experiment they were about[)] he told him the stranger would trie an Experiment of making lead ductile as wax, or rather as butter but that he might not apprehend he was deceived he wisht him to send for a Crucible and some lead which being done and the lead put in fusion he put into an ______ of lead about ______ of a bright powder and a litle while took the Crucible from the fire and when they Judged it was cold he expecting no such thing was not a litle surprised to find instead of lead Gold which after all the trialls that could be made was found true Gold of ______ finenesse.[28]

Clearly, Burnet's notes recount the very same incident as told in Philoponus's speech. The *Dialogue* fragment, however, is far more detailed, vivid, and intricate. Consonant with its preminent status as Boyle's own eyewitness account of the projective transmutation of lead into gold, fragment 10 is by far the longest account in the dialogue.

The importance of this event for Boyle should not be underestimated. The meeting with this "forreigner," whom Boyle learns to be the disciple of a

[27] Boyle, *Producibleness,* 160.

[28] Hunter, *Robert Boyle by Himself,* 30; blanks left in the original MS.

French adept, was a momentous event. Its importance to Boyle is witnessed by the substantial amount of space he devotes to it in the *Dialogue* as well as by its prominent recounting in the "Burnet Memorandum," not to mention the brief allusion in *Producibleness.* Although Boyle never entirely dismissed the possibility that the transmutatory Elixir existed, he does express some reservations in published works and confesses to Burnet (in the "Burnet Memorandum") that he "was long very distrustfull of what is comonly related concerning the Philosophers Stone."[29] By Boyle's own account, it was this very incident that gave him "convincing satisfaction in that matter."[30] The drama of the event comes through very clearly in the *Dialogue* account. The reader can almost see the projector's smugly smiling countenance replying to Boyle's shocked expression when he empties the crucible and holds in his hand, instead of the lead he put in or the mercury he expected it to be turned into, a lump of solid yellow gold.

Echoes of this event seem to have resounded even after Boyle's death. In 1706, fifteen years after Boyle had died, Johann Michael Faustius, a medical doctor of Frankfurt am Main and member of the Academia naturae curiosorum, published an edition of Eirenaeus Philalethes' extremely popular alchemical work *Introitus apertus ad occlusum regis palatium* (Open entrance to the shut palace of the king). In the preface to the work, Faustius endeavored to ascertain some biographical details regarding the highly celebrated but equally anonymous Philalethes. Faustius visited England and there heard an account of how Boyle "was once visited by chance by a disciple of a certain Adept who was afterwards imprisoned by the French. This disciple, while Boyle watched, converted lead into gold."[31] The alchemist's disciple claimed that the elusive Philalethes was living in France, and at Boyle's behest agreed to carry letters to him from Boyle. Unfortunately, the letters never reached their destination, for the young man, Faustius reports, fell from his horse in the course of the journey and died.

Faustius's story has a strong resonance with the account tendered by Philoponus—both tell of a disciple of a French adept whom Boyle meets by chance and who transmutes lead into gold before him. Even if it does not refer to the very same event, its transmission by Faustius nonetheless bears witness to the survival of stories about Boyle's chrysopoetic interests and experiences. Faustius's comment about the imprisonment of the French adept echoes yet another account of Boyle's alchemical experiences, this

[29] Boyle's previous uncertainty about the Stone was expressed, for example, in "Imperfection of the Chymists' Doctrine of Qualities," in *Mechanical Origin,* 35, and *Excellency of Theology, Campar'd with Natural Philosophy* (London, 1674), 134.

[30] Hunter, *Robert Boyle by Himself,* 30.

[31] *Philaletha illustratus . . . sive Introitus apertus,* ed. Johann Michael Faustius (Frankfurt am Main, 1706), sigs. c2–c3. "Erat aliquando, ut Togatus iste Heros [Boyle], à discipulo cuiusdam Adepti, postea apud Gallos in vincula conjecti forte inviseretur, hic ubi inspectante Boylaeo, Plumbum in Aurum convertisset . . ."

time recorded by Constantijn Huygens in his diary on 20 June 1689. There Huygens notes that his brother Christiaan had been told by Boyle himself that a man had once visited him and made an ounce of gold from lead (using a bright red powder); Boyle later heard that the alchemist had been arrested in France.[32] The stories diverge in terms of who exactly was imprisoned (the adept, according to Faustius, or his disciple, according to Huygens) and whether or not anyone died in falling from a horse, but such differences may be attributable to garbling during transmission; whoever recounted the story to Faustius fifteen years after Boyle's death was at least temporally removed from its origin.

Philoponus's account has a further significance—it provides sufficient information to enable us to link it to a published account of a contemporaneous projection, allowing for the identification of the projector's master. In fragment 10 Boyle recounts that he later learned "of two severall transmutations that our Traveller had made in one towne, of both which operations they were assur'd by the Persons themselves in whose sight & favour they were made, whereof one whose name I know is at present a learned professor of Phisick in a university not very far of[f]."[33] This "learned professor" is likely to be Boyle's friend Edmund Dickinson (later to be named the executor of Boyle's chymical papers), who was an M.D. and Linacre Lecturer at Oxford. In 1683, Dickinson wrote a lengthy letter to an adept of his acquaintance named as Theodore Mundanus, requesting answers to several questions regarding the making of the Philosophers' Stone. This letter, together with Mundanus's reply (translated into Latin from the the original French), was published in 1686.[34] In his own letter, Dickinson states that he had long been interested in transmutatory alchemy but had

[32] Maddison, *Life,* 176; quoted from *Historische Genootschap te Utrecht,* Nieuwe Reeks (1876), no. 23: "Broer was 'smergens bij mrs Boyle geweest, die hem onder anderen vertelt hadde, dat een man bij hem geweest hadde, die met een poeder dat root en claer was, een once goudt gemaaeckt hadded von loot; dat hij gehoort hadde, dat die man in Vranckrijck gearresteert was."

[33] Boyle, *Dialogue on Transmutation,* appendix 1, 268.

[34] *Epistola Edmundi Dickinson ad Theodorum Mundanum, Philosophum adeptum* (Oxford, 1686); the letter is dated at London, July 1683 (92). The French original is preserved in British Library Sloane MS 3629, 202–29.

There is some uncertainty over when Dickinson moved to London from Oxford. His biographer (and grandson) William Blomberg sets it at 1675 in *An Account of the Life and Writings of Edmund Dickinson* (London, 1739), 84; Frank cites the date 1678 in *Harvey and the Oxford Physiologists,* 70–71, while the *DNB* and *BC* state 1684. This last date is certainly not correct, for Dickinson was resident in London in 1680 when he helped the exiled J. J. Becher upon his arrival there (see the dedication to Becher's *Tripus hermeticus fatidicus* [Frankfurt, 1689]). Dickinson bought Thomas Willis's London house upon Willis's death in 1675, which provides the first date (although there is no evidence that Dickinson relocated immediately); the last date, given initially in *Biographica britannica,* 2d ed., 5 vols. (London, 1778–93), 5:175–79, on 176, probably comes from an erroneous date of 1684 for Willis's death. Regardless of Dickinson's abode at the time of Boyle's writing, he might still be correctly referred to as an Oxford professor. On Dickinson, see the *New DNB* article by Principe.

been skeptical of its validity. But all these doubts, he writes, were removed by "that excellent demonstration which your excellency exhibited two years ago."[35] Mundanus's reply gives much more detailed information about the event. He states that he met Dickinson originally at Oxford in the mid-1660s, at which time he noted his aptitude and interest in chymistry. In 1679, however, he passed through England again, met Dickinson in London, found him still interested in chymistry, and, in order to excite his industry toward chrysopoetic studies, showed him two projections.[36]

Several points of correspondence suggest that the two transmutations performed by Mundanus before Dickinson are the *very same* as those cited by Boyle in the Philoponus fragment. First, Boyle's description of a "learned professor of Phisick in a university not very far of[f]," who must have had an interest in transmutatory alchemy, clearly fits the profession, Oxonian connection, and interests of Dickinson. Second, Mundanus is a Frenchman, like Boyle's projector. Third, *two* separate transmutations are mentioned by both Mundanus and Boyle. Fourth, the timing is correct—Mundanus explicitly dates his transmutations to 1679, within the period when Boyle was writing the *Dialogue,* and in sufficient time for him to insert the cryptically brief and allusive reference to the event in the delayed publication of *Producibleness,* which was printed in late 1679 and offered for sale in January 1680.[37] Mundanus's memory was perhaps slightly faulty, for clinching evidence (to be discussed below) argues that the projections before Boyle and Dickinson occurred instead in early 1678. In that year, in the context of another of his alchemical contacts, Boyle makes *explicit reference to an alchemist whom he calls Mundanus.*

The only difficulty in this identification is that Boyle describes his projector as a "servant & young Disciple of an Adeptus," whereas Mundanus states that he had been "an Adept philosopher for more than forty years," in which case (unless he were impossibly precocious) he must have been at least sixty.[38] Nonetheless, the difficulty is not insurmountable. For the notable similarities, especially in light of Boyle's later mention of Mundanus by name, are sufficiently striking to countenance the speculation that Boyle's "Traveller" was Mundanus's apprentice, and the two traveled together

[35] Dickinson, *Epistola,* 4; "illustris ea demonstratio, quam vestra excellentia biennio jam elapso coram exhibuit, omnem ansam dubitandi mihi praecidisset." This would give a date of 1682 for the transmutation, but that is contradicted explicitly by Mundanus; see below.

[36] Ibid., 148–49. "Quando autem Angliam redeunti, anno millesimo sexcentesimo septuagesimo nono, iterum mihi nomen tuum & fama Londini occurrisset . . . *duas istas Projectiones* paulo post coram teipso praestiti." Dickinson's interest in chymistry at Oxford is described in *Olai Borrichii Itinerarium,* 3:26–27, 36, 46–47, 50.

[37] Fulton, *Bibliography,* 29. It might be objected that the imprimatur was given in 1677, but that would not preclude the insertion of a paragraph here and there, as was Boyle's custom, even as late as while a work was being printed. I think it possible that the delay in publication of this chymical work may in fact be due to Boyle's alchemical experiences of 1677–78, the most dramatic of which was this transmutation of lead into gold.

[38] Boyle, *Dialogue on Transmutation,* appendix 1, 267; Dickinson, *Epistola,* 149.

through England; Boyle met the disciple, and Dickinson the master with whom he had previously been acquainted. Boyle, who discovered from the disciple his master's name recognized it at once from having seen it "in printed Gazetts where he was mention'd as a very rich man famous for extraordinary knowledge & who is now rais'd . . . to very eminent dignities."[39] This high rank accords with Dickinson's repeated address to Mundanus as "your excellency." Indeed, Boyle uses the high status of this adept (whether or not he is rightly identified with Mundanus) to combat criticisms based upon the questionable character of alchemical projectors, and Boyle immediately contrasts him with a thieving projector. This virtuous adept also supplies an implicit contrast with Wenzel Seyler, whose immorality is highlighted by Becher.[40]

Of course, it is almost superfluous to point out that "Mundanus" is a pseudonym and not the genuine name that Boyle and Dickinson presumably knew.[41] It is also worth noting that "Mundanus" (world-citizen) can also be rendered as "Cosmopolite," a title used by several alchemists, including, most particularly, Michael Sendivogius and Eirenaeus Philalethes. This similarity of monikers may bear upon the origin of the Philalethes connection in the version of the story that Faustius recounts.

Philoponus's speech gives us as detailed a description of a chrysopoetic projection from as reliable a source as we are ever likely to find. It ties Boyle fully into the tradition of projection witnesses, along with Helvetius, the accounts in Hoghelande, and others. Yet it was by no means the only transmutation by projection that Boyle witnessed. The "Burnet Memorandum" recounts that prior to this incident, Boyle had witnessed a somewhat botched attempt at transmutation.

[39] Boyle, *Dialogue on Transmutation,* appendix 1, 267.

[40] Becher, *Magnalia naturae.*

[41] Furthermore, there is the question of the identity of one Becket who was also present at the projections, whom Mundanus calls his friend (148) and through whom the letter to Dickinson was conveyed (1), and who is possibly the "Dr. H. B. Philochymicus" listed as the translator of Mundanus's letter into Latin (145).

In Mundanus's reply to Dickinson, he mentions "the philosophical or romantic parable called *The Journey of Philaretus to the Mount of Mercury* which came to me with your most learned letter," which, though anonymous, Mundanus suggests was Dickinson's (212–13). Dickinson does not mention this work in his letter to Mundanus, and no such work was found among Dickinson's papers by his grandson William Blomberg (*Life,* 192–93; although Blomberg owned his grandfather's papers, as he published a work on the Grecian games, "from the Doctor's own manuscript"). Philaretus is, however, a pseudonym used by Boyle in his early writings, and while "lover of virtue" is hardly a name unique to Boyle, it is intriguing to speculate that this might have been a Boylean work sent anonymously by Dickinson to the adept whose powder performed projections before them both. The only scrap of evidence for this (admittedly imaginative) notion is an alchemical manuscript (datable to the 1680s) in the Boyle archive (presented in appendix 2, text E) that is endorsed in barely legible pencil "Philaretus speaks." This text does not fit into the dialogue (where Philaretus is not a character anyway), and is the only example of Boyle's use of Philaretus outside of projects conceived after 1650. The fragment of text is, unfortunately, not noticeably "parabolic."

One brought him [Boyle] a grain pretended to be the powder of projection but that person putting it unskillfully into the lead the flame scorched his hand so that the shaking it made him lose it[;] somewhat of it stuck to the paper but so litle that the eye without help of a Microscope could not discern it[;] that [particle] he [Boyle] carefully gathered and put to lead in fusion which being so litle could not have great Operation but it stopt the fusion of the lead which made him conclude that a greater quantity might have produced some more considerable change.[42]

The fact that this minute particle of the Elixir "stopt the fusion of the lead"—that is, caused it to solidify—would have suggested that transmutation to gold was beginning, as gold has a higher melting point than easily fusible lead. Boyle's description in fact echoes one of van Helmont's own accounts of transmutation. While van Helmont freely confesses that he is not an adept, he (like Boyle and Helvetius) nonetheless claims to have been given minute portions of the Stone by generous adepti. In "*Vita aeterna,*" van Helmont writes that he was given a quarter of a grain (1/2400th of an ounce in his Flemish system of weights) of the Stone, which, when projected upon eight ounces of boiling mercury, caused "the mercury immediately to stand solid from its flux."[43] This account of van Helmont was well-known, owing not only to the notoriety of its author but also to its precise information regarding weights and proportions (e.g., the Elixir transmuted about nineteen thousand times its weight of lead) and the fact that it was reprinted as part of the introduction to Helvetius's account in *Vitulus aureus;* Boyle himself cites it explicitly in his *Origine of Fixtness.*[44]

Yet there is still a third account of Boyle's witnessing a transmutation, told thirdhand by Jean Jacques Manget in the preface to his collection *Bibliotheca chemica curiosa.* There Manget recounts his meeting with "a most reverend man, illustrious by his great merit, and now [1702] a bishop in England." This "illustrious man," judging by Manget's description and his intimacy with Boyle, was surely Gilbert Burnet, who at the stated time of this meeting (1685) was in exile and did, in fact, pass through Manget's native Geneva.[45] The story told by Burnet to Manget runs thus:

A certain unknown man, dressed quite shabbily, once visited the illustrious Boyle and, after some discourse on chemical operations, asked him to have his servants (that is, the aforesaid Boyle's) bring some antimony with certain other quite com-

[42] Hunter, *Robert Boyle by Himself,* 29.

[43] Van Helmont, "Vita aeterna," in *Ortus medicinae,* 743; "confestim totus hydrargyrus cum aliquanto rumore, stetit à fluxu . . ." Van Helmont's other account of performing a projection occurs in "Arbor vitae," ibid., 793. Note here how the Stone functions like rennet in "curdling" mercury into gold.

[44] Helvetius, *Vitulus aureus,* 828–29; the account from van Helmont's *Arbor vitae* is also excerpted here; Boyle, *Of the Mechanical Origine or Production of Fixtness* (London, 1675), 15–18.

[45] Gilbert Burnet, *Dr. Burnet's Travels* (Amsterdam, 1687); the title page claims these letters to have been written to the "Honourable *R. B.* Esq." Burnet's stay in Geneva in 1685 is mentioned on 5.

mon metallic materials, which being then (by good fortune) found in the most noble Boyle's own laboratory, he threw them into a crucible and placed the crucible in a furnace adapted for the melting of metals. After he had done so, and the contents of the crucible were seen to have melted, the unknown man then showed a little powder to the aforementioned servants, which they threw into the same crucible, and immediately thereafter he left, ordering that the crucible should be kept in the furnace until the fire had quite gone out, and promised that he would return in a few hours. But he indeed forgot his promise and was seen neither that whole day nor the following day, nor on any thereafter, so Mr. Boyle, taking up the crucible from the furnace, saw within it a truly golden material that, after subjecting it to all assays, he was forced to declare good gold, although it was slightly inferior in weight. And this is that material of which I (not without admiration) saw a portion in the possession of that illustrious man who kindly communicated this account to me, as I said at the outset.[46]

This story bears some resemblance to Helvetius's account—the visit of a shabbily dressed stranger who promises to return but does not do so. Also, the fact that Burnet carried this fragment of transmuted gold around with him recalls Boyle's own statement in the Philoponus account regarding the piece of transmuted gold that he "carr[ies] always about me with that concerne for it that such a rarity deserves."[47] Both also recall Philetas's initiation of the *Dialogue* by exhibiting a piece of alchemical gold he carried to a society meeting. As this event told to Manget is not mentioned in the "Burnet Memorandum" along with the other transmutation accounts, it presumably occurred after that document was dictated. Hunter has dated the memorandum as prior to Burnet's exile (1685–1688); the inclusion in the memorandum of the projection experience now identified with Mundanus sets a terminus a quo for the document at 1678.[48] Thus since Burnet's interview with Manget is explicitly dated to 1685, this projection by the "shabby stranger" must have occurred sometime between 1678 and 1685.

A further indication of Boyle's witness of projection occurs in the Royal Society Journal Book, where mention is made of the efforts to repeal the Act against Multipliers. The reader will recall that the repeal of this act, owing to Newton's reference to it in his letter to John Locke, was one of those long-visible signs of Boyle's alchemical involvements. The part of the statute enacted under Henry IV in 1404 that dealt with transmutation reads, "that none from thenceforth should use to multiply gold or silver, or use the craft of multiplication; and if any the same do, they incur the pain of felony."[49] The act was repealed in 1689, and according to Newton's testimony, the repeal was initiated by Boyle himself. This Royal Society Journal Book entry, first noticed by Hunter, gives independent confirmation of Boyle's in-

[46] *BCC*, 1:[iii].
[47] Boyle, *Dialogue on Transmutation*, appendix 1, 267.
[48] Hunter, *Robert Boyle by Himself*, xxv–xxvi.
[49] Quoted in I Gul. & Mar. Ch. 30; the act of repeal is reprinted in *Works*, 1:cxxxii.

volvement. The record reads, "Inducement to the Parliament to Repeal the Act of Henery 4th against Multiplication of Metalls was from the Testimony of Mr Boyle and the Bishop of Salisbury who affirm to have seen projection, or the transmutation of other metalls into Gold."[50]

It should also be noted that in all of these instances of Boyle's witness of transmutation, Bishop Burnet plays a role—either as the person to whom such accounts are told and by whom they are recorded, or as the person by whom it is recounted, or, as the note regarding the repeal of the gold-making statute implies, as a cowitness to projection itself. This involvement of Burnet not only evidences the position of the bishop as a close companion to Boyle, with whom he enjoyed "close and entire friendship" for over thirty years, but also develops further the link between Boyle's alchemical activities and issues of morality and spirituality that is made clearly in the "Burnet Memorandum" and is provocatively explored by Hunter.[51] Burnet's value as a trustworthy relater of experiences should also be considered. The important spiritual aspects of Boyle's alchemy will be fully taken up in chapter 6, where the reader will note that Boyle's alchemy was not entirely "rationalist" but rather included important supernatural aspects.

Witnessing Projection

The instances of Boyle's direct observation of chrysopoetic transmutation are surely the "more immediate arguments" that convinced him of the real existence of the adepti, as he suggests in the preface to *Producibleness.*

> what I have heard from divers very credible eye-witnesses, *and perhaps some more immediate arguments* [emphasis added], strongly incline me to thinke that there may have been, and may yet be, some such men [as the adepti], and whatever be to be thought of what they call the *Philosophers Stone,* I confess my self convinc'd by what I have seen, that there are in the World as difficult *Arcana* as diverse of those which have been (perhaps not all of them justly) derided under the name of Chymicall *non-entia.*[52]

All of these transmutations witnessed by Boyle himself point to the fact that Boyle was convinced of the reality of the Stone and of alchemical transmutation, not primarily by reading or by theoretical considerations, but rather by his own direct experience. We can thus hear Boyle speaking clearly in the *Dialogue* through the character of Zosimus when he replies to Erastus that "I have never said that I accept the Philosophical Stone because I can clearly explain how its properties and operations can be performed, but rather because I have been convinced by sufficient testimony that such oper-

[50] Royal Society Copy Journal Book, 7:213–14; quoted in Hunter, "Alchemy, Magic and Moralism," 405.

[51] On Burnet's friendship with Boyle, see H. C. Foxcroft, ed., *A Supplement to Burnet's History of My Own Time* (Oxford, 1902), 464–65; see also Hunter, "Alchemy, Magic and Moralism."

[52] Boyle, *Producibleness,* [x].

ations have actually been carried out." As mentioned at the start of this chapter, it is primarily upon experimental accounts rather than upon theory that the argument of the *Dialogue on Transmutation* rests. While recognizing the difficulty of arguing negatively from a fragmentary document, one can at least note that in the surviving text, very little of the argument comes from theoretical considerations. The ratiocinations of fragment 3 regarding the shape and activity of corpuscles are intended only to weaken Erastus's argument of the *impossibility* of alchemical transmutation by projection, not to provide a mechanism for its action or a proof of its reality. If Boyle's certainty of the truth of alchemical claims was based on his own eyewitnessing, it is not unreasonable for him to attempt to convince others by recounting these significant experiences.

The topic of witnessing is treated in detail in Steven Shapin and Simon Schaffer's *Leviathan and the Air-Pump*. There the authors rightly point to the importance of witness testimony in Boyle's writings and in the practices of the early Royal Society in terms of ajudicating experimental "matters of fact." According to these authors, this "social technology" for the production and validation of "knowledge-claims" was strengthened by a complementary "literary technology," a prose style specifically and strategically developed by Boyle for the description of experimental narratives, able to make "virtual witnesses" of readers. By packing descriptions with a wealth of circumstantial detail (time, place, persons present, performer, etc.), Boyle could make "virtual witnesses" of readers unable to be present or to replicate the experiment. These virtual witnesses then could add their "social validation" of the matters of fact to that of direct witnesses.[53] In spite of Shapin and Schaffer's own ability to compel assent from many of their readers, their portrayal of the witnessing strategy of Boyle and the Royal Society is open to some objections.

The notion of Boyle's "literary technology" has been called into question elsewhere by the revelation of less contrived and more supportable origins for Boyle's prolixity, complex sentence structure, wealth of circumstantial detail, and overly qualified prose.[54] But more problematic in light of the present study is the way Shapin and Schaffer consider probation by witness accounts novel to Boyle and the Royal Society in the context of Restoration England, and how they distinguish it from the usages of "the alchemists" in particular.[55]

One problem is the uncomplicated way in which the authors invoke alchemy as a monolithic entity inherently opposed to the "modern science"

[53] Shapin and Schaffer, *Leviathan*, 55–78; the "strategies of validation" retailed here appeared first in Shapin, "Pump and Circumstance."

[54] Principe, "Virtuous Romance," 395–96; see also "Style and Thought of Early Boyle"; Hunter, *Robert Boyle by Himself*, lxxiv–lxxv; Hunter and Davis, "The Making of Robert Boyle's *Free Inquiry*," 220 (where one of Boyle's contemporaries notes that Boyle's writing style reproduced his speech patterns, "with many circumlocutions"); Sargent, *Diffident Naturalist*, 300.

[55] The contrast between the practices of Boyle and those of the alchemists is made repeatedly in Shapin and Schaffer, *Leviathan*, 39, 55–59, 70–72, and elsewhere.

Boyle and the Royal Society were constructing. They reinforce their linkage of the Royal Society's program with Restoration social order by heaping together as their opposite "the knowledge claims of alchemical 'secretists' and of sectarian 'enthusiasts' who claimed individual and unmediated inspiration from God, or whose solitary treading [*sic*] of the Book of Nature produced unverifiable observational testimony."[56] But this grouping of alchemists with enthusiasts is not historically accurate. While most alchemists did in fact acknowledge their arcana as divine gifts, few expected dramatic divine inspirations after the manner of the "sectarians." Additionally, quite a number of Royal Society fellows—including Boyle himself, as we shall see in chapter 5—were themselves (when it came to chrysopoeia) "alchemical secretists." Thus the asserted division of the fellows from "the alchemists" is less than clear.

More pertinently, the question here is one of probation (or "validation"), and alchemical workers did not, unlike sectarian enthusiasts, rely upon the putative divine origin of their knowledge(-claims) to convince ("compel assent"). Instead, they too turned to the accumulation and retelling of experiential evidence, most notably codified in the tradition of the transmutation history. Transmutations and air-pump experiments have in common the fact that they are not accessible to everyone—both the Philosophers' Stone and the air-pump are rare commodities. Further, just as some who did acquire air-pumps encountered difficulties in their use, so too some who obtained a bit of the Stone encountered, like Helvetius, problems in its projection. Therefore, in similar ways knowledge of the "matters of fact" must come predominantly from eyewitnesses fortunate enough to have been present at the exhibition of the phenomena.

Importantly, the use of circumstantial experiential accounts in the presentation of such "matters of fact" is both less novel to Boyle's air-pump and less complicated than Shapin and Schaffer claim. While the legal tradition (as noted by Kargon, Sargent, and Barbara Shapiro) provides clear precedents to probation by witness accounts, the transmutation history presents perhaps a more surprising example of the continuity of Boyle's methods of probation with earlier traditions.[57] The accounts provided by Boyle in the *Dialogue*, while they are typical of his style, also strongly resemble those in the classical examples of Hoghelande, Helvetius, van Helmont, and Seyler.[58] The probatory importance of such accounts was a matter of course

[56] Ibid., 39.

[57] Robert H. Kargon, "The Testimony of Nature: Boyle, Hooke and Experimental Philosophy," *Albion* 3 (1971): 72–81; Rose-Mary Sargent, "Scientific Experiment and Legal Expertise: The Way of Experience in Seventeenth-Century England," *Studies in the History and Philosophy of Science* 20, (1989): 19–45, and *Diffident Naturalist*, 42–61. Barbara Shapiro, "The Concept 'Fact': Legal Origins and Cultural Diffusion," *Albion* 26 (1994): 227–52.

[58] The transmutation history continued as well long after Boyle. Throughout the eighteenth century, the last generations of alchemical apologists turned out several collections of transmutation accounts—for example, Güldenfalk's *Sammlung von mehr als hundert wahrhaften Transmutation-geschichten*.

among alchemical writers, and Boyle writes and deploys them in an identical fashion.

A further example of such accounts is the renowned case of Nicolas Flamel and his wife Pernelle. This account, published in 1612, again has a literary format full of circumstantial details regarding times, dates, places, and witnesses similar to that which Shapin and Schaffer imply is novel to the Royal Society and Boyle's program.

> The first time that I made projection, was upon *Mercurie,* whereof I turned halfe a pound, or thereabouts, onto pure *Silver,* better than that of the *Mine,* as I my selfe assayed, and made others assay many times. This was upon a Munday, the 17 of *January* about noone, in my house, *Perrenelle* only being present; in the yeer of restoring mankind, 1382 . . . I made *projection* of the *red stone* upon the like quantity of *Mercurie,* in the presence of *Perrenelle* onely, in the same house, the *five and twentieth day* of *Aprill* following, the same yeere, about five a *clocke* in the *Evening.*[59]

Furthermore, recent work has shown the existence of accounts of experimentation in early- and mid-seventeenth-century agricultural literature that likewise fulfill all the criteria cited by Shapin and Schaffer for "virtual witnessing," and it is likely that other loci in other fields could also be adduced.[60] The style for the delivery of creditable accounts did not suddenly appear as a novel (and socially motivated) development in the Royal Society's Restoration England.

One aspect of witnessing that is extended in Shapin's *Social History of Truth* is the link of credibility with the witnesses' standing in English aristocracy. This is, on the one hand, a perhaps unnecessary qualification, for at no period would anyone be surprised at the notion that some witnesses are inherently more creditable than others on account of skill, experience, or a history of veracity. In the *Dialogue,* the antagonist Erastus recognizes the differences in witnesses, and his comments do not involve the social status of the witnesses but rather their skill. "If I found that he who gives the report were distinguished by no little probity and experience, and knew what he said with sound knowledge, I confess that I would then easily credit his words, even without ever seeing his gold."[61] Even if we admit *arguendo* the necessity of high social status for witnesses, transmutation histories not infrequently list such persons as observers or inspectors of a projection. Helvetius was physician to the prince of Orange, Seyler's transmutations

[59] Nicholas Flamel, *Exposition of the Hieroglyphicall Figures,* trans. Irenaeus Orandus (London, 1624; reprint, New York: Garland Publishing, 1994), 13. In spite of the medieval date (correct for the lifetime of Flamel, a real Parisian notary), the book containing this transmutation history was actually written ca. 1600 as has been shown by Robert Halleux, "Le mythe de Nicolas Flamel, ou les méchanismes de la pseudépigraphie alchimique," *Archives internationales d'histoire des sciences* 33 (1983): 234–55.

[60] Kathleen Whalen, "Robert Boyle, Experimental Reports, and Agricultural Literature," forthcoming.

[61] Boyle, *Dialogue on Transmutation,* appendix 1, 239.

(and earlier examples performed at the imperial court) were witnessed by noblemen, and so forth.

Having shown the *existence* of such accounts in alchemical literature and the similarity of their structure and deployment there with Boyle's, I find it useful also to demonstrate how the *epistemological and moral status* of these accounts was viewed by alchemical writers. A useful example comes from the chrysopoetic writings of George Starkey, alias Eirenaeus Philalethes. The Philalethes tracts constitute the most popular and influential alchemical corpus of the second half of the seventeenth century.[62] The earliest published of the Philalethes tracts, the *Marrow of Alchemy* (1654–55), deals explicitly and at length with the importance of transmutation histories. Starkey describes the format of his treatise in such a way that much of his description, with a slightly rearranged order, is equally applicable to Boyle's *Dialogue.* Starkey, like Boyle, shows the truth of alchemy foremost "by Testimonies, and those of such who were themselves professedly Adepti, and also of such who did not pretend to the Art, so that an Art confirmed by the testimony of its own Sons, and Strangers also to it, all being men of undoubted Credit, is not questionable, but by unreasonable Cavillers."

Here Starkey, a common immigrant from the American colonies, summarizes in the preface to his account of the Philosophers' Stone the characteristic hallmarks that the sociological school has posited as novel to the "validation strategy" of the Royal Society's gentry—testimony and the creditability and multiplicity of witnesses. But Starkey continues, again in a mode notably similar to Boyle's *Dialogue,* "In the next place [the author] adjoynes firm Reasons, proving the probability of what Artists by their Art do promise; and thirdly, brings in his own experimental testimonies, concerning what hee with his eyes had seen, and with his hands handled in this particular, upon which account, as an Ocular witness, he might write with confidence, and certainty."[63]

The verse text of the *Marrow* likewise makes clear the importance of witnessing. It notes first the importance of the credibility of such witnesses: "All Ages, Countrys, Nations eke afford / Us store of Testimonies, men of worth / For skill and learning"; and then cites the *moral* necessity of believing accounts offered by such witnesses.

> . . . many witnesses of it are found,
> Of credit good, who if it were not thus,
> Must all b'adjudged false, nor is it sound
> To censure those who have the Art affirm'd
> As false, that so they may be juglers term'd.
> For by this rule there is not any thing
> May credence gain but what our selves do know,
> This would the world into confusion bring, . . .

[62] On Philalethes and his corpus, see Newman, *Gehennical Fire.*
[63] Philalethes (Starkey), *Marrow of Alchemy,* advertisement to pt. 2, [i–ii].

> Methinks if I a thing affirm'd should reade
> Or hear from one, against whom there doth lie
> No just exception, 'twere a shameful deed,
> To tax the truth of what I hear, . .[64]

This *moral* necessity of believing accounts is found likewise in Boyle's *Christian Virtuoso;* however, the pursuit of this topic into the domain of Boyle's theology, as illuminating as it is neglected, would take us too far afield from the issues at hand. In passing, I will point out that Boyle's willingness—indeed, *obligation*—to believe experiential accounts is intimately tied to the necessity of believing the eyewitness accounts of the Apostles regarding the miracles of Christ—this crucial link was glossed over by Shapin and Schaffer because they failed to acknowledge that Boyle's chief allegiance in life was not to class or country or social order but rather to biblical Christianity. Boyle's overriding concern seems to have been that undermining belief in testimonial accounts, especially in those which seem incredible, simultaneously undermines the miraculous proofs of the Christian faith offered by the testimonial accounts of the Gospels.[65]

The important point at present is that the structure, status, and deployment of Boyle's experiential accounts—whether in the *Dialogue* or elsewhere—is neither novel nor divergent from that of a significant portion of traditional alchemical writers, nor in fact from that of various other traditions. As the format and status of witness accounts did not suddenly appear or change notably with Boyle and Royal Society, the notion that these formats were constructed to further social order in the Restoration settlement is undercut.

The dichotomy that Shapin and Schaffer draw between Boyle and the Royal Society and the "alchemical secretists" reproduces the rather casual use of alchemy as a foil to "modern" science current in the older positivist historiography. The transmutation histories upon which so much of late-sixteenth- and seventeenth-century alchemical apologetics rest are not in fact significantly different from the later or contemporaneous accounts typical of the "New Science" either in technical format, emphasis on creditability, or epistemological status. Such a similarity of style, format, and status binds together the alchemical tradition and early modern science; experiential witness accounts are a locus of continuity, not of disjunction.

Alchemical Contacts

The accounts of transmutations studied above were all important experiences that had a profound effect upon Boyle. They convinced him of the reality of the alchemical adepti and their transmutatory Philosophers' Stone;

[64] Ibid., pt. 1, 4–5.

[65] See the proposition framed in Boyle, *Christian Virtuoso*, 60, and how this proposition is used to undergird the authority of Scripture.

they provided important testimony for his *Dialogue* and later for his drive to repeal the Act against Multipliers, and they weighed sufficiently heavily on his mind to occupy a substantial part of his biographical interview with Bishop Burnet. But Boyle had many other contacts with alchemists both in England and on the Continent, which, though perhaps less dramatic, help to fill out our picture of Boyle within a context of alchemical practitioners. The surviving Boyle correspondence is full of letters from alchemists promising or requesting different things. Some present Boyle with receipts or manuscripts, others dedicate books to him, while yet others ask his advice on alchemical processes. Erasmus Rothmaler dedicated to Boyle a manuscript entitled *Consilium philosophicum vel potius Harmonia philosophica* (The philosophical counsel, or rather, the philosophical harmony), which survives among the Boyle Papers.[66] Interestingly, the copy bears an endorsement in Boyle's hand reading, "Mr. Rothmaler's Booke that I had from himselfe," continued in John Warr's hand, "by way of Gift as I understood it, which I bequeath to Mr Newton the Mathematian of Cambridge."[67] Apparently this bequest was never delivered.

Several published alchemical texts bear dedications to Boyle. *The Philosophicall Epitaph* of one W. C., most probably William Chamberlyne, was dedicated to Boyle in 1673 and contained (among other things) an English translation of Helvetius's *Vitulus aureus.* George Starkey's *Liquor Alchahest* was posthumously dedicated to Boyle in 1675 by the publisher J. Astell, who there takes upon himself the Boylean title of Pyrophilus. Starkey had himself dedicated his *Pyrotechny Asserted and Illustrated* to his friend and collaborator Boyle in 1658. Likewise, Johann Seger von Weidenfeld addressed the 1685 English edition of his thoroughly alchemical *De secretis adeptorum* to Boyle. The mysterious Y-Worth, whose alchemical collaborations with Newton have recently been investigated by Karin Figala, dedicated his *Spagyric Physician* to Boyle (posthumously) in 1692.[68] None of these authors would thus have addressed their works to Boyle if he were opposed to their contents.

Johann Joachim Becher had considerable alchemical interaction with Boyle.[69] Becher left Holland for England in 1679 and seems to have had

[66] BP, 23:307–473.

[67] Ibid., 309.

[68] Karin Figala, "Zwei Londoner Alchemisten um 1700: Sir Isaac Newton und Cleidophorus Mystagogus," *Physis* 18 (1976): 245–73; Figala and Ulrich Petzold, "Alchemy in the Newtonian Circle: Personal Acquaintances and the Problem of the Late Phase of Isaac Newton's Alchemy," in *Renaissance and Revolution: Humanists, Scholars, Craftsmen, and Natural Philosophers in Early Modern Europe,* ed. J. V. Field and F.A.J.L. James (Cambridge University Press, 1993), 173–92.

[69] On Becher, see Smith, *The Business of Alchemy,* and *Johann Joachim Becher (1635–1682),* ed. Gotthardt Frühsorgen and Gerhard F. Strasser (Wiesbaden: Harrassowitz, 1993); neither book focuses on Becher's natural philosophy and chymistry, which would yet benefit from a comprehensive treatment. For an introduction to Becher's chymistry, though dated in approach, see J. R. Partington, *A History of Chemistry,* 3 vols. (London: Macmillan, 1961), 2:637–52.

much contact, hitherto virtually unnoticed, with Boyle in England. Becher's 1680 *Magnalia naturae,* already mentioned as containing the tale of Wenzel Seyler, bears a title-page reference to Boyle as the reason for its publication. Later, in 1689, Becher's posthumously published *Alphabetum minerale,* which constitutes the third part of his *Tripus hermeticus fatidicus* (The prophetic hermetic oracle), bears a dedication to Boyle. There he cites Boyle as being "not the least of the esteemers of my *Physica subterranea.*"[70] A manuscript copy of this work was originally among the Boyle Papers; it was listed in a catalog compiled by Thomas Birch's assistant Henry Miles in the early 1740s. It was probably discarded when it was recognized as not being one of Boyle's compositions.[71] This is quite a pity, for it was presumably the copy Becher presented to Boyle in the early 1680s with the request that it appear with Boyle's name as dedicatee.

Although the *Alphabetum* manuscript is not extant, several other Becher manuscripts survive among the Boyle Papers. The most interesting of these is the *Concordantia purgationis,* a work summarizing and comparing methods for preparing the Mercuries of metals (particularly silver) for use in confecting the Philosophers' Stone. This text was appended to the *Alphabetum* in the 1689 Frankfurt edition and was republished in 1719.[72] This work, as well as the *Alphabetum,* may well have been written especially for Boyle himself; it fits in neatly with the topic of metalline Mercuries that occupied Boyle's thought for forty years, as will be described in detail in the following chapter. Additionally, there exists also a shorter (and apparently unpublished) work entitled "Dr. Becheri theoria seu opinio singularis de metallorum generatione et tractatione" (Dr. Becher's theory, or a singular opinion on the generation and treatment of metals) in the same (yet unidentified) hand as the previous manuscript.[73] The same hand also wrote out lengthy notes and extracts from Becher's 1680 work *De arenaria perpetua.*[74] All of these manuscripts were probably given to Boyle by Becher himself while he was in England, where he remained until his death in 1682.

Alchemical Pretenders

Beyond these positive or, at worst, neutral interactions, Boyle also experienced a few somewhat more negative contacts with alchemical practitioners. He was well aware, as early as his first natural philosophical compositions (ca. 1650), that chymistry included not only a few adepti but also

[70] Becher, *Tripus hermeticus fatidicus,* dedication to Boyle, 99–101. "Scio, Te Physicae meae subterraneae aestimatorem esse non postremum" (101).

[71] Becher, *Alphabetum minerale,* in *Tripus hermeticus fatidicus,* 98–149. Miles's list dated 10 February 1742/43 is at BP, 36:141–43, 163–64 (bound out of order); *Alphabetum minerale* is item 116 (fol. 142) and marked in Miles's shorthand "laid aside, not his."

[72] BP, 19:57–77; Johann Joachim Becher, *Opuscula chymica rariora* (Nuremberg, 1719), 150–82.

[73] BP, 29:1–10.

[74] BP, 31:295–397.

a significant proportion of "Pretenders to that excellent Science," whom he later referred to more bluntly as "cheats."[75] Boyle seems to have encountered his share of cheats as well as "true adepts." The surviving Boyle correspondence testifies to the range of alchemists who contacted Boyle—some were sincere seekers after the Stone, but others seem to have been the kind of cheats whom Boyle contrasted with the adepts.

A straightforward instance of this unwelcome kind of an encounter occurs in a series of letters from one Gottfried von Sonnenburg written to Boyle from Frankfurt am Main in 1684–1685. After a brief letter of self-introduction, Sonnenburg sent Boyle a lengthy narrative recounting how he, by accidental sight of a secret manuscript bequeathed by an adept to his son, discovered the method of producing the Philosophers' Stone.[76] Ordinarily this would have been sufficient good fortune to render a man happy for life, but when Sonnenburg set about the process, although all the signs were good, he found that he could not afford the coals and maintenance of the furnace to bring the work to completion over nine months or more of continuous and carefully regulated heating. Indeed, he claims, he has not the money to carry this "true and infallible process" to conclusion, for "poverty is the thing that suppresses the intellect." As a result, Sonnenburg asks Boyle in a cover letter if he can obtain for him a "good man . . . devoted to the reformed religion," possibly from among the members of the Royal Society, who would be willing to pay him seven thousand pounds for the process. As part of the bargain, Boyle would get a commission of sorts—the revelation of the whole process gratis. Apparently, Boyle did not leap upon this opportunity. Six months later, on 3 May 1685, Sonnenburg wrote again to Boyle to express his amazement that Boyle had neither responded to him nor found him a buyer.[77]

Now clearly Boyle saw the obvious problems with Sonnenburg's proposal—even if one were convinced of the possibility of producing the Philosophers' Stone (as Boyle was), Sonnenburg's process, for all his protestations regarding its origin with an authentic adept, was still untried. Furthermore, Boyle was hardly likely to act as a broker for such a process and for such a sum as seven thousand pounds, thus putting his own reputation very much at risk. But Sonnenburg's last letter provides an additional reason for Boyle's "Pythagorean silence." Sonnenburg, thinking that Boyle might be dead or grievously ill, made inquiries about him of "a certain person from Holland"; this informant told Sonnenburg that "some imposter had most wickedly deceived" Boyle, and as a result Boyle was likely to be highly suspicious of similar offers. "That monster, whose intentions were polluted by sordid greed," had apparently promised Boyle some revelation of alchemical mysteries but left Boyle disappointed, "perhaps after having

[75] Boyle, "Essay of the Holy Scriptures," BP, 7:1–94, on 55; *Sceptical Chymist,* [xiii].

[76] Gothefredus à Sonnenburg to Boyle, 24 July/3 August 1684, BL, 5:112; Sonnenburg to Boyle, 15/25 October [1684], 103; Sonnenburg's prodromus, 104–9.

[77] Sonnenburg to Boyle, 3 May 1685, BL, 5:110–11.

paid out large sums of money."[78] Sonnenburg invokes this incident to protest that he is different and should not be forced to suffer because of the evil actions of another, and, as he points out, *he* is willing to give Boyle the secret for *free,* provided, of course, that Boyle extracts the seven thousand pounds from somebody else.

Sonnenburg's offer is illustrative of the kind of alchemically colored schemes that were prevalent in the seventeenth century, and that many princes and others in power or possessed of wealth (like Boyle) were likely to encounter. A similar attempt appears in an unsigned letter that begins rather abruptly, "Give me or a mutual friend three thousand pounds . . . ," in order to help some adept's disciples and to retrieve some hidden documents. The writer promises Boyle "very brief experiment" that will restore Boyle's money "very quickly." [79] But Sonnenburg's mention of Boyle's bad encounter with an imposter, which was apparently fairly well-known if it was related by an unnamed Dutchman, reveals more fully Boyle's active interest in alchemical secrets and his willingness to pay handsomely for them. There could be no better illustration of Boyle's intense eagerness to gain arcane alchemical knowledge than the curious affair in which he was engaged from 1677 to about 1680 involving the elusive Georges Pierre des Clozets; his patron, the supposed patriarch of Antioch; and the Asterism, a secret society of alchemical adepti. The story is long and convoluted, involving a large number of people and all sorts of remarkable tales and adventures. This bizarre encounter was briefly studied by the late R.E.W. Maddison but deserves a much closer look within the context of the present study.[80]

Georges Pierre and the Asterism

Boyle's interaction with Georges Pierre began with a visit from Pierre in London sometime in the summer of 1677. A letter dated 25 June 1677 signed by "Georges, patriarch of Antioch," and sent to Pierre (then copied by Pierre and sent to Boyle) orders Pierre to go to England and to visit Boyle, show him "the projection," and tell him that "it will not be long before God allows him [Boyle] to enjoy the happiness of being a true philosopher."[81] Whatever Boyle saw and heard from Pierre that summer greatly impressed him, for he soon began an extensive correspondence with Pierre; thirteen letters from Pierre survive, and these make reference to an equal number of letters from Boyle of which we presently have no direct knowledge. Boyle's

[78] Ibid.

[79] BP, 44:55; endorsed on verso "Arcana arcanè custodito [*sic*]."

[80] Maddison, *Life,* 166–76; a study of Boyle's interaction with George Pierre was to form the eighth installment of his "Studies in the Life of Robert Boyle, F.R.S." published in *NRRS* from 1951 to 1965. He began a manuscript on the subject in 1968 but made little progress, probably owing to the lack of data and the incorporation of all the material into his *Life* published in 1969. The manuscript is currently with his papers in the archives of the University of Kent.

[81] George, patriarch of Antioch, to Georges Pierre, 25 July 1677, BP, 25:337.

esteem of Pierre is clear not only from the sheer number of letters and gifts sent to him but also from Pierre's own words: "You had promised me, Sir, not to pay me any more compliments, yet your letter is still filled with them."[82]

Boyle's introduction to Pierre in mid-1677 closely precedes the 1678 publication of *Anti-Elixir,* opening the possibility that Georges Pierre may be the "Foreign Virtuoso" who gave Boyle that substance. The brief interval between their meeting and publication fits with the hurried writing and publication of *Anti-Elixir* (see p. 225), and such a gift would explain Boyle's immediate high esteem of Pierre. The patriarch of Antioch would then correlate well with the stranger's "Eastern Patron" from whom he received the powder.[83] While the correspondence states that Pierre was sent to England by the patriarch specifically to visit Boyle, *Anti-Elixir* claims that the meeting was accidental; but since Boyle was still involved with Pierre while *Anti-Elixir* was being printed, he may not have wished to reveal the whole truth about Pierre's mission.

In early 1678, Boyle received a letter dated 24 December 1677 at Constantinople from Georges du Mesnillet, who identifies himself as the patriarch of Antioch and the leader of a society of alchemical masters. In this letter, the patriarch acknowledges Georges Pierre as his agent and informs Boyle that he is to be nominated for membership in a secret society of adepti at their next general assembly.[84] The patriarch thanks Boyle for the gifts he has sent him through Pierre, and promises gifts in return. Pierre and du Mesnillet both declare that Boyle's genius in natural philosophy has been made known to this secret society through his books, and reports of his piety, generosity, and humility have likewise reached them; therefore, he is worthy to be inducted and have all the major arcana revealed to him.

Throughout the first half of 1678 letters passed at least weekly between Boyle and Pierre, as did the flow of numerous gifts from Boyle to Pierre and to the patriarch. In December Boyle sent Pierre a large telescope, assay balances, and a globe; in February, he sent copies of the New Testament (presumably the Turkish one he had helped finance in 1666) for the patriarch and a case of one hundred glass vials for Pierre. Pierre then asked Boyle to send the patriarch gifts for the Turkish court: jackets, fine fabric ("for the sultana queen mother, eight rods of flesh-colored moiré, eight of gold-colored moiré, and eight of flame-colored moiré"), and a chiming clock ("more than three feet high"), among other things. Letters from both Pierre and the patriarch promise Boyle gifts in return, or at least bills of exchange or cash from du Mesnillet's banker—Gaspard Cassati of St. Mark's Square,

[82] Pierre to Boyle, 13/23 December 1677, BL, 4:106; "vous maviez promis Mr de nuser plus de Complements cependent vostre lettre en est encore remplie."

[83] Boyle, *Dialogue on Transmutation,* appendix 1, 282. Note that the "Eastern patron" of the draft and Latin translation becomes "Eastern *Virtuoso*" in the published *Anti-Elixir.*

[84] BL, 3:6–7. The name *L'Asterism* does not appear in the correspondence until May 1678, up until which time reference is made only to an *assemblée générale.*

Venice. Pierre reckons that the gifts requested for the patriarch and the Turkish court will cost £50, but Cassati will send him two hundred pistoles (£115) to cover the expenses.[85] As Boyle's side of the correspondence is missing, we cannot tell whether these gifts or reimbursements actually arrived, with the exception of some fruits and cheeses that Pierre sent Boyle in December 1677.[86] The patriarch promises Boyle "two excellent pearls" and several shipments of gifts including gold brocade, satin, Persian cloth, a bed ("of incorruptible wood"), Chinese porcelain, silk carpets, two gold ingots, and a perpetual lamp in rock crystal.[87] In order to facilitate this exchange of gifts from the East, Pierre first advises Boyle to have himself "received into the India Company [*Compagnie des Indes*], and the quicker the better."[88] Boyle had already been appointed to a committee of the East India Company in April 1677. More strikingly, in April Pierre advises Boyle "to continue what you have begun, to do what is necessary for you to belong to the Company of Turkey."[89] Thus, apparently in accordance with Pierre's advice, Boyle requested entry to the company and was in fact made a freeman of the Turkey Company the following month on 15 May 1678.[90] While Boyle's membership in the company has long been known, it has previously been attributed to his interest in receiving natural historical reports from distant locales; this newly revealed motivation places it in a wholly different light.

While the fate (or existence) of these costly gifts is unknown, it is clear from the surviving letters that the patriarch's greatest gift did not make its way to Boyle. For in expressing concern that the two pearls travel safely to him, the patriarch writes that "they must not be lost like the blessed powder that I entrusted to Mr. Pierre. This powder was sent to Mr. Des Mulens when he was in Italy with the son of My Lord Halifax; I do not doubt that he projected it somewhere or other."[91] By *pulvis benedictus,* the patriarch

[85] On the Turkish New Testament, see Maddison, *Life,* 111. This and all subsequent conversions from French currency to pounds sterling are based upon the rate of 1.73 pistoles = £1, as mentioned in a letter relating to Pierre at BL, 1:102.

[86] Pierre to Boyle, 17/27 January 1678, BL, 4:107.

[87] Pierre to Boyle, 18/28 February 1678, BL, 4:127; Patriarch to Boyle, 24 December 1677, BL 3:6.

[88] Pierre to Boyle, 7/17 March 1678, BL, 4:109; "Je vous conseille de vous faire recevoir dans la Compagnie des Indes, et le plus tost est le meiux."

[89] Pierre to Boyle, 1/11 April 1678, BL, 4:112*v.* "Je vous conseille de poursuivre ce que vous avez commencé pour faire ensorte que vous soyez de la Compagnie de turc[quie]"; the letter is torn away at the end of the line; thus the final four letters are conjectured. On the Turkey Company, see Alfred C. Wood, *A History of the Levant Company* (London: Oxford University Press, 1935).

[90] Maddison, *Life,* 134. The Turkey Company records read "Upon the motion of Mr Deputy Governor [John Buckworth] the Freedome of this Company was conferred Gratis upon the Honble Robert Boyle, Esq. who tooke the Oath." This occurred at the meeting of the general court 15 May 1678; Public Record Office, State Papers 105/154, fol. 43*v.*

[91] Patriarch of Antioch to Boyle, 24 December 1677 (N.S.?), BL, 3:7; "non est necesse ut sicut pulvis benedictus quem D[omi]no petro confideram, amittantur, qui pulvis mittebatur ad

undoubtedly means a sample of the Philosophers' Stone, "projected" by the intermediary rather than sent on to Boyle. Pierre mentions the powder as well, saying that he gave it to a Mr. Audierne "who gave it into the hands of Mr. de Meulans."[92] This des Mulens or de Meulans is certainly the Du Moulin who was in fact tutor to Henry Saville (1660–1688), eldest son of George Saville, Lord Halifax (1633–1695). The two were in fact in Italy during 1677–1678, as the patriarch states. This Du Moulin is possibly related to the family of French Protestants that included Peter (1601–1684) and Lewis (1606–1680), with whom Boyle was familiar, but the exact relationship is at present unclear.[93] The tutor Du Moulin is almost certainly not Peter du Moulin as Maddison suggests, for at that time he would have been in his late seventies and thus unlikely to be traveling about Europe as a tutor.[94]

In January 1678, in fulfillment of one of the requirements for nomination, Boyle sent two copies of a collection of at least twenty medical receipts to Pierre. Boyle was apparently worried that he had not sent a sufficient number, testifying to his eager concern over making a good case before the Asterism. But Pierre reassures him on 18/28 February that "you do not need any more receipts for the assembly; do not feel concerned about it."[95] Pierre adds that he appended (with du Mesnillet's blessing) a few of his own recipes under Boyle's name—"a few tinctures of antimony . . . and a fixation of mercury"—sealed the packet and sent it to Philalette, the adept who will act as Boyle's sponsor at the assembly.[96] While it would be intriguing to link this "Philalette" with the elusive alchemical author and wandering adept Eirenaeus Philalethes, it is clear from related documents that he is Petrus

D[omin]um Des mulens tunc temporis in Italia cum filio Mylord halifax, non dubito quin alicubi illum proiecerit."

[92] Pierre to Boyle, 13/23 December 1677, BL, 4:106; "le sr audierne qui est presentement en angleterre, Cestoit a luy que je donné la poudre et cest luy qui la mit entre les mains de mr. de meulans."

[93] Peter had no children, but his brother Louis had one son who would be of the appropriate age for a tutor; see Henry Wagner, *Pedigrees of the Du Moulin and De L'Angle Families* (London, 1883), 2–5. The tutor's letters to Lord Halifax survive among the Devonshire House MSS and Spencer MSS; see H. C. Foxcroft, *The Life and Letters of Sir George Saville,* 2 vols. (London, 1898), 1:116 and 131 n. 1. Lord Halifax was also Boyle's neighbor, having built and taken up residence in 1673 in Halifax House on the northwest corner of St. James's Square, very close to Lady Ranelagh's house on Pall Mall where Boyle was living at the time.

Letters from Lewis Du Moulin survive at BL, 2:137 and 139; letters from Peter Du Moulin are at 141, 143, 145, and 147. Further, Boyle commissioned Peter to prepare the *Devill of Mascon* (Oxford, 1658), an English translation of Francois Perrault's account of spirit activity; Boyle's letter to Peter was published as an introduction to the volume. Peter dedicated his *Parerga* (Canterbury, 1670) to Boyle.

[94] Maddison, *Life,* 168.

[95] Pierre to Boyle, 18/28 February 1678, BL, 4:127–128*v,* on 127*v;* "vous naviez bien plus besoingt de receptes pour lassemblée que cela ne vous inquiete pas."

[96] Boyle was himself apparently interested in the receipts that Pierre sent under his name, for he asked for them in his next letter to Pierre; see Pierre to Boyle, 7/17 March 1678, BL, 4:109.

Philalepta, a member of the Asterism referred to elsewhere as being "from Canterbury in the county of Kent."[97]

On 7/17 March 1678, Pierre tells Boyle that the alchemical masters have assembled "near Nice," and that he should prepare himself "to receive the greatest of all advantages."[98] A week later, Pierre informs Boyle of some of the goings-on at this remarkably international assembly. An adept named Sabitieli showed a powder that could coagulate water into a transparent stone. A "Polish philosopher" caused plants to flower and set fruit in the space of two hours. Pursafeda, "the Chinese gentleman," exhibited flasks that he said contained a developing homunculus, a five-month-old foal, and a fox. The Asterism is awaiting three adepti "from the banks of the Ganges." Pierre reports that Philalette gave a speech to the assembly on Boyle's "excellent merits" and compared him to the mythical founder of alchemy, Hermes Trismegestus.[99]

This same letter contains an enormously interesting reference to a question Boyle had apparently asked in his previous letter. Pierre writes that "I do not know whether this Mundanus whom you mention has been in England," and further, "I am sorry that you have not seen this Mundanus of whom you told me."[100] This seems to clinch the Mundanus-Dickinson-Boyle connection that was postulated above in regard to Boyle's momentous witness of transmutation. It seems that Boyle was trying to learn more of Mundanus from Pierre, perhaps hoping that Mundanus was a member of the Asterism. This reference dates Boyle's witness of transmutation to 1678 rather than the date of 1679 given by Mundanus in his letter to Dickinson. The fact that Boyle "has not seen" him would also support the postulation that Boyle met the disciple while Dickinson met the master. Again, it would be most enlightening to have Boyle's letter to Pierre on this matter. At present, however, the reference to Mundanus in Pierre's letter suffices to show the high degree of interconnection among Boyle's alchemical exploits, and the drama of his alchemical annus mirabilis of 1677–78 when he encountered so many alchemical experiences.

About this same time Boyle received a remarkable document signed by Philalette and ostensibly from the secret assembly itself. The first half of the document, dated "at Herigo, in the county of Nice, on the 24th of February," records that Boyle's books and "a collection of receipts" were presented to the Asterism by Philalepta (Philalette), that the assembled adepti found both them and Boyle's character outstanding, and that they have therefore accepted Boyle's nomination along with those of two other

[97] BP 40:91.

[98] Pierre to Boyle, 7/17 March 1678, BL, 4:109*v;* "preparez vous a recevoir le plus grand de tous les avantages."

[99] Pierre to Boyle, 14/24 March 1678, BL, 4:110–11.

[100] Ibid., 110: "je ne scay pas sil ce mondanus dont vous me parlez avroit passé en angleterre," and 110*v:* "ce mundanus dont vous me parlez, je suis en douleur que vous ne lavez vue."

applicants—Giovanni Baptista Baldi, a nobleman of Bologna, and Peter Wark, a commander from Warsaw. (Pierre would later inform Boyle that twenty-seven people were proposed, but of these only three were accepted.)[101] The official declaration then stipulates that the three successful nominees present themselves before the assembled adepti at Herigo on 6 May. Herigo (as we learn from Pierre) is the castle of Don Gabriel, a cousin of the duke of Savoy, who was received into the Asterism two years previously at their assembly in Otranto.[102] The second half of the document, dated 7 March 1678 and entitled "transcript from the journal-book of the Cabalistic Society of Philosophers in an ecumenical and general meeting at Herigo," carries a message from du Mesnillet, Philalepta, and Domingo Alvares (of Zaragoza) that nominates Boyle to the post of treasurer for France, England, and Spain, reminds him to appear on 6 May, and allows him the option (in the event of illness) of sending one of three men in his place as a proxy—Johannes Baptista de Montclar (a canon from Lyons), Francis Claudius Picolomivisti (from the duke of Savoy's presidium in Chambéry), or Georges Pierre.

Another letter and mandate arrived at about this time from the Asterism; Pierre refers to it in his letters of 1/11 and 11/21 April, noting that it is from Philalette himself at Herigo, even though he "made a mistake in addressing it from Caen and not Herigo."[103] This letter and document survived until the 1740s when Henry Miles sorted the Boyle archive; he listed it in his inventory as "1678—from Caen with a Mandate from Philosophers' Assembly—in very uncouth English, not very intelligible Signed by a hand I cannot read."[104] Unfortunately, Miles seems to have destroyed this valuable item along with another valuable part of the correspondence from Herigo mentioned below.

On 14/24 March Pierre states that he has received orders to appear on Boyle's behalf if Boyle should so designate him. But apparently Boyle requested special considerations, which Pierre tells him will be debated on 7 April. This request from Boyle may explain a further letter from Herigo signed by Sephrozimez, who identifies himself as an adherent of the patriarch, the leader of the monastery on Mount Sinai, and one of the unsuccessful nominees at the Herigo assembly.[105] He introduces himself as the person who transmitted the patriarch's 24 December letter to Boyle, and thus is identifiable with the "certain monk" whom the patriarch then mentioned as the courier of a letter to Boyle.[106] Indeed, the Patriarch's December letter,

[101] BP 40:91; Pierre to Boyle, 14/24 March 1678, BL 4:110–11, on 111.

[102] Pierre to Boyle, 1/11 April 1678, BL, 4:112.

[103] Pierre to Boyle, 11/21 April 1678, BL, 4:114.

[104] BP, 36:144.

[105] Sephrozimez to Boyle, 6 April 1678 (N.S.), BL, 5:76. The leader of St. Catherine's monastery on Mount Sinai bears the title of Archbishop and is autonomous.

[106] Patriarch of Antioch to Boyle, 8 February 1678, BL, 3:8. The "Superior of Mt. Sinai" is mentioned as speaking with du Mesnillet in Pierre's letter of 17 May 1678, 4:117.

which Pierre says the patriarch "employed a monk to write . . . to which he had time only to append his signature, being occupied with the business of the Seraglio," is in the same handwriting as this letter from Sephrozimez.[107] Sephrozimez tells Boyle that he now has until 6 June to appear either in person or by proxy; if he chooses one of the approved representatives, he should send whichever of the three he chooses six hundred livres (about fifty-three pounds) to help defray the travel expenses. In either case, Boyle or his proxy must go to Aix-en-Provence and the Asterism will pay for their travel thence to Herigo.[108]

Pierre's next letter reiterates what Boyle must do to accept the invitation to membership. He must write letters of thanks and acceptance to the assembly and an address (in Latin) that will become part of their records. He must choose who will go as his representative, and who will accompany him, and must supply the representative with a letter of recommendation and six hundred livres. Pierre explains that the journey is long and expensive, for if he were to go, he would have to travel 350 leagues past 260 tolls (this distance must be for a round trip, assuming that Aix-en-Provence is the destination). The representative will "carry out all the necessary functions" at Herigo, and Boyle can give all necessary satisfaction of allegiance personally at a planned colloquium of adepti to be held in London in the fall, perhaps in Boyle's own house.[109] Pierre informs him that the assembly also offers the option of Boyle's going to Lyons on 30 June, when "six of the principals to whom you must swear allegiance" will be there.[110]

The vow of allegiance Boyle must make to the society troubled him, as did vows in general. Boyle's "great (and perhaps peculiar) tenderness in point of oaths" is now well recognized, and it was on account of this peculiarity that he declined the presidency of the Royal Society (which required swearing an oath of office) when it was offered to him in 1680.[111] Apparently Boyle expressed such reservations to Pierre, for the patriarch's agent tried to reassure him on more than one occasion that it would involve nothing "against mankind and the interests of a true Christian toward his Savior Lord, or against his king and republic."[112] As might be expected of Boyle's "scrupulosity," his unease did not abate so easily, and in another letter Pierre

[107] Pierre to Boyle, 18/28 February 1678, BL, 4:127–8*v*, on 128*v*; "il me dit quil vous a fait escrire par un moine et quil neut que le temps de signer a cause daffaire audedans du serail." On the meaning of *seraglio*, see n. 141.

[108] Here again, the exchange rates mentioned in BL, 1:102 are employed, viz. 11.25 livres = £1. The letter mentions "Aquenses" as the rendezvous point, which could be either Aix-en-Provence or Aix-en-Chapelle (Aachen), but the locality of Herigo in the province of Nice strongly suggests the former identification.

[109] Pierre to Boyle, 1/11 April 1678, BL, 4:113; 3/13 May, BL, 4:115, specifies November for the "London colloquium."

[110] Pierre to Boyle, 11/21 April 1678, BL, 4:114.

[111] Hunter, "Conscience of Robert Boyle," 153–57.

[112] Pierre to Boyle, 1/11 April 1678, BL, 4:112–13, on 112*v*; "contre lhomme et linterest dun veritable Chrestien envers son seigneur savueur ny contre son Roy et La Republique."

again tries to reassure him that the vow "will hold you to nothing but your service, and will not lead you into any temptation, but will serve as a spur for you to love and serve God the more perfectly, and in another fashion than you have done hitherto."[113] In addition to his ordinary aversion to oaths, Boyle might well have feared that making a vow to a corporate body overseas might be, or at least appear, treasonable. As it turns out, Boyle persisted in his refusal to acquiesce in a vow of any sort, and an exception seems to have been made for him.[114]

Boyle sent the money along with a "Latin expression of faith" and the necessary letters for the assembly; Pierre acknowledges their receipt on 13 May. But Pierre has encountered a new difficulty—consequences of his relationship with Mr. le Moine. Le Moine has appeared from the beginning of the correspondence and seems to have functioned as an intermediary between Boyle and Pierre; clearly he was known to Boyle and resident in England for at least some of the correspondence. As far as can be made out from the letters, Pierre showed le Moine something he should have shown only to Boyle, and as a result the Asterism has sentenced him to three months in prison. Thus Pierre asks Boyle for a letter of support to beg "these gentlemen to let me off my prison sentence that has been brought on me by Mr. le Moine."[115] Boyle obliges and writes at least one letter for which Pierre thanks him—"all that you write to the Asterism on my behalf is very fine, and I hope for a happy outcome."[116]

The Pierre affair, which has admittedly been odd with its Antiochian patriarch in the Turkish court and Chinese alchemists wandering about with embryonic homunculi in flasks, now becomes yet more bizarre. Pierre writes on 20/30 May 1678 that he has had to delay his departure for Herigo because of serious injuries from an exploding cannon—"my lower jaw is broken, and I had a hole in my forehead that was packed with two ounces of linen wadding."[117] Pierre had asked for such cannons ("two feet long, using half-pound balls, and weighing about twenty pounds") as a gift from Boyle to fulfill a request for them by du Mesnillet, who was to use them as a present to the ladies of the Turkish court![118] Boyle wisely (considering the tense relations between France and England in 1678) declined to send

[113] Pierre to Boyle, 11/21 April 1678, BL, 4:114; "ne vous obligera a rien qu[']a servir et ne vous induira dans aucunnes tentations mais vous scervira comme deguillon pour aymer et servir dieu plus parfaictement & dune aultre maniere que vous ne lavez fait cy devant."

[114] Pierre to Boyle, 20/30 May 1678, BL, 4:119; the text is not entirely clear, but writing from Herigo, Pierre states, "vous avez este audessus des lois en ce rencontre car nen ser[vi]t[ez] vous ne f[er]ez point de novitiat et avez tout le pouvoir de Nostre patron de vostre Coste."

[115] Pierre to Boyle, 3/13 May 1678, BL, 4:115–16, on 116 *v;* "menvoyez une lettre non latine en vostre langue vulgaire par laquelle vous priez ces mrs de mescompter de la prison voyla ce que mr le moyenne ma attiré."

[116] Pierre to Boyle, 20/30 May 1678, BL, 4:119.

[117] Ibid.; "Jay eu la maschoire Inferieure cassée et un trou au frond ou lon mettoit 2 once des Charpies."

[118] Pierre to Boyle, 7/17 March 1678, BL, 4:109; "2 pettits Canons de fonte de 18 poulces ou de 2 piedes de lange pour les dames du serail, et qui fuseau de Calibre de demy livre du balle, de la pesanteur denvirons 20 lbs."

them.[119] Thus Pierre had them cast in France, but when at the request of a friend he fired one, it burst and killed four people. Pierre, not surprisingly perhaps, was sued and had to pay more than seven thousand livres (£622) in damages. Although this little incident delayed his departure for the assembly at Herigo, he assures Boyle that his traveling companions have arrived and they will leave the next day.

On 15/25 June, Pierre writes to Boyle from Herigo. There, in Don Gabriel's chamber, he and "Pyrocles" were visited by a deputation of the "Cosmopolites."[120] These officials announced with much pomp that they now "reveal the learned secrets of their school of transmutational philosophy" to Boyle. Pyrocles accepts in Boyle's name the letters of membership and "a great chest locked with three keys," which holds a "book of life" containing

> the true interpretation of all our emblems and all our formulations, which have been employed by the inhabitants of the chemical mountain to hide their foliated earth from unbelievers, the sworn enemies of God, and the allegories, parables, problems, types, enigmas, sayings of nature, fables, portraits and figures of the foster children of Nature.[121]

"Pyrocles," which seems to be an alias for Philalette, is supposed to relay messages between Boyle and the Asterism, and then to visit Boyle in London, presumably bringing with him the "great chest."[122]

In spite of this joyous event, Pierre is still in trouble and due to be locked up, for which he asks Boyle's help again, including an advance of eight hundred livres (£71). Another letter from this high point in the Pierre affair existed until the 1740s. Henry Miles listed it in his catalog simply as being in "a mystic strain from Herigo, June 1678." As this date corresponds to the date of 6 June given by Sephrozimez for Boyle to appear himself or by proxy before the assembly, it was presumably a document regarding his admission to the Asterism; as Miles notes that it was in English (the only language he

119 War between France and England was widely expected, and Pierre comments on the situation thrice (13/23 December 1677, BL, 4:106*v*; 14/24 March 1678, 4:111; 1/11 April 1678, 4:112). Additionally, the Turkey Company records for 1678 are full of their fears of a "breach with France," and the consequent disruption of shipping, a concern that arises throughout the Pierre correspondence. See State Papers 105/154, fols. 35*v* (12 January 1678); 105/114, 104–13: "War with France (now daily expected to bee openly declared)," on 109–10, 8 February 1678. On 20 March 1678, Parliament passed an act forbidding trade with France for three years.

120 Pierre to Boyle, 15/25 June 1678, BL, 4:125.

121 Ibid.; "pour le sr boille . . . que lasterisme a ouvert aujourdhuy les scavants secrets de son Escolle philosophalle Transmutatoire." "la veritable Interpretation de toultes nos Emblesmes et de tous nos stilles, desquels les habitants de la montagne chymique se sont servis pour cacher leur terre feuillée aux Impies Ennemis Jurez de dieu, & des doctes nourrisons de la Nature, leurs allegories, parabales, problesmes, Types, Enigmes, dires naturals, fables, pourtraictes & figures."

122 Although an alias for Philalette, itself an alias, may seem peculiar, Pierre also uses the aliases Theophile and Ignace for the patriarch. The clearest proof that Pyrocles is Philalette is that Pierre first asks Boyle to write a letter to Philalette regarding Pierre's prison sentence (7/17 May, BL, 4:117) and then thanks Boyle, saying that "all that you have written to Pirocles for the Asterism on my behalf is very fine" (20/30 May, BL, 4:119).

could read competently), it was also presumably from Philalette, who would write to Boyle (as Pierre states) in English.[123] Unfortunately, this letter, like the aforementioned "Mandate," was probably destroyed by Miles; the rest of the Pierre correspondence seems to have survived thanks only to Miles's ignorance of foreign languages. Both lost items were in "very uncouth English" and were therefore readable by Miles, who found them either irrelevant to or inappropriate for Boyle's eighteenth-century image. The rest of the Pierre documents are in Latin or French and were therefore unintelligible to Miles, and so were spared, presumably on the premise that they might contain something of value. This may explain why there exist very few alchemical letters in English but dozens in Latin and French. Modern historians can reflect thankfully on the poor language instruction afforded eighteenth-century Dissenting ministers; had Miles read Latin or French, we might have no record at all of Boyle's alchemical correspondence.[124]

A month later, in July 1678, Pierre writes to say that the Asterism has ordered him to travel to Bordeaux and thence into Spain and Portugal.[125] But now uncertainties seem to have arisen on Boyle's end, for after another month, Pierre (allegedly from Bordeaux) tries to reassure Boyle and quell his "strange fears."[126] It seems that one Mr. la Marche, who according to Pierre was employed as a scribe for the assembly but dismissed for disloyalty, wrote a letter to Boyle through a Mr. Carbonnel and aroused his suspicions. But Pierre protests that la Marche is a scoundrel, for he not only wrote this disquieting letter to Boyle but after his dismissal grew suspicious of the assembly and wrote of it to the king (presumably Louis XIV), who sent spies and "dispersed most of the Cosmopolites." Worse yet, presumably at the bidding of the nervous king, the "guard of the magazine in the castle of Herigo, having been bribed by money or other means, laid a fuse that blew up the walls, the bastions, and the superb ramparts and more than thirty of our masters."[127] Pierre asks Boyle not to be "overcome by sadness" at this news of the effects of la Marche's "pernicious designs," but this was apparently the last time that Boyle heard from Pierre; at least no later letters survive.

It was possibly during this time of "strange fears" and Pierre's disappearance that Boyle received a short note in French from a corporate body signed only as "your humble and obedient servants, B. R."[128] The writers

[123] Pierre to Boyle, 17/27 January 1678, BL, 4:107.

[124] Clearly the papers were not purged as thoroughly as the letters (their volume is greater by almost an order of magnitude), or the English fragments of the *Dialogue* might not have survived.

[125] Pierre to Boyle, 14/24 July 1678, BL, 4:121.

[126] Pierre to Boyle, 10/20 August 1678, BL, 4:123. An early (and quite unsuccessful) attempt to transcribe this letter out of Pierre's cramped handwriting occurs at BL, 8:37.

[127] Pierre to Boyle, 10/20 August 1678, BL, 4:123; "le garde magazin du chasteau de herigo ayant esté gaigné par argent ou altre [*sic*] part mit une mesche qui a fait saulter les Murs les Bastions et le superbe bastiment et plus de 30 de nos maistrez."

[128] B. R. to Boyle, undated, BL, 6:76.

note that "it is with great distress that we have learned of the complaints you have concerning us," but strive to reassure Boyle and urge him to be patient, for he "will have cause to praise us in the future." The tone, anonymity, and text all suggest that this missive originates with the Asterism. As proof of their goodwill, they promise to send "through the person who has served us until now"—presumably Pierre—"a menstruum proper for everything that will lose neither weight, quality, nor virtue."[129] The description of this solvent fits that of the eagerly sought-after Helmontian alkahest—a universal solvent (in which Boyle was greatly interested) capable of dividing all bodies into their principles, and then able to be distilled from them with no loss of strength—or at least something quite like it. (Boyle was very keen to obtain this solvent; besides numerous references to the alkahest in the *Sceptical Chymist,* Boyle's ca. 1680 catalog of his tracts on the "Hermetic Philosophy" mentions one entitled "Of the liquor Alchahest and other Analizing Menstruums," and Boyle also mentions this tract in his preface to *Producibleness.*)[130]

While Boyle never (to our knowledge) heard from Pierre again, he did receive further intelligences regarding him. In September 1678 he wrote a letter of inquiry to one Descqueville, a colleague of Pierre's in Caen, whom Pierre had mentioned to Boyle in January as a "good gentleman" and under whose name he sent a bill of exchange to Boyle.[131] Descqueville's response primarily regards an opiate water that Boyle had sent to Pierre to be given to Descqueville; Descqueville believes that the water was adulterated or exchanged for a worthless liquor, and wishes to verify its appearance and provenance with Boyle. He returns a "memoir" with his letter, claiming it to have been given to him by Pierre, who said it was written by Boyle and accompanied the opiate. This returned memoir is probably the mutilated French manuscript signed "Boyle" that survives in the archive. The handwriting is not identifiable with any of Boyle's known amanuenses, and the contents—which describe the use and cost of the opiate—do not sound like genuine Boyle. Thus it is probably, as Descqueville feared, a forgery.[132]

This forgery puts Pierre in a bad enough light, but Descqueville's letter contains further damning statements. Descqueville asks about the bill of

[129] Ibid.; "nous vous promettons de vous envoyer part ecrit par la personne qui nous a servi Jusques a present un menstrue propre a toutes choses qui ne perdra ni son poids ni sa qualite ni sa vertu."

[130] RSMS, 198:143*v;* Boyle, *Producibleness,* [xii]. Some of this tract may have survived until the 1740s; Miles's catalog of the Boyle Papers (BP, 36:141) mentions five manuscripts on menstrua: "#64. preface to tract about Menstruums by way of L[ette]r to a friend p. 2"; "#66. Of Kinds multiplication and use of Menstruums pages 17"; "#67. Of Menstruums pages 21"; "#68 Of Kinds multiplication of Menstruums pages 30"; and "#69 heads disc[oursing]. about Menstruums 2 pages." None of these is extant.

[131] Descqueville to Boyle, 22 September 1678, BL, 2:119; Pierre to Boyle, 17/27 January 1678, 4:107.

[132] BL, 1:102. The lower part of the sheet has been cut away; several lines of text are missing.

exchange, claiming that "it was my money," and wants to know for what purposes it has been spent (the forged memoir mentions the eight pounds Desqueville sent and itemizes the costs involved in making the opiate). He then relates that Pierre "since his journey to England last year [1677, when he met Boyle] has not left this town." His only excursions have been to nearby Bayeux where there lives a "girl whom he got with child and for which the order of his arrest has been given." Pierre himself left Caen a week previously and had been telling people for quite some time that "he would be making a great journey in September for you [Boyle] and for Monsieur le Chevalier du Mesnillet."[133]

No surviving letters from 1679 deal at all with Pierre, but in 1680 Boyle received two further notices of Pierre that partly or wholly contradict the negative views expressed by Descqueville. One correspondent named as De Saingermain, who was on very good terms with Boyle (he visited Boyle several times, corresponded with him from before 1680 until at least 1685, carried gifts from Boyle to Pierre Bayle, visited Leeuwenhoek, and collaborated with Boyle on chymical processes), wrote to Boyle in February 1680 in response to his request for information "concerning Mr. raxppx." Once this code is cracked (see chapter 5), Mr. raxppx turns out to be Mr. Pierre, and the balance of De Saingermain's coded epistle is clear. De Saingermain has found that Pierre is in fact "intimate with the most illustrious figures of this century, he has the honor of being disciple to certain of them," and, more interestingly, "he is not lacking in powder [rlntpx = poudre]."[134] This *rlntpx* is certainly a reference to the transmuting powder or Philosophers' Stone. De Saingermain adds that "those here [Paris] gave me a very complimentary account of him," and he offers to transmit letters to Pierre on Boyle's behalf. A copy of an unsigned letter to an unnamed "Madame" regarding transmutation, which mentions Georges Pierre very favorably, may be a result of De Saingermain's inquiries.[135]

De Saingermain also tells Boyle about a gold ducat recently whitened at "our Academy"—"one could thus always extract and replace the color" and the volatile tincture extracted could be "put to good use in the transmutations of metals." He ends by appending another receipt for the decolorization of gold (tLb = d'Or), which the reader will recognize as an example of the transplantation of golden tincture I discussed in the preceding chapter. This coded recipe recurs in uncoded form elsewhere in the Boyle Papers,

[133] Descqueville to Boyle, 22 September 1678, BL, 2:119; "lontemps quil nous eust dit a tous leur quil faysoit dans le mois de Septembre Un Voyage pour Vous et pour monsieur le chevallier du Mesnillet."

[134] D[e] S[aingermain] to Boyle, 26 February 1680, BL, 6:41–42. On deciphering this code, see Principe, "Robert Boyle's Alchemical Secrecy," 66–67. For more on De Saingermain, see chapter 5.

[135] BL, 6:22–23*v;* the letter is dated from Caen, 11 February 1677, which was before Boyle met Pierre, and recommends Pierre to the addressee for his merit and knowledge of transmutational arcana. The handwriting is very similar to De Saingermain's.

among a collection of chymical receipts that presumably came from De Saingermain.[136]

De Saingermain's next letter (May 1680) shows his solicitude in finding a secure means of transmitting letters from Boyle to one "Mr. Aristander."[137] In view of the secrecy and delicacy with which this transmission of letters is handled, and the pains taken to "oblige him [Aristander] to reply to [Boyle] immediately," it seems that Aristander (hardly a likely French name) is Pierre. Boyle was still attempting to contact him, and De Saingermain has found Pierre's trail and a mutual friend, one Monsieur Verpy at Paris, willing to transmit letters securely to him. This helpful service, however, would prove futile, for by the date of de Saingermain's letter it was too late for Boyle to get any more responses from Pierre, as we shall see very shortly.

On another front, about the time of Pierre's disappearance, Boyle wrote to la Marche, the "traitor and scoundrel" whom Pierre condemned, but received no reply. Several years later, in 1682, however, la Marche wrote to Boyle asking for his opinion of a chymical receipt and seeking employment in England. Boyle responded in a letter of 15 June 1682 (the only known letter from Boyle relating at all to the Pierre affair) in which he dismissed the receipt and then told la Marche that he would not find employment in England. But Boyle continued in an unusually strong tone that he was not inclined to help la Marche in any way because "have some reason not to take it very kindly, that when I wrote to you three or four years ago, about Mr. Pierre des Closets, and seemed much concerned to receive an answer, (and therefore, if I mistake not, wrote more than once) you never vouchsafed me a line of answer."[138] Boyle adds that la Marche might redeem himself somewhat if he helps a certain person (whose name is left blank in the copy) "to recover all those papers of mine, that are to be found among those of your great friend monsieur Pierre, or wherever you know them to be." No further communication with la Marche is known.

The conclusion to the story of Georges Pierre des Clozets comes in "an exact copy of that concerning Mr. Pierre which is contained in a letter from Caen to a person in London, 17 July 1680."[139] While this letter is presumably the result of Boyle's inquiries, its author and addressee are unknown. The letter itself bears the endorsement (in the manner of an address) "Monsieur Le Mênillet, &c" in Boyle's own hand. Coupled with this endorsement is the puzzling evaluation of Henry Miles (partly in his shorthand): "in Cypher, & not his [Boyle's]"; it is unclear to what "cypher" Miles is referring, for there are none in the letter, unless the letter originally enclosed another document that Miles discarded.

According to this unsigned document, "the famous Monsieur Pierre des Clozets" returned to Caen in late 1679 or early 1680 from a yearlong jour-

[136] BP, 27:287–89.

[137] De Saingermain to Boyle, 1/11 May [1680], BL, 2:117–18.

[138] Boyle to De La Marche, 15 June 1682, *Works,* 6:60.

[139] BL, 6:43–44.

ney of which he told "marvelous things." He entered the town with great ostentation "on a horse worth sixty louis, with a servant no less well mounted, not to mention another packhorse, laden with the richest clothing and other furniture and very rare and precious curiosities." He stayed for a while "at Fontenay Abbey working on chymical preparations," and waiting for the order of his arrest from Bayeux (stemming from his having "abus[ed] a young lady of quality") to be lifted. Once this was done, he displayed himself around Caen in great splendor and opulence of clothing, carriage, and equipage. He then, "to flout gossip and let it be seen that his condition was not of false brilliance," purchased an estate in Bretteville costing fourteen thousand livres (£1,244), which he paid in gold. That spring he set about building and planting on the site, but he soon became ill with an inflammation of the lung and died in May 1680. His family inherited the estate and continued in mourning to the date of the letter. The missive closes with words that perfectly sum up the mystery of this whole (mis)adventure: "Such was the end of this man, whose character has been so little known, and after his death we know even less than when he was alive."[140]

What is one to make of this curious affair? It is easiest to dismiss it all as a confidence trick, as Maddison did in 1969; Desqueville's letter appears to support this conclusion. But merely denominating Pierre as a con artist seems not only too facile but somewhat beside the point, and smacks too much of the dismissive spirit that once rejected all of alchemical thought as unworthy of investigation. The interest of the Pierre affair does not revolve exclusively about whether or not Pierre was a charlatan. Rather, we should be ready to learn what Boyle's involvement in this affair tells us about Boyle's attitude toward alchemy, secrecy, and a number of ancillary issues. It seems judicious, however, to attempt to ascertain, as far as is now possible, where (if anywhere) the truth lies in Pierre's correspondence with Boyle. Investigation of the names and places cited in the correspondence raises a fair number of correlations that seem to argue for a greater complexity to the whole affair than the tendency to dismiss it all as a simple hoax would allow. The affair involved many people, some presumably quite well known to Boyle, and Pierre seems oddly well informed about Boyle's friends, colleagues, contacts, and even neighbors.

One of the main characters in this business is Monsieur le Chevalier Georges du Mesnillet, patriarch of Antioch. Is this patriarch of Antioch only a self-validating fiction of Georges Pierre? Descqueville's letter, even while indicting Pierre, refers to du Mesnillet in the same breath with Boyle as if he were known to the writer. Further, the 1680 letter about Pierre's death seems to be addressed to du Mesnillet. If du Mesnillet is a real person, who is he? To begin, it is quite unlikely that a man with so French a name would be patriarch of Antioch or that a Levantine churchman would bear the title of

[140] Ibid.; "telle fut la fin de cet homme, dont on a si peu connu le caractère, qu'apres sa mort on en sçait encor moins que lors qu'il estoit vivant."

"Chevalier." Even less likely is the proposition that a Christian cleric would have the authority to discharge affairs in the seraglio of the Turkish sultan Mehmed IV.[141] Indeed, the patriarchate of Antioch (which by this period was actually resident in Damascus) was led by Neophytos, who occupied the see from late 1672 to 1682, having temporarily deposed Cyril V.[142] Even though there were rival patriarchs representing Oriental churches not in communion with the Orthodox Church of Constantinople (e.g., Andrew Akidgean, consecrated in 1662 as Syrian Catholic patriarch; and the Maronite patriarch from 1670 to 1704, Stephanus Dovaihi Ehdenensis), the obvious Frenchness of the name du Mesnillet suggests that he bears an old crusader title.[143] Indeed, the Latin Church continued to name Latin patriarchs of Antioch (who never set foot in the Levant) long after the demise of the crusader states. The titular patriarch for 1678, however, was no du Mesnillet but rather one Alessandro Crescenzi.[144]

It is noteworthy also that when Boyle is named to the post of treasurer of England, France, and Spain, the nominators are Philalepta from Canterbury, du Mesnillet, and Domigo Alvarez from Zaragoza; thus du Mesnillet seems to be representing France. The correspondence claims also that the "patriarch" had a sister in France and an uncle who was *premier président* of the Parlement of Dijon—unlikely connections for an Oriental prelate.[145] During the time of the Pierre correspondence, the Dijon Parlement was led by Nicholas Brulart, marquis de La Borde (1627–1692), who held the post of *premier président* from 1657 to 1692.[146] I have been unable, however, to trace his nephews at this time.[147]

[141] Readers should note that the term *seraglio,* which now connotes the harem in particular, did not have so exclusive a meaning in the seventeenth century, at which time it was often used synonymously with the Turkish palace or court, the Grand Porte. Thus a patriarch writing from the seraglio need not evoke an image as incongruous (not to say shocking) as modern definitions would seem to render it.

[142] *Dictionnaire d'Histoire et de Géographie Ecclésiastiques,* ed. A. Baudrillart (Paris, 1924), vol. 3, cols. 643–44.

[143] Giuseppe Simone Assemani, *Series chronologica patriarchum Antiochiae,* Catalogue of Maronite Patriarchs (Rome, 1881), 39–40; *Catholic Encyclopedia,* s.v. "Antioch."

[144] *Géographie Ecclésiastiques,* col. 625; the author's source is "l'Index des sieges tituliares déposés aux archives de la Congrégation consistoriale au Vatican." Crescenzi gained the title around 1675, which is the same date mentioned by du Mesnillet as the start of his patriarchate, but that is presumably a coincidence.

[145] Patriarch of Antioch to Boyle, 8 February 1678, BL, 3:8; Pierre to Boyle, 7/17 March 1678, 4:109.

[146] M. De la Cuisine, *Le Parlement de Bourgogne,* 2d ed. (Dijon and Paris, 1864); a list of the members and officials of the Parlement is recorded in 3:415–29; on the post of *premier président,* see 1:97–104.

[147] Beyond these leads, the pedigree of the Du Moulin family (which is implicated in this business at least insofar as one of its members was to be the carrier of the portion of the Philosophers' Stone destined for Boyle) provides some intriguing but inconclusive relationships. The Du Moulins of Boyle's acquaintance were descended from one Denis Du Moulin, who, after the death of the wife who bore him a single son, took orders and became first the archibishop of Toulouse, then the bishop of Paris, and then in 1445 was named Latin patriarch

It is reasonable to conclude, then, that du Mesnillet (real or not) bore the title patriarch of Antioch as a sign of his primacy in the Asterism. The historical preeminence of the see established by St. Peter at "Antioch, that stately City where the disciples of our Saviour did first derive their title from their master's name," would render its patriarchate suitable as a sign of primacy.[148] The difficulty in simply making du Mesnillet a Frenchman is that he represents himself as writing from Constantinople and Antioch, requests gifts for the Turkish court, and has Pierre ask for letters from Boyle to English agents in various Levantine cities. It might be advanced that there exists a code of secrecy—of which Boyle was cognizant—throughout the correspondence, and in this case the names of Eastern cities are used as cover names for French cities. This supposition is indeed supported by the fact that Pierre's letter of 7/17 May 1678 is dated at *Athens* even though in the text he states explicitly that he is writing from Paris. Such a notion might be extended into a romantic reading of the affair, with everything being a code for something else, and might suggest that Turkey is really France, that the Great Lord is not the sultan but Louis XIV, that the *Porte* in Constantinople is the court at Paris, and perhaps even that the Greek Church is really constituted of French Protestants. Caen was after all an important Protestant city, and Boyle dispensed advice to the patriarch on the Greek Church. One might then suggest that the Asterism is a group of conspirators, and that alchemy is a cover for a subtext of political or denominational intrigue. However titillating such an imaginative reading might be, there is no evidence to support it, and Boyle's notable aversion to political involvement argues strongly against it. Also, in spite of Caen's Protestant tendencies, Pierre and his family were Catholics. Furthermore, our attempt to make du Mesnillet a resident of France with a lofty title runs into the problem that Boyle apparently *believed* his residence was in Constantinople, for he sent him his Turkish New Testament and joined the Turkey Company so that he could obtain his gifts more easily, indicating that he did in fact expect them to come from the East. If there were a coded subtext capable of rendering the details of this episode more plausible, Boyle himself would have to have been aware of it.

---

of Antioch. This relationship is rather proudly displayed in the autobiography of Pierre du Moulin the elder—father of Peter and Louis—who also called his ancestor a cardinal, even though Denis actually declined that dignity offered to him in 1440 by the antipope Felix V. Now, while this may seem too distant to have importance in the Georges Pierre mystery, it seems less unrelated when we note that the same autobiography avers that when Pierre Du Moulin came to England with his sister Esther in 1615, he was befriended by a fellow expatriate who eventually married his sister and whose name was du Mesnillet. At present, the children (if any) from this union are not known (du Mesnillet is not mentioned at all in Wagner, *Pedigrees*). Esther eventually returned to France; there she married René Bouchart and settled in Caen, Georges Pierre's hometown, where she died in 1641.

[148] The quotation about Antioch derives from Acts 11:26 and is the opening line of Boyle's original version of *Theodora,* St. John's College (Oxford), MS 66A; see Michael Hunter, "A New Boyle Find," *British Society for the History of Science Newsletter* 45 (October 1994): 20–21.

The castle of Herigo presents further clues and problems; no place by that name has yet been identified. The lord of the castle of Herigo, Don Gabriel, "cousin to the duke of Savoy," is a real person. (Maddison mistranscribed this name as the meaningless "Doubabiel.")[149] Don Gabriel, prince of Savoy, was the great-uncle of the then-child duke of Savoy, Victor Amadeus (reigned 1675–1730), and was in command of the troops of Savoy in their war against Genoa in 1672 and in the "Salt Wars" in Mondovì in 1681–1682.[150] *Cousin,* a very fluid term of kinship in the seventeenth century, might refer either to Don Gabriel (great-uncle to the duke) or to one of his descendants who shared his name and title.

Caen was an important intellectual center in the late seventeenth century, containing the only organized French provincial scientific society of the seventeenth century, the Académie de Physique, and a society of men of letters and an important university.[151] One of Caen's chief lights, Pierre-Daniel Huet (1630–1721), had some interest in chrysopoeia and even records his own secondhand transmutation history, and the local historian Gabriel Vanel devotes considerable space to alchemical goings-on in his lengthy study of Caen; unfortunately, neither makes mention of Georges Pierre, his associates, or any apparently related alchemical affairs.[152] Thus Georges Pierre has seemingly gone unnoticed in the published annals of Caen.

My exploration of the Archives départmentales in Caen, however, provided some surprising results, and Georges Pierre des Clozets can now take on a more concrete existence. He lived in the parish of St. Jean at Caen, was the son (probably the eldest) of Guillaume Pierre, sieur des Clozets, and Elisabeth Beaugendre, and had three brothers—Jean Jacques, the priest at St. Jean (d. 14 October 1689); Robert, a doctor of medicine at the University of Caen; and Guillaume (1657–21 January 1713).[153] The birthdate of Guillaume, the youngest son, suggests that Georges was born in the early 1650s, putting him in his late twenties during his commerce with Boyle. The

[149] Maddison, *Life,* 168 and 172.

[150] Don Gabriel was one of ten illegitimate children of Duke Charles Emmanuel of Savoy (1562–1630); see Samuel Guichenon, *Histoire généalogique de la Royale Maison du Savoie,* 5 vols. (Turin, 1778–1780), 2:446, where he is described as a "Prince de grande espérance & que donne tous les jours nouvelles & glorieuses preuves de son courage & de la passion qu'il a pour les intérêts de la Coronne de Savoie." Also, Guido Amoretti, *Il Ducato di Savoia dal 1559 al 1713,* vol. 3, *Dal 1659 al 1690* (Turin: Daniela Piazza Editore, 1987), 88–119 and 151–68.

[151] David S. Lux, *Patronage and Royal Science in Seventeenth-Century France* (Ithaca: Cornell University Press, 1989); Harcourt Brown, *Scientific Organizations in Seventeenth-Century France* (Baltimore: Johns Hopkins University Press, 1934); Katherine Stern Brennan, "Culture and Dependencies: The Society of Men of Letters of Caen from 1652 to 1705" (Ph.D. diss., Johns Hopkins University, 1981).

[152] Pierre Daniel Huet, *Les origins de la ville de Caen et des lieux circonvoisins,* 2d ed. (Rouen, 1706); *Commentarius de rebus ad eum pertinentibus* (Amsterdam, 1718), 223–26; Gabriel Vanel, *Une grande ville aux XVIIe et XVIIIe siècles,* 3 vols. (Caen, 1910), 3:317–42.

[153] All archival study on the Pierre affair was made possible by a grant from the American Philosophical Society. These data on the des Clozets family were gathered from Archives départmentales du Calvados (ADC), St. Jean Parish records, mircofilms 5 MI 1 R18 and R19; register of marriage contracts, 8E 4236 fol. 1001; and land contracts, 2E 757.

archives also verified all the claims made by the 1680 letter to Boyle—Georges did in fact die in 1680 and was buried on 26 April 1680 in the parish of St. Jean.[154] Moreover, the claim that Pierre returned to Caen as a very rich man also appears to be true. He, or rather his family, did actually buy a large estate in Bretteville-sur-Odon, a southern suburb of Caen, on 12 November 1679 for the sum of fourteen thousand livres, exactly as Boyle's anonymous informant claims. The history of the des Clozets' ownership of the estate is recounted in a contract of sale dated 22 April 1711, made by Georges's brother Robert to whom the estate had passed.[155] Where did Pierre obtain his sudden wealth? The sum total of Boyle's gifts to him amount to only a fraction of what was spent on the estate alone. Perhaps Boyle was not the only wealthy man to send cash and gifts to Pierre to facilitate enrollment in the Asterism.

Perhaps more oddly, even Pierre's seemingly outrageous story of the fatally exploding cannon receives some support from local records. According to Pierre, one of the casualties of the detonation was Charles le Mor, "who begged me to fire them," the young son of the pewterer on rue St. Jean to whom Boyle directed his letters for Pierre.[156] Surprisingly, the St. Jean parish register records Charles le Mor's burial on 30 May 1678, the same day as Pierre's letter informing Boyle of the accident. While there is of course nothing in the parish record to confirm that le Mor was killed by a badly cast cannon, why would Pierre bother to fabricate so bizarre a story and to weave into it an actual death?[157] If le Mor really was killed by Pierre's cannon, then what was Pierre doing with cannons in the first place?

Some of Georges Pierre's associates mentioned in his letters can also be traced. Both Pierre and Descqueville mention Mr. de Carbonnel, who seems to have been in London, as Boyle was supposed to contact him; he may well be the same de Carbonnel noted as a traveler in a contemporary account of travelers from Caen.[158] He may be one of the brothers Daniel and Thomas de Charbonnel listed in the archives of the French Protestant Church of London, who relocated from Caen to London in 1676.[159] Maddison sug-

[154] ADC, St. Jean Parish burial records, 5 MI 1 R18; Georges Pierre, No. 58 for 1680: "Aujourd'hui 26 Avril 1680, a eté par moi Simon de la Vigne p. curé de S. Pierre de Caen par permission de Sieur curé de S. Jean [Jean Jacques Pierre, Georges's brother], inhumé le corps de Georges Pierre decédé le jour d'hui."

[155] ADC, 2E 757; cf. BL 6:43–44. The sale price in 1711 was 15,000 livres. Robert is here cited as a doctor of medicine at the university.

[156] Pierre to Boyle, 20/30 May 1678, BL 4:119.

[157] ADC, St. Jean Parish records, 5 MI 1 R18, Charles le Mor, No. 47 for 1678. Charles's father (or possibly eldest brother) is listed as "Michel Le Mor, master pewterer" in the register of marriage contracts, 8E 2688, fol. 999 (22 December 1689). It is perhaps significant that Charles le Mor is one of the few deceased *not* buried by Jean Jacques Pierre; could this imply that as Georges's brother he was too closely related to the fatality itself to perform the priestly office for le Mor?

[158] Michel de Saint-Martin, *Le livret des voyageurs de la ville de Caen* (Caen, 1677), and *Supplement au livret* (Caen, 1678), both cited in Vanel, *Grande ville,* 2:291.

[159] French Protestant Church Archives, MS 20, entry for 2 August 1676; MS 230, fols. 22 and 26*v*. There are several Carbonnels from Caen listed at later dates.

gested that "Mr. le Moine," who acted as an intermediary between Pierre and Boyle, was Etienne le Moine, a notable theologian who was born in Caen in 1624, studied under Pierre du Moulin, received a D.D. from Oxford in 1676, and died in Leiden in 1689 where he had taken up the chair of theology.[160] Although Etienne le Moine matches the le Moine of the Pierre correspondence in terms of his connection to Caen and his residence in England during the time of the correspondence, I rather doubt that so important a churchman would have acted as intermediary in this affair.[161] More probably, Pierre's le Moine may be the one mentioned in an alchemical context in a letter from Fatio de Duillier to Isaac Newton; Fatio states he knows le Moine because "I have seen him at Mr. Boyle's."[162] This person might be the "Monsieur Le Moyne, docteur en Medecine de Caen," who registered at the French Protestant Church of London on 7 March 1675; his connection to Caen and standing as a physician accord well with the le Moine of the correspondence.[163]

In sum, the tale of Robert Boyle's involvement with Georges Pierre must be one of the more outlandish stories in the history of science. It involved a substantial number of people and twists of plot that would seem excessive even in a bad novel. Where the truth lies remains hard to ascertain. In spite of the lingering questions, however, one thing is quite clear: the affair powerfully asserts Boyle's great eagerness to acquire alchemical knowledge. Not only was he convinced of the existence of adepti, he was anxious, perhaps to the point of gullibility, to join what he believed to be their assembly. He fulfilled all the obligations he could, and even was fearful that he had not executed them sufficiently well to ensure that his case for admission was strong. Boyle wrote weekly letters to Pierre, laid out considerable sums of money in gifts (some quite unsolicited), and allowed himself to be directed by Pierre to join the Turkey Company in order to receive the patriarch's gifts more expeditiously. All these facts attest to his enthusiastic interest. His interest was presumably heightened by the transmutation he witnessed at the hands of Mundanus's disciple, which occurred contemporaneously. If the Pierre affair is no more than a confidence trick, it was Boyle's overwhelming ardor to obtain access to adepti and their alchemical secrets that clouded his discretion, enabling him to overlook the inconsistencies and improbabilities in Pierre's letters. Furthermore, if Pierre was just a con man, his voyage to England to swindle Robert Boyle with alchemical promises must argue that Boyle's avid interest in chrysopoeia was widely known.

Ultimately, Boyle was disappointed in his aspirations, as his letter to la

[160] Anthony à Wood, *Athenae oxonienses,* 2 vols. (London, 1692), 2:875. Maddison, *Life,* 168.

[161] On Boyle and French Protestants, see M. E. Rowbottom, "Some Huguenot Friends and Acquaintances of Robert Boyle (1627–91)," *Proceedings of the Huguenot Society of London* 20 (1959–1960): 177–94.

[162] Fatio de Duillier to Newton, 4 May 1693, *Correspondence of Isaac Newton,* 3:265–67, on 266.

[163] French Protestant Church Archives, MS 20, entry for 7 March 1674/75.

Marche indicates. But this one bad experience did not extinguish his enthusiasm for alchemy or his willingness to entertain the claims of other would-be adepti, although it may have made him a bit more wary, as Sonnenburg found, to his disappointment. Boyle could still fall back upon his experiences of projections (and the piece of alchemical gold "I yett have by me & carry always about me") to verify for himself the truth of chrysopoetic alchemy. He did not turn his back on the Noble Art because of one disappointing experience; he had known all along that the ranks of alchemists were filled with both cheats and adepts, the former predominating. Indeed, Boyle's interaction with Pierre was not his only unsuccessful venture with a promiser of alchemical secrets. A short tract penned by his onetime operator Ambrose Godfrey Hanckwitz gives a detailed account of Boyle's patronage of another alchemical scheme. Godfrey's remembrances also shed new light on Boyle's relationship with at least one of his operators.

Godfrey Hanckwitz and the "Crosey-Crucian"

Ambrosius Gottfried Hanckwitz (1660–1741), known after his arrival in England simply as Ambrose Godfrey, is one of the few of Boyle's operators we know by name, and one of even fewer who went on after their association with Boyle to construct successful independent careers. Godfrey began working for Boyle sometime in the late 1670s, and his chief claim to fame in Boyle's eyes was his conspicuous success in preparing extremely pure samples of phosphorus in considerable quantities far more readily and efficiently than anyone else could do. The sections dealing with white phosphorus in Boyle's tract *Aerial and Icy Noctiluca* (1681/82) rest in great measure upon Godfrey's experimental labors and discoveries. After leaving Boyle's service sometime in the 1680s, Godfrey set up a trade in chemical preparations, most notably in white phosphorus, and founded a pharmaceutical house that remained in business until the present century. He was elected a fellow of the Royal Society in January 1730. Godfrey was not only a competent entrepreneur but also, and foremost, a highly skilled preparative chymist.[164]

Around the beginning of the eighteenth century, Godfrey composed a short tract entitled "An Apology and Letter touching a Crosey-Crucian [*sic*]." The work was never published, nor perhaps was it intended to be. In the nineteenth century, Joseph Ince, a writer for the *Pharmaceutical Journal*, examined this manuscript pamphlet, which was then preserved along with a huge volume of notebooks, correspondence, papers, diaries, and memorabilia constituting a Godfrey archive. He published some lengthy extracts from Godfrey's "Apology" in the *Journal* but thereafter, with damnable

[164] See Maddison, "Boyle's Operator: Ambrose Godfrey Hanckwitz, F.R.S.," *NRRS* 11 (1955): 159–88, and *New DNB* entry by Principe. The work of another German, Frederick Slare (Schloer), was also crucial to Boyle's studies of phosphorus, although Slare soon discontinued work on this substance while Hanckwitz developed his career from it. On Slare, see Marie Boas Hall, "Frederic Slare, F.R.S. (1648–1727)," *NRRS* 46 (1992): 23–41.

carelessness, actually seems to have lost very nearly the entire cache of Godfrey family documents; he blithely lent away the original manuscript of the "Apology" to the pharmacist Daniel Hanbury, F.R.S. (1825–1875), and it has never been seen again.[165] The fragments that survive by way of quotations in Ince's article provide the modern reader with another glimpse of Boyle's interest in alchemical practitioners and, on a wider scale, make some unexpected comments on Boyle's character.

According to Godfrey, Boyle came into contact with a certain German who claimed to know the method of confecting the Philosophers' Stone. Boyle paid for his passage to England and set Godfrey as his supervisor, presumably on the grounds of both Godfrey's competence in chymistry and his ability to communicate with the visitor in his native German.[166] The would-be alchemist was housed in Godfrey's lodgings on Chandos Street, where Godfrey lived (with his wife) until 1693. The German alchemist must have arrived sometime around 1680 (the year Godfrey successfully prepared phosphorus for Boyle), for Godfrey states he was then "a poor young beginner, and a servant too."[167] Since there is no mention of Godfrey's first son (named Boyle Godfrey in honor of his father's master) even when Godfrey complains of his cramped quarters and narrow means, it is likely the event took place before Boyle Godfrey's birth in 1683. Boyle entrusted Godfrey with paying a salary to the émigré alchemist, observing and witnessing his experiments, and reporting on their progress. It is interesting to note how Boyle "farmed out" this German into the lodging of one of his operators rather than setting him up in his laboratory in the Pall Mall house. One wonders, then, how many ancillary laboratories there were in London (or elsewhere) carrying out Boyle's studies.

Godfrey, by his own statements, was as much a believer in chrysopoeia as Boyle and was therefore eager to see results. Unfortunately, no progress at all was made, and Godfrey's wife became increasingly annoyed at the stranger's presence in their quarters. When the alchemist lamented that he missed his family, Godfrey reported this to his master and Boyle paid to

[165] Joseph Ince, "Ambrose Godfrey Hanckwitz," *Pharmaceutical Journal*, ser. 1, 18 (1858): 126–30, 157–62, 215–22; "The Old Firm of Godfrey," *Pharmaceutical Journal*, ser. 4, 2 (1896): 166–69, 205–7, 245–48. Both Maddison and I have attempted to recover Godfrey items from the archives of the Royal Pharmaceutical Society. Maddison found several items, but the Pharmaceutical Society's archive has since lost many of them again; the society seems to have inherited Ince's attitude toward early documents. The society's collection seems to be in disarray, and adequate access is denied to knowledgeable scholars (I was not even allowed to see a putative catalog sequestered in a back room).

[166] Maddison ("Hanckwitz," 161n) suggested that this German might be J. J. Becher. Indeed, Becher's origin, his family (wife and grown daughter), and the time of his arrival do seem to coincide with those of Hanckwitz's charge. On the other hand, it is not clear why Becher would be called a Rosicrucian, and Becher's writings do not present the strong religious overtones that Hanckwitz mentions as crucial to his German would-be alchemist. While the alchemist of Hanckwitz's story had a falling out with Boyle, Becher dedicated books to Boyle throughout his time in England right up until his death in 1682. It seems, therefore, that Becher is not a likely candidate for our "Crosey-Crucian."

[167] Ince, "Hanckwitz," 159.

have his wife and grown daughter brought over as well. This made matters worse for Godfrey's domestic harmony, for he writes that all three "were lodged . . . in my own bedchamber at Chandos Street, along with me and my wife all in one room."[168] But Boyle shortly thereafter paid for separate lodging for the alchemist, where he soon set up what Godfrey called "some fiddling furnace, enough to make one laugh."[169] Godfrey laments bitterly of being caught between Boyle, who was all this time "continually eager and desirous of news," and the promising projector, who was continually eager and desirous of more money. Godfrey set up in his own house three processes according to the instructions of the alchemist: "the resuscitation of plants, the separation of sulphur in animals, and the mercurification of lead."[170] One wonders whether the preparation of the phosphorus was done here on Chandos Street as well. If so, it is truly appalling to imagine Godfrey (and his long-suffering wife) withstanding the stench of the ingredients—human urine and feces—and the worse stench, dangers, and toxicity of the product phosphorus in their poor-rate lodging.

Although the projector continued making fine orations on the unity of alchemy and religion, declaiming "against the three antichristian religions as he called it, viz. Popedom, Lutherdom, and Calvinism," no alchemical progress was forthcoming. Then Boyle, who in Godfrey's words "paid everyone to a tee, yet knew well enough how to bestow money, how to limit his gifts and not be flush," began to decrease the would-be alchemist's salary. This caused the alchemist to fly into a rage at Godfrey and provoked some very unpleasant exchanges, including several with the alchemist's wife, "a terrible bawling creature," who "followed once [Godfrey's] wife in the street with scolding, spitting at her, and exclaiming at her, though, thank God, all in German, that the people understood her not." The alchemist then circumvented Godfrey and went to address Boyle directly, offering a letter and some curiosity in a box, but Boyle, having been informed of Godfrey's misfortunes with the man, "excused them civilly" and refused to see him.[171] After another bout of reviling Godfrey, the charlatan departed, leaving Godfrey to philosophize over "meddling with Rosicrucians and madmen. This is the fruits of meddling with process-mongers."

Conclusion

Robert Boyle had many contacts with alchemists. Indeed, his certainty regarding the reality of projective transmutation stems from his eyewitness of

[168] Ibid.

[169] Ibid., 160.

[170] Ibid.

[171] A "Dr. Moulins" plays also into this drama; he relayed the problems Godfrey was having to Boyle.

transmutations in the 1670s and 1680s. The most striking of these was the conversion of lead into gold at the hands of a French alchemist who has here been tentatively identified as a disciple of one Theodore Mundanus, an adept known personally by Boyle's friend Edmund Dickinson. This momentous event receives a prominent place in the *Dialogue on Transmutation,* where it and other experiential accounts are deployed by the Lapidists to carry their cause, squarely in the tradition of the transmutation histories popular in seventeenth-century alchemical apologetics.

Boyle was solicitous to learn as much as possible about performances of transmutation, especially those of Wenzel Seyler, "the famous friar" of Vienna, about whom he quizzed the imperial ambassador. His avid interest in traditional alchemical topics made him the object of numerous dedications of alchemical publications as well as the target of several alchemical schemers. I have treated his dealings with Georges Pierre and the Asterism at length, and whether or not the affair was a confidence trick, it clearly displays the avidity of Boyle's pursuit of alchemical knowledge—to the point of (what seems to us) gullibility. Yet even the unsuccessful outcome of his interaction with Pierre was not sufficient to dampen Boyle's enthusiasm for his alchemical quest—he continued to patronize promising projectors, to receive alchemical information, and to inquire into chrysopoeia.

While Boyle frequently received confirmation of alchemical claims, even directly from the hands of adepti, he did not receive thence what he certainly wanted even more—the practical knowledge of the secret preparations of their arcana. To see the effects of the Philosophers' Stone was tremendous in itself, but to be able to prepare and to command that precious Elixir oneself would be far more wonderful and enlightening. The alchemical literature repeatedly stressed how the adepti would never divulge the secret of their Stone openly—indeed, the second half of the *Dialogue* spends much time bemoaning the adepti's reticence to speak candidly—and Boyle's experiences verified this policy. If Boyle was to learn the secret of the Stone (or of any other *arcanum maius,* such as the alkahest), he would have to employ the other two sources (failing direct divine revelation) remaining—the cryptic books of the adepti and his own laboratory experimentation. The following chapter will document Boyle's approaches to these twin sources of knowledge.

CHAPTER V

Boyle and Alchemical Practice

WITHIN ALCHEMY there exist side by side two traditions of equal importance: the textual and the practical. The textual tradition contains two facets—reading the texts of others and writing one's own. In order to learn the preparation of the Philosophers' Stone, the aspiring adept had to scrutinize the written legacy of earlier generations of adepts, to attempt to solve their riddles and enigmas, and to piece together the scattered intimations of "right practice" buried in their books. Sir Isaac Newton presents an extreme case of addiction to the textual tradition with his voluminous collections of alchemical quotations and his *Index chemicus;* a majority of Newton's "million words of alchemy" are copied extracts from respected authors. Newton's exercises aimed at building up a fuller understanding of the secret Art by laying together innumerable comments from a multitude of authors, each of whom reveals (according to common belief) only a piece of the whole.[1] In writing their works, alchemical authors align themselves along defined methods of concealing and revealing. The preparation of the Stone was a secret, never to be revealed in its entirety or in plain language; hence the enigmatic, allegorical, and obscure texts of chrysopoeians. But chrysopoeia is intimately tied to the laboratory as well; it is not a mere textual tradition. The Stone has to be prepared by art and the fire, and thus it requires the development and mastery of manual skills and the knowledge of chymical operations and techniques. Practical experimentation is crucial to alchemy.[2]

To complete the study of how fully Boyle placed himself within transmutational alchemy, I need to show his devotion to and activity in both the

[1] Richard S. Westfall, "Isaac Newton's *Index chemicus,*" *Ambix* 22 (1975): 174–85; more generally on Newton's alchemy, see Betty Jo Teeter Dobbs, *The Foundations of Newton's Alchemy, or "The Hunting of the Greene Lyon"* (Cambridge: Cambridge University Press, 1975), and *Janus Faces of Genius* (Cambridge: Cambridge University Press, 1991).

[2] I here disagree sharply with Marco Beretta, who concludes his *Enlightenment of Matter* with an appendix on alchemy riddled with astounding errors of both fact and interpretation, where he broadly states that alchemy should not be regarded as an experimental endeavor whatsoever (331). See the fuller criticism in Newman and Principe, "Etymological Origins." I also respectfully dissent from the thesis of Brian Vickers, "Alchemie als verbale Kunst," in *Chemie und Geisteswissenschaften: Versuch einer Annäherung,* ed. Jürgen Mittelstrass and Günter Stock (Berlin: Akademie Verlag, 1992), 17–34, on 17: "die Alchemie war keineswegs eine empirisch-praktische Kunst mit handwerklicher Tradition." More investigations of laboratory alchemy occur in Lawrence M. Principe, "'Chemical Translation' and the Role of Impurities in Alchemy: Examples from Basil Valentine's *Triumph-Wagen,*" *Ambix* 34 (1987): 21–30; "Apparatus and Reproducibility in Alchemy"; William R. Newman, "Alchemy, Assaying, and Experiment," in *Instruments and Experimentation in the History of Chemistry* (Chicago: University of Chicago Press, 1998).

textual and the experimental. I have already shown that Boyle was an alchemical author, not only writing the *Dialogue on the Transmutation of Metals* but also at least planning several additional works on the subject, not to mention the published signs of his chrysopoetic interests in *Anti-Elixir* and his *Philosophical Transactions* paper on the incalescence of mercury with gold.[3] But it is necessary now to explore how Boyle responded to the alchemical literary traditions. More explicitly, how extensively did Boyle read chrysopoetic authors? Did he attempt to puzzle out their "mystie speech"? And did he himself adopt their cryptic methods of communication, contrary to standard portrayals of his commitment to "openness" and "public knowledge"? This inquiry leads directly into the subsequent investigation of Boyle's own alchemical experimentation—his attempts to follow the chrysopoetic processes described by the adepti in metaphorical or allusive guise. Since Boyle has rightly gained renown as an *experimentalist,* his pursuit of chrysopoeia and the Stone must have manifested itself in experimental programs.

Reading Alchemy

The substantial number of alchemical manuscripts among Boyle's surviving papers clearly witnesses his study of chrysopoetic authors. In addition to the works of Becher and Rothmaler mentioned in the previous chapter, the Boyle archive contains texts from almost all the chrysopoetic authors considered important in the seventeenth century. For example, there exist over seventy pages of works (in the hand of Boyle's copyist Robin Bacon) by the fifteenth-century Sir George Ripley, including a sixteenth-century English version of his *Marrow of Philosophy* and "fragmenta" concerning the Elixir.[4] The *Concordantia chymica* of Alexander van Suchten, an important source for the Philalethean works, appears in two different translations, as do the *Epistolae philosophicae, Statua philosophorum incognitorum,* and other works attributed to Michael Sendivogius.[5] The multiple copies suggest that Boyle may have been comparing manuscripts to arrive at more accurate texts. Eirenaeus Philalethes' *De metallorum metamorphosi* exists in an early copy in the hand of a scribe of the Hartlib circle.[6] Basil Valentine appears as well in a partial French translation (in Oldenburg's hand) of the

[3] See his catalog at the start of chapter 4; Boyle's paper on incalescence will be dealt with at length below.

[4] BP, 30:1–72.

[5] Ibid., 14:47–74 and 34:1–152; 31:399–554 (in Robin Bacon's hand) and 34:238–323 (in Frederick Clodius's hand). These works are most probably spurious additions to the Sendivogian corpus; Zbigniew Szydlo, *Water Which Does Not Wet Hands: The Alchemy of Michael Sendivogius* (Warsaw: Polish Academy of Sciences, 1994), considers them authentic, but he gives no solid argument for his case.

[6] BP, 44:1–20; see also Newman, *Gehennical Fire,* 151–63, and 274 (MS 27).

Natürliche und Übernatürliche Dingen and *Stein Ignis.*[7] There exist also extracts from the *Testamentum novissimum* of the pseudo-Lull, biographical materials concerning him, and a catalog of sixty treatises attributed to him, as well as two sets of transcripts and extracts from the tenth book of the *Archidoxis* of Paracelsus dealing with the alkahest and the Philosophers' Stone.[8]

Lesser known works are also present, such as two contemporaneous accounts of making the Philosophers' Stone. One is "A Diary and Practicke given me by Mr. Oughtred," which first recounts (in typically enigmatic language) the preparation of the matter for the Stone and then presents a running account of its progress during ten months of prolonged heating. This document apparently circulated fairly widely in manuscript, for another copy (in a similar if not identical hand), endorsed "Oughtred to Mr. Thomas Henshaw from whose manuscript I coppied it June the 6 166-," exists in the British Library.[9] Another is the mysterious booklet *The Art Begun* (*Ars coepta*), dated 1681 and written in the hand of Boyle's scribe Robin Bacon. Directions involving gold and mercury reveal the general alchemical nature of the text, but its specifics remain unknown since the entire document is in cipher.[10] The fact that these texts can be dated on the basis of handwriting from the 1650s (Clodius's hand and that of a Hartlib circle scribe) to the late 1670s and 1680s (Bacon's hand and "hand A") shows that Boyle's interest in collecting and studying chrysopoetic works stretched over his entire career.

Since Boyle's library was dispersed after his death and no catalog of it exists, we cannot calculate its proportion of alchemical books as has been done for Newton's.[11] Nonetheless, Robert Hooke records on 21 March 1693 that he "saw neer 100 of Mr. Boyles high Dutch Chymicall bookes ly exposed in Moorfeilds on the railes."[12] How many of these would fall into the category of chrysopoeia is open to some question, but if they represented

[7] BP, 26:182–92.

[8] BP, 44:56–93, 41:56–63, 8:117; 14:166–77 and 39:75.

[9] BP, 30:445–53; British Library Sloane MS 2222, fols. 136–41*v*. The Sloane copy continues for several entries beyond the BP version, which is apparently incomplete. The last digit of the date in the Sloane MS is illegible. William Oughtred (1575–1660) was a mathematician and divine, whom John Aubrey calls "a great lover of Chymistry" and who once told John Evelyn that he was very close to finding the Stone; *Aubrey's Brief Lives,* ed. Oliver Lawson Dick (London: Secker and Warburg, 1950), 222–25. Oughtred also tutored Sir William Backhouse, who was Elias Ashmole's "Father in the Art" (see *DNB*). Thomas Henshaw (1618–1700), also a student of Oughtred's, was a founding member of the Royal Society, and French undersecretary to Charles II, James II, and William III. He is cited as an expert in traditional alchemical matters by Ashmole, *The Way to Bliss* (London, 1658), preface. See Stephen Pasmore, "Thomas Henshaw, F.R.S.," *NRRS* 36 (1982): 177–82.

[10] BP, 11:296–309; see Principe, "Robert Boyle's Alchemical Secrecy," 67.

[11] Richard S. Westfall, "Alchemy in Newton's Library," *Ambix* 31 (1984): 97–101. RSMS 23 was identified as the catalog of Boyle's library, but Hunter argues effectively against that attribution (*Guide,* xxi–xxii).

[12] Fulton, *Bibliography,* v.

a statistical sampling of seventeenth-century German chymical publications, we can assume that a substantial majority dealt with traditional alchemy. A small fraction of Boyle's library, preserved by accident, reinforces the impression that Boyle's library was well stocked with chrysopoetic texts. Seven manuscript volumes inadvertently bundled up with Boyle's papers upon his death were rediscovered in the Royal Society library in 1994. These volumes include the sixteenth-century *Clangor buccinae,* the *Enarratio methodica trium Gebri medicinarum* attributed to Philalethes, a collection of seventeen various alchemical tracts, and a laboratory notebook of George Starkey's that records some of his labors to make the Stone and the alkahest.[13] This high proportion (four out of seven) of treatises on chrysopoeia suggests not only that there were a large number of such titles in Boyle's library, but also that these particular volumes were in use around the time of Boyle's death (as they were packed up with his working papers instead of sold with his library), thus reinforcing our sense of the longevity of Boyle's interests.

One component of the Boyle Papers that has, perhaps curiously, remained largely unexploited is the large quantity of experimental notes and results. These collections, often stretching for nearly one hundred pages of continuous entries, record a diverse array of material; experimental processes and trials are interpersed with memoranda and reading notes. Many of these entries are dated, and their patchwork appearance in the hands of many different scribes testifies to their use as continuous records of Boyle's activities. As such, they are useful indicators of both the textual and the experimental in Boyle's chrysopoeia.

These accounts themselves stem from Boyle's habit of keeping collections of "Promiscuous Thoughts" from a very early age. The earliest example, entitled "diverse peeces" and only recently discovered, dates from his stay at Geneva while he was a student in 1642, and contains short essays, exercises, and memoranda.[14] Examples from the later 1640s contain rhetorical elements and romantic excerpts in the manner of commonplace books.[15] The sudden appearance of scientific and medical receipts in such collections (to the exclusion of all else) in 1650 has recently been noted for the first time by Hunter.[16] The entries of the 1660s–1680s differ again from those of the early and mid-1650s. While the earlier collections testify to Boyle's sudden interest in matters medical and natural philosophical, their contents are almost wholly derived from other workers; they are the collections Boyle made of receipts communicated to him from people who were themselves experimentalists,

[13] Lawrence M. Principe, "Newly-Discovered Boyle Documents in the Royal Society Archive: Alchemical Tracts and His Student Notebook," *NRRS* 49 (1995): 57–70, on 59. There is also the claim (but without substantiating evidence) that Boyle's library contained a copy of the medieval *Liber de distinctione mercurii aquarum* of Morienus; see Arthur E. Waite, *Lives of the Alchemistical Philosophers* (London, 1888), 57.

[14] Principe, "Newly-Discovered Boyle Documents," 59–63.

[15] Principe, "Virtuous Romance," 381.

[16] Michael Hunter, "How Boyle Became a Scientist," *History of Science* 33 (1995): 59–103, on 66–67.

and as such they continue the commonplace format of his earlier collections. Collections after the mid-1650s, however, show Boyle's own work and thus his gradual maturity into a self-sufficient experimentalist.

Boyle's long-acknowledged significance and preeminence as an experimentalist should have underscored the promise of such documents to provide a more direct, vivid, and (perhaps) reliable picture of Boyle's experimental endeavors and methodology than his more filtered and polished published tracts. Several factors have, however, militated against this approach. First, the disorder and volume of his papers buried these accounts from view; fortunately, this problem has been remedied by the cataloging of the papers.[17] A perhaps greater difficulty is the very diversity and disorder of the entries themselves. The scattered gatherings of laboratory results do not constitute anything resembling a coherent whole of the sort readily amenable to historical study. There is little or no continuity between entries. Simply put, the researcher must wade through a very large volume of miscellaneous material and be able to recognize items pertinent to his present study. But this very diversity and lack of cohesiveness reveal the complexity of Boyle's own interests and activities. Throughout his maturity, Boyle habitually had very many projects going at once, with various hands to various tasks. Results reported to him by his operators and by others were set down in these collections together with excerpts and memoranda from whatever he was reading at the time. To evaluate these entries here would lead far afield; I only point out that scholars interested in seventeenth-century experimental practice should examine such first-generation documents directly rather than relying predominantly upon filtered, published forms the applicability of which for their purposes seems (in some cases) to be undermined by the very studies for which they are deployed.

These records, when quarried for details pertinent to Boyle's alchemy, provide a further dimension and vividness to Boyle's chrysopoetic pursuits. Some reveal Boyle's attempts to puzzle out the meanings of respected authors; in one case an entry dated 20 August 1678 reads

> The name Adrop is well handled by Arnaldus [of Villanova]. According to Ripley & some others it signifies the ⊕ [vitriol] of our ♄ [lead] (that is venus disolvd very well in our 🜅 [spirit of wine].[18]

Some laboratory records provide vivid examples of Boyle's possession and study of alchemical substances. For example, among a collection of "hydrostatick tryals," that is, density determinations, stand experimental testimonies of materials mentioned in published writings. Amid the lengthy lists (dated late 1689 to early 1690) of materials tested occur the "Mercury from Saturn [lead]," "Mercury coagulated with the fumes of lead," and "the spoil'd gold." This last entry recurs elsewhere (dated May 1690) as

[17] Hunter, *Guide.*

[18] BP, 28:337.

"gold deprav'd by the Antielixir" and shows a density considerably less than that of gold, as is recounted in the *Anti-Elixir* tract. Clearly Boyle was still interested in the "degradation" of gold he performed in 1677–1678, and continued to test the debased metal[19]. In another collection Boyle writes of assaying a "gold coagulated from mercury," a topic he wished to examine in his projected but unwritten *Diffident Naturalist.*[20]

Writing Secrets

These written records of Boyle's experimental activities, however, do not easily disclose their chrysopoetic content. For Boyle almost always masks chrysopoetic and spagyric material with a profusion of codes. I have elsewhere shown in detail how in private papers and communications dating from the 1650s to his death, Boyle employs no fewer than nine different codes, ciphers, and nomenclators to veil his alchemical activities and processes.[21] It was this persistent use of codes that prevented John Locke, Daniel Coxe, and Edmund Dickinson from sorting Boyle's chymical papers after his death; they complained to the executor John Warr, Jr. that they needed "the key or keys of the Chemical terms, without which they could order nothing."[22] One can gain an understanding of the executors' befuddlement by perusing laboratory notes like the following:

> Dissolve ℥ [ounce] iv of clean negerus in a s. q. of Baqla made of 2 parts of strong spirit of Or. and one part of wel rectify'd Polyralb. Dissolve also ℥ [ounce] i of Paz in a s. q. of our simple Baflia.[23]

By careful analysis and comparison of contexts I have deciphered most of these codes. In some cases Boyle simply replaces words with their Greek or Hebrew equivalents: for example, *cassiteros* and *theion* for *tin* and *sulphur,* or *zahab* and *melech* for *gold* and *salt.* In some instances Boyle invents barbarous Hebraic terms from a Semitic triconsonantal root to provide words for substances not named in classical Hebrew. In other cases Boyle employs meaningless words like *durca, ormunt,* and *albeda* to stand in the place of substances or operations. In these latter cases, the code word always begins with the letter alphabetically following the first letter of the

[19] BP, 21:191–218, "The XVI Century [of experiments]," on 214–15; ibid., 219–257, "The XVII Century," on 239 and 242; Ihde ("Alchemy in Reverse," 57) suggested that the operator faked the results of the anti-Elixir projection in order to please Boyle; this is clearly refuted by these records of the "deprav'd gold" from the 1690s. A partial chemical explanation of the anti-Elixir is presented in appendix 1, conclusion, 284n. p.

[20] BP, 27:19; dated 27 May 1659; on *Diffident Naturalist,* see "Aurum mercuriale," BP, 36:13.

[21] Principe, "Robert Boyle's Alchemical Secrecy."

[22] From a letter from John Warr, Sr., to John Warr, Jr., dated 16 July 1692. Quoted by Henry Miles in British Library Additional Manuscript 4314, fol. 90.

[23] BP, 17:48.

word (always in Latin) to be replaced; thus words beginning with *a* are replaced with code words beginning with *b*, and so forth. Hence *metallum* is replaced by *nidorum*, and *plumbum* by *quatrum*.[24] Boyle adds to the complexity of his systems by hybridizing words, so that, for example, *polira* (*oleum*) and *albeda* (*vitriolum*) combine to give *poliralb* (*oleum vitriolis*). Crossing or nesting different encoding systems further expands Boyle's secret vocabulary. For example, *Jupiter*, which very commonly stands for tin in chymical books, can be further encoded as *kursella*. Similarly, *solis* (or *sun*), which is gold's astrological equivalent, can be itself translated into Greek or Hebrew, providing *helios* or *shemesh* as additional code words for gold. Thus Boyle constructed a huge vocabulary in order to protect the secrecy of his alchemical experimentation.

Even Boyle's reading notes are sometimes encoded. There exists a lengthy set of closely spaced notes in the hand of Boyle's amanuensis Robin Bacon entitled "from Parrac his fourth tome."[25] These notes manifest several of Boyle's encoding systems at once—using Hebrew, Greek, and even French, Italian, and German equivalents, plus the alphabetical code—and deal exclusively with finding and treating the true matter of the Philosophers' Stone, the interpretation of the classical alchemical authorities, and the making of metalline Mercuries. Many begin with the phrase "he sayes," which clearly distinguishes such reading notes from Boyle's own experimental records. Curiously, a space is left in the manuscript after almost every code word, as if Boyle had ordered Bacon to prepare the manuscript in such a way that Boyle himself could later return to it to write in the plaintext terms. The implication is that Boyle did not want Bacon or other scribes or readers to know the true nature of the notes he was taking. Indeed, a large number of Boyle's alchemical receipts and notes—from the 1650s to the 1680s—show two hands. In most cases, the dictated text taken down by the amanuensis is of less chrysopoetic interest than the text as corrected later; in one case a receipt for turning tin into copper is corrected in Boyle's own hand to describe instead the transmutation of silver into gold.[26] This method of note taking suggests that Boyle was protecting his knowledge by keeping the totality of a process out of the hands of any one person, guarding against the recognition of their alchemical value and subsequent theft. The method thus fulfills much the same function as his codes.

[24] Two small scraps of the keylist to this code survive at BP, 28:333, and were briefly noted by Maddison, *Life*, 171.

[25] BP, 26:33–36*v*. The identity of "Parrac" is unclear; one might assume it to be a simple substitution code like the one cited below, in which case *Parrac* might decipher into *Beccer* (Becher), and the "fourth tome" might be the fourth book (second supplement) of his *Physica subterranea*. This explanation, however, does not accord with the contents of the notes. No locus for such notes exists in the *Physica*, and the sentiments do not accord with Becher's. For example, when commenting on the veracity of various authors, the notes read, "He laughs at Ph as a plagiary of Suttons works" (fol. 33*v*), which presumably refers to Philalethes and van Suchten, but Becher, far from laughing at Philalethes, held him in high esteem.

[26] BP, 25:166.

Just as these encoding systems could hide the true nature of Boyle's notes from unwanted inquiry, they could also safeguard correspondence. For example, in Boyle's dealings with Georges Pierre in 1678, Pierre writes in great exasperation that a copy of all Boyle's choice processes intended for the Asterism has found its way "to Clomarez, the apothecary's boy who lives with Lefebure opposite your house."[27] Boyle then apparently suggested that they employ the code which he had been using for years to protect his own alchemical preparations, for in his next letter Pierre encourages Boyle to "send the chymical code that you have told me about, and we shall enjoy ourselves, as I shall send you some processes."[28] Pierre's letter of 3/13 May in fact uses Boyle's codes to transmit a process for the prized product *Potabile Banarum* (or, in plaintext, *potable gold*), which begins "extract the Negirus from Banasis and Dicla . . . ," that is, "extract the Mercury from antimony and copper . . ."[29]

Similarly, Boyle's correspondence with De Saingermain uses simple two-line nomenclators to encode the parts of his February 1680 letter dealing with inquiries about Pierre and an alchemical recipe for the extraction of the color from gold.[30] De Saingermain obligingly writes out in full the coding system he suggests

A N G E L U S b c d f H
i L m o p q r t x y z &c

A word with no doubled letters (the keyword) is first written out and then followed by the balance of the alphabet in two lines of equal length; each letter can then be ciphered into the one above or below it. Boyle copied out De Saingermain's ANGELUS code in one of his notebooks (now RSMS 194) and used it to decipher (or to cipher) Laosso = Pierre.[31] The ANGELUS code does not work for Saingermain's February 1680 letter, but assuming that the code there works on the same principle and that the "*Mr. raxppx*" decodes to *Mr. Pierre,* this other nomenclator may be shown to be based on the keyword *AMOUR*.

A M O U R b c d e f g h
i k l n p q s t x y z &c

Some of Boyle's secret correspondence still defies decipherment (in the absence of the coding key), as in the chrysopoetic correspondence with a French alchemist known only as Monsieur Le Green, who uses numbers for the names of people, operations, and substances.[32] Le Green in fact agrees

[27] Pierre to Boyle, 18/28 February 1678, BL, 4:127.

[28] Pierre to Boyle, 14/24 March 1678, BL, 4:110. Boyle apparently did not send the code immediately, for Pierre asks again for "your table of codes" on 11 April.

[29] Pierre to Boyle, 3/13 May 1678, BL, 4:115.

[30] De Saingermain to Boyle, 1/11 May [1680], BL, 2:117–18.

[31] RSMS, 194:4*v;* the identical text occurs again on 9*v.*

[32] BL, 3:150–55; 6:78, 80–81. R.E.W. Maddison ("A Tentative Index of the Correspon-

to send secrets relating to the Stone by letter to Boyle only "because there is no one but you who has the key to these numbers."[33]

These private schemes to discourage uncontrolled access and thievery are paired with less obvious methods of concealment in published writings. The *menstruum peracutum* and its ability to render gold volatile and to transmute some of it into silver (described in chapter 3) provides a good example of both Boyle's experimental pursuit of the Stone in the 1650s and 1660s and his method of writing about it. The process, as I pointed out in chapter 3, draws its origin from one of Basil Valentine's most cryptic works on making the Stone, and as such, this "gold process" showcases not only Boyle's reading of alchemical authors but also his experimental attempts to re-create their results. Further, while Basil's text is intentionally obscure, Boyle's description of his own work shares some of that obscurity. Prior to the publication of a full account in *Origin of Formes,* Boyle makes no fewer than six allusions to this "gold process" using the *menstruum peracutum* where he rather offhandedly mentions no more than that he knows a "certain substance" of his making which can volatilize or "destroy" gold.[34] When Boyle does describe his gold process at length, he carefully crafts a subtext dealing with the confection of the Philosophers' Stone to supplement the main text dealing with corpuscularianism; by a judicious use of canonical phrases Boyle signals the chrysopoetic import of the process. For example, in describing the dissolution of gold in *menstruum peracutum,* Boyle writes that gold is "wont to melt as it were naturally . . . without Ebullition (almost like Ice in luke-warm water)."[35] While Boyle's simile seems unremarkable to most readers, a contemporaneous seeker for the Stone would at once recognize its enormous importance. Classical descriptions of the preparation of the Stone repeatedly use this very simile of dissolving "naturally like ice in warm water" quite specifically in order to describe the action of Philosophical Mercury on gold. The dissolution of gold in Philosophical Mercury is the first key step in making the Stone. Sendivogius, in his *Novum lumen chemicum* (1604), advises the seeker for the Stone to find "a certain humidity that dissolves gold without violence or

---

dence of the Honourable Robert Boyle, F.R.S.," *NRRS* 13 [1958]: 128–201, on 184) erroneously attributes the latter two letters to Renaudot, reading the characters at the bottom of the 17 August letter (BL, 3:78*v*) as REDT, which he considered an abbreviation for RE[nau]D[o]T. These characters, however, are not Roman and have not yet been identified. The handwriting of BL 6:78 and 80–81 is identical to that of BL 3:150–55, which is signed "Le Green."

[33] Le Green to Boyle, 19 September 1680 (N.S.), BL, 3:150–55, on 153; "ie dit beaucoup a cause quil ny a que vous que aves lexplication de ces nombres."

[34] Boyle, "Unsuccessfulness of Experiments," in *Physiological Essays,* 62–63; "A History of Fluidity and Firmnesse," in ibid., 146; *Sceptical Chymist,* 39–40, 222, 407, and see also Boas [Hall], "Early Version," 161. The full account is in *Origin of Formes and Qualities,* 349–73, and is partially repeated in "Mechanical Origin of Volatility," in *Mechanical Origin,* 45–47.

[35] Boyle, "Unsuccessfulness of Experiments," in *Physiological Essays,* 63.

sound, indeed so gently and naturally as ice melts with the aid of warm water."[36] Similarly, the extremely popular *Wasserstein der Weisen* (1619) notes that in the true philosophical solvent, gold dissolves "like ice in some warm water."[37] Likewise, in the *Introitus apertus ad occlusum regis palatium,* Eirenaeus Philalethes remarks repeatedly on how gold "melts in our Mercury like ice in lukewarm water."[38] Boyle's use of the phrase could not be accidental; he was well read in the alchemical literature, and he uses this highly specific phrase in an appropriate context. The use is deliberate, and its meaning could not be missed by alchemically attuned readers; it proclaims to them (and to them alone) that Boyle believes that he has successfully prepared the eagerly sought-after Philosophical Mercury, the first step in making the Philosophers' Stone.

Several times Boyle uses the expression "to destroy gold" in the context of the gold process. Noting the fixity of gold, Boyle asserts that it "can as little be Volatiliz'd as Destroy'd"; yet he goes on to explain that he *has* volatilized it, and further that he has "really destroyed even refined gold." By "destruction" Boyle presumably means his transmutation or debasement of gold into silver, a process I have elsewhere explained chemically.[39] The significance of these twin claims is made clear by his juxtaposed (but not explicitly connected) quotation of an axiom he attributes to Roger Bacon that "it is easier to make gold than to destroy it" (*facilius est aurum construere, quam destruere*).[40] If we take these comments together, Boyle implies that since he has in fact destroyed gold (the more difficult task), the making of gold, an easier process, should follow readily. The volatilization of gold, the most fixed of the metals, was a well-known alchemical desideratum. Indeed, the difficulty and desirability of the volatilization of gold was well enough known to show up decades later in the Scriblerus triumvirate's play *Three Hours after Marriage,* where Dr. Fossile responds skeptically to a character (disguised as a traveling Polish alchemist) who asserts that he has volatilized gold: "Have a care what you assert. The Volatilization of Gold is not an obvious Process. It is by a great elegance of speech called *fortitudo fortitudinis fortissima.*"[41]

Some of Boyle's other published writings employ a well-documented method of alchemical secrecy known as the "dispersion of knowledge"—

[36] Sendivogius, *Novum lumen chemicum,* 584, also 586–87.

[37] Johann Ambrosius Siebmacher, *Wasserstein der Weisen* (Frankfurt and Leipzig, 1760), 44.

[38] Philalethes, *Introitus apertus,* 682; see also 678.

[39] Principe, "The Gold Process"; Boyle, *Origine of Formes and Qualities,* 369; *Sceptical Chymist,* 407.

[40] Boyle, *Sceptical Chymist,* 177; *Origine of Formes and Qualities,* 363. West, "Notes on the Importance of Alchemy," cites the two quotations in the *Sceptical Chemist* (407 and 177) but does not connect the early intimations of the gold process with its full exposition in the *Origine of Formes and Qualities.*

[41] John Gay, Alexander Pope, and John Arbothnot, *Three Hours after Marriage,* ed. Richard Morton and William M. Peterson (Painesville, OH: Lake Erie College Press, 1961), 32.

the dividing of a single important process into several pieces and scattering them disconnectedly through a largely unrelated text. This "dispersion" is found in writers from Jābir to Philalethes.[42] In this way Boyle "reveals and conceals" the manufacture of a special solvent (for extracting the tincture from glass of antimony) in *Usefulnesse of Experimental Naturall Philosophy* (1663).[43] A somewhat different "secrecy by silence" appears in Boyle's later *Philosophical Transactions* paper on an "incalescent mercury" (1676), which I will treat in detail below. While this paper has been cited as an example of Boyle's openness even in alchemical matters, it is not, in fact, open at all. It is actually equivalent to chrysopoeian tracts in terms of explanatory content, for while it describes the *effects* of a peculiar substance, it explicitly keeps the preparation and origin of this material in the strictest secrecy.

Boyle's Attitude to Secrecy

Significantly, Boyle's codes and allusive or dispersed descriptions occur *only* in chrysopoetic and spagyric contexts. This fact indicates that Boyle invested alchemical matters with a status different from that of his other studies. This situation impinges directly upon Boyle's attitude to secrecy—an aspect of his thought that has often been oversimplified. It is a mistake to presume that Boyle straightforwardly objected to secrecy and strove to make even alchemical matters public.[44] He did not. We saw in chapter 2 how the *Sceptical Chymist* explicitly excuses adepti for "writing darkly and aenigmatically" about their grand arcana even while holding those who would teach the general principles of natural philosophy to a standard of openness and clarity. Boyle's allowances for alchemical secrecy are renewed in the 1680 appendix *Producibleness of Chymical Principles* where, as we saw, he not only condones secrecy for chrysopoetic arcana but actually claims to understand the basis and meaning of at least some of the adepts' metaphorical speech. Anti-Lapidists in the second part of the *Dialogue on Transmutation* treat the writing style of the adepti harshly, but this criticism centers on hypocrisy and arrogance rather than on secrecy itself.[45]

Much of the argument in favor of Boyle's supposed dedication to public knowledge and rejection of secrecy comes not only from those passages of the

[42] Paul Kraus, *Jabir ibn Hayyan: Contribution à la histoire des idées scientifiques dans l'Islam,* in *Memoires présentés à l'Institut d'Egypte* 44, 2 vols. (Cairo, 1943), 1:xxvii–xxxiii; Maurice Crosland, *Historical Studies in the Language of Chemistry* (New York: Dover Publications, 1978), 36–40, cites an excellent example of dispersion in Glauber. See also Newman, *Gehennical Fire,* 125–35, for examples in the Philalethes tracts.

[43] Principe, "Robert Boyle's Alchemical Secrecy," 67–69.

[44] "Boyle had no use for secrets, real or imaginary." Boas [Hall], *Robert Boyle on Natural Philosophy* (Bloomington: Indiana University Press, 1965), 278; see also 86. Shapin and Schaffer, *Leviathan,* 71, 78.

[45] Boyle, *Dialogue on Transmutation,* appendix 1.

Sceptical Chymist that I have already shown to have exempted adepti, but also from Boyle's early essay entitled "An Invitation to a Free and Generous Communication of Receipts in Physick."[46] This work, Boyle's first (and anonymous) publication, takes the form of a letter written to a fictitious possessor of secrets named Empiricus. Therein Boyle makes numerous appeals to morality and Scripture to convince the addressee that valuable medical receipts should be made public for the benefit of all mankind. Here Boyle includes even the "Chymists" and "that great Elixir" in his call for openness, with none of the distinction he employs in later writings. But the "Invitation" must be understood in context. Its conception dates from 1647, and it was sent in 1649 to Samuel Hartlib, who had requested it. These dates fall within Boyle's moralist period, before he had any serious interest in natural philosophy. More specifically, the "Invitation" belongs to a set of moralistic epistolary fictions Boyle wrote in 1647–1648. These conceits, strongly influenced by the contemporary French epistolary tradition, all employ an elevated and highly rhetorical moralistic tone.[47] It is anachronistic to set this very early work alongside Boyle's later scientific interests and writings without recognizing its origin in the mind of a very different Boyle—one whose recognized activity and connection with the Hartlib circle were not those of an independent experimentalist.[48] Boyle's mature attitude to secrecy cannot be gleaned from a youthful work whose motives are far removed from the natural philosophical activities that characterized his later life.

In terms of traditional alchemy, Boyle not only made allowances for the secrecy of others but indulged in it himself, often employing traditional methods of concealment. Such secrecy is not subsumable under the headings that have previously been marked out as exceptions for public knowledge, namely, protection of proprietary and trade value or obligations to informants, another kind of secrecy that Boyle was obliged to maintain in certain cases, but for very different reasons.[49]

Experimental Chrysopoeia

The foregoing section makes it clear that Boyle was deeply involved in the literary aspects of chrysopoetic alchemy—reading the authorities, writing

[46] Boyle, *Invitation to Communicativeness,* in *Chymical, Medicinal and Chyrurgical Addresses,* ed. Samuel Hartlib (London, 1655), 113–50; M. E. Rowbottom, "The Earliest Published Writing of Robert Boyle," *Annals of Science* 6 (1950): 376–89; R.E.W. Maddison, "The Earliest Published Writing of Robert Boyle," *Annals of Science* 17 (1961): 165–73; Michael Hunter, "The Reluctant Philanthropist: Robert Boyle and the 'Communication of Secrets and Receits in Physick,'" in *Religio Medici: Medicine and Religion in Seventeenth-Century England,* ed. Ole Peter Grell and Andrew Cunningham (London: Scholar Press, 1996), 247–72.

[47] Principe, "Virtuous Romance," 385.

[48] Hunter, "How Boyle Became a Scientist," 62–65.

[49] Golinski, "Chemistry in the Scientific Revolution," 382–86; "A Noble Spectacle," 29–30.

his own texts, maintaining secrecy, and employing the veiled speech typical of chrysopoetic authors. But for Boyle, experimentalism was the key to knowledge of the natural world—the Book of Nature took precedence over the books of men, and alchemy was no exception. Unlike Newton, whose alchemy consisted predominantly in textual studies, Boyle gave preeminence to laboratory activities.[50] Having already shown Boyle's "gold process" to be an early example of his pursuit of the Philosophers' Stone, I will now try to uncover some of his other transmutational experimentation.

Boyle's correspondence provides glimpses of his pursuit of experimental chrysopoeia. For example, one of the letters from Georges Pierre thanks Boyle "for telling me in your letter of the progress that your disciple is making . . . if the gold is found to be good, let me know." Clearly someone in Boyle's employ is working on a chrysopoetic preparation, and Pierre notes that the Asterism will be pleased to learn of Boyle's "interest in the metallic [tincture]."[51] Long after he went looking for Pierre on Boyle's behalf, De Saingermain continued to visit and to correspond with Boyle. Their correspondence from 1685 contains a record of experimentation to make the Philosophers' Stone that De Saingermain was carrying out under Boyle's direction. In January, De Saingermain writes of the "plan that we agreed upon when we spoke face-to-face," asks for further directions, and assures Boyle that a due portion of the product will be sent to him when it is finished. Another of Boyle's colleagues apparently had this same process underway, presumably in London, for De Saingermain refers to Boyle's "expert friend" and asks for news of his progress "on the powder." In April, De Saingermain announces that his work on "our experiment" is completed (even though its difficulty was compounded, he complains, because Boyle's friend had not given all the details accurately), and thus "it only remains therefore, Sir, to give a dose of this purgative to our fugitive servant." De Saingermain asks for details of its administration, and whether "your German has seen some of the fugitive servant halted in quantity by this medicine."[52] The "fugitive servant" refers to common mercury in reference to its volatility (it "flees" the fire) and the office of the mythological Mercury as messenger for the gods. Its "purgation" and "halting" refer to its transmutation into gold, when it is purged of its superfluities to become the per-

[50] The contrast between Newton and Boyle is studied in Lawrence M. Principe, "The Alchemies of Robert Boyle and Isaac Newton: Alternate Approaches and Divergent Deployments," in *Canonical Imperatives: Rethinking the Scientific Revolution,* ed. Margaret J. Osler (Cambridge: Cambridge University Press, 1998).

[51] Pierre to Boyle, 14/24 March 1678, BL, 4:110–11; "Je vous suis sensiblement obligé de me marquer dans la vostre le progres de vostre disciple . . . si lor se trouve bon vous me le ferez scavoir . . ."; "marques de vostre curiosité sur la metallicque."

[52] De Saingermain to Boyle, 14 April 1685, BL, 6:48; "il ne reste donc plus Monsieur qu'a donner une dose de cette purgation a nostre Serviteur fugitif"; "sil a veu du serf fugitif arresté en quantité par cette medecine." He refers to Boyle's associate as "vostre All.," which I interpret as "vostre Allemand" judging from Saingermain's later use of "portes d'Allem." clearly in reference to the borders of Germany; see Saingermain to Boyle, 25 December 1685, BL, 6:50.

fect, noble metal, and halted from its running nature by conversion into solid gold. What became of this experiment we do not know, for De Saingermain's last letter (Christmas, 1685) finds him in Amsterdam as a ruined exile, having fled France and unable to return, presumably owing to the revocation of the Edict of Nantes on 18 October 1685.[53]

This transmutational experiment shows not only another example of Boyle's laboratory pursuit of chrysopoeia but also its *collaborative* nature. Boyle is directing (and perhaps financing) alchemical studies carried on simultaneously by both De Saingermain in Paris and an anonymous German (Hanckwitz?), presumably in London. While Boyle clearly saw the necessity of maintaining secrecy in regard to chrysopoeia, he nonetheless maintained a "circle" of trusted collaborators.

Letters seeking Boyle's advice on the Stone reinforce the collaborative nature of his experimental pursuits. A striking example is provided by John Matson, a chymist in Dover. Unfortunately, almost all the letters from Matson to Boyle met the same sad fate as the Asterism items sent by Philalette—they were destroyed by Henry Miles in the 1740s. An early inventory lists several letters from Matson, but now only a single one survives.[54] It probably owes its survival to oversight, for it was included in a large bundle of chymical receipts, mostly in French and Latin, that Matson sent to Boyle, and thus was probably not seen by Miles. This letter, dated from Dover 18 May 1676, thanks Boyle for "favorable assistance in my proceedings" and requests further instruction. While Matson does not explicitly state the nature of his "proceedings," his reference to gentle heating "not to exceed the induring one's hand on the inner cover" "until the Coulours varry, & from thence to fixation & whitnesse," identifies his goal as the Philosophers' Stone. The latter expression refers to two stages in the preparation of the Stone—the *cauda pavonis,* or peacock's tail, where the digesting material exhibits many transient colors; and the first fixation into the White Stone, a halfway point in the process of confecting the complete Red Stone, where the immature Stone can transmute metals only to silver and not yet to gold. Matson's reference to gentle heating recapitulates the axiom "Let never thy Glasse be hotter then thow may feele / And suffer styll in thy hand to holde" enunciated by George Ripley and oft repeated down to Boyle's day.[55] Matson also expresses the need for secrecy in this matter and tells Boyle that he is ready to discharge his operator who is "a litle too cuning barely to performe what I direct without prying further." Matson even declares that "I thinke I should with Reluctancy accept of such a secret my selfe, if by the same means it should come to the hand of an indiscreet person." The secrets of chrysopoeia are privileged and of great power; both Boyle and his corre-

[53] De Saingermain to Boyle, 25 December 1685, BL, 6:50; "Vous scaves asses, Monsieur, LImpossibilité quil y a de Retourner En france; toutes nos familles Sont perdues et ruinee de fonds En Comble. J'ay une partie de mes parens en prison, une autre vagabonde . . ."

[54] Hunter, "Dilemma of Biography," 131.

[55] Ripley, *Compound of Alchymie,* 138.

spondents recognized and strove to fulfill the obligations of secrecy. Matson promises that "soe soone as [the operator] is gone," he will send Boyle a fuller account of his work.

There is a similar letter from one Le Febvre that expressly asks for Boyle's advice on a specific point in making the Stone. After a lengthy digestion where Le Febvre sees the succession of colors classically associated with the Stone—"several mixed colors, then white, then citrine, and after this a brownish red," he obtains a hard, fixed mass. When he cuts this with a knife, he reports to Boyle, "the middle and point that had touched the substance and been driven through it were turned immediately into a golden color."[56] While this result is very favorable, Le Febvre is concerned that his digested material is not "of easy fusion," which is "a property announced in the books"; so he turns to Boyle in the hope that the latter will "let me know how I can make this substance melt easily." The process of increasing the fusibility of the Stone is known as "inceration," from *cera* (wax); it is necessary that the Stone be as fusible as wax so that in projection it can melt and penetrate the metals like oil soaking into paper and cause transmutation.[57]

Both Le Febvre and Matson turn to Boyle for guidance in their confection of the Stone; clearly, both knew of Boyle's interest in the matter and recognized that he had expertise in practical chrysopoetic processes. A general sense of Boyle's pursuit of the Stone is clearly provided by a remarkable document presented in appendix 2 (text F). There Boyle cites his various attempts to prepare the Philosophers' Stone and records how he pursued this goal by numerous "Tryals made upon different matters." This document is only a fragment of an original text that, judging from the references in the surviving portion, outlined the methods and materials with which Boyle had worked. The extant section is written as an explanation of Boyle's manifold approaches to the Stone, for he notes that though " 'tis said to be the unanimous doctrine of the Adepti, that there can be but One true matter of the grand Elixir," he nonetheless experimented with a variety of starting materials and processes, and had his opinion that the Stone might be prepared by a number of ways "confirm'd . . . by the authority or confessions of some Hermeticks now alive." Clearly, this document refers to many years of experimental efforts. Boyle also states that although these efforts may not have afforded him the Stone, they still provided him some illuminating phenomena and good medicines.

We have already encountered a few of these varied "Tryals" Boyle carried out on his own or in conjunction with others. Now I wish to turn to the approach to the Stone on which Boyle was active for the longest period of

[56] Le Febvre to Boyle, undated, BL, 3:148; the letter is unsigned, and the identification of the author comes from the endorsement "Le Febvre's letter" on the back. "La pointe et le milieu que avoit touché et passé par violence à travers de la matiere furent teinte à l'heure mesme dune veritable couleur d'or."

[57] On inceration, see, for example, Ripley, *The Bosom Book,* 138–40.

time, a project that superbly showcases the textual, the experimental, and the social dimensions of Boyle's chrysopoetic enterprise.

Boyle and the Philosophical Mercury

Rather than amassing diverse examples of Boyle's alchemical experimentation recorded in correspondence and notes, I will focus instead on a substance that Boyle believed to be the crucial Philosophical Mercury needed for the Philosophers' Stone. The trail of this single process covers forty years—from Boyle's earliest natural philosophical studies right up until his death—thus vividly representing the steadfastness of his alchemical quest. In terms of the number of years this material attracted his attention and the number of experiments and people it involved, the quest for the correct preparation and deployment of the Philosophical Mercury rivals in importance Boyle's vastly better known studies of the air.

The importance of the Philosophical Mercury in traditional alchemy rests upon its indispensability for preparing the Philosophers' Stone, for it is the solvent able to dissolve gold "radically." Here again I must note the diversity of alchemical schools, for there existed a range of opinion as to what this Mercury was and how it was to be prepared and used. I have already shown how in the late 1650s and early 1660s Boyle implied allusively that his *menstruum peracutum* was Philosophical Mercury; this marks Boyle's brief and early exploration of the so-called *via humida,* or "wet way," which stipulates the use of watery solvents to dissolve gold in preparation for making the Stone. This wet way had many adherents and was often attributed to Basil Valentine, the writer from whom, in fact, Boyle adopted the process. But Boyle much more often adhered to the "Mercurialist" method. The Mercurialist school follows the *via sicca,* or "dry way," maintaining that the Philosophical Mercury is a metallic solvent prepared from common mercury, "the water which does not wet the hands." A brief consideration of the Mercurialist doctrines relating to the preparation and employment of the Philosophical Mercury will facilitate the identification of the Philosophical Mercury in Boyle's writings and his adoption of Mercurialist theories. Fortunately, an analysis of the Mercurialist doctrines as expressed by the foremost exponent of that school, Eirenaeus Philalethes, has already been presented by Georg Ernst Stahl, and more recently by William Newman; here it will suffice to recount the main points illuminating the significance of the Boylean texts to follow.[58]

The Mercurialists' Philosophical Mercury

The process for preparing the Philosophical Mercury can be traced back through at least a score of important alchemical authors, and its theoretical

[58] George Ernst Stahl, *Philosophical Principles of Universal Chemistry,* trans. Peter Shaw (London, 1730), 401–16; see also Newman, *Gehennical Fire.*

foundations located as far back as the thirteenth-century writings of the pseudo-Geber. During Boyle's lifetime, the Mercurialist school attracted many adherents including Sir Isaac Newton and Johann Joachim Becher, and Stahl notes that it has "the greatest number of Votaries." Four chief exponents of this school are useful for outlining its main tenets—Gaston DuClo, or Claveus, whom Boyle cites approvingly in the *Sceptical Chymist,* Alexander van Suchten, Jean Collesson, and Eirenaeus Philalethes.[59]

The main tenet of the Mercurialist school held that Philosophical Mercury was to be prepared by the purification of common mercury. This purgation must remove not only external, visible impurities but also its "internal superfluities." The first are removed by straightforward means of purification—washing and grinding with salt and/or vinegar, distillation, combination with sulphur to form cinnabar (mercuric sulphide) followed by "revivification" with reducing agents, and so forth. The internal "impurities" are considered more intrinsic, being carried from the first formation of the mercury in the earth, not separable by ordinary means and yet not an essential part of the mercury; their removal required a "more Philosophical" manipulation. These impurities render common mercury useless for the Philosophical Work, because as DuClo writes, "it is necessary first to purge it and free it from excessive coldness and moistness" and incorporate other materials with it to increase its hotness.[60] Collesson refers to the liberation of common mercury from "its phlegmatic nature . . . and from a black excrementitious earth that was not part of its natural composition."[61] Van Suchten proposes cleansing common mercury from these debilitating impurities with an alloy of "martial regulus of antimony" (antimony metal reduced from the native sulphide ore by iron) and silver. This same agent is prescribed throughout the Philalethean corpus under disguised names, as Newman has shown.[62] Both Philalethes and van Suchten claim that besides cleansing common mercury, the antimonial alloy impregnates it with a "volatile Gold," otherwise called the "Mercury of Iron," which gives the prepared product its alchemically significant properties.[63]

Once common mercury is rightly purified, it becomes more active and penetrating and is then called "animated" or Philosophical Mercury.[64] This

[59] Stahl, *Principles,* 401; on 2 and 396 Stahl cites three of these Mercurialists approvingly. I hope to publish in due course an examination of the most important schools of seventeenth-century chrysopoetic alchemy and their origins; my preliminary contribution to this effort is "Gaston DuClo."

[60] DuClo (Claveus), *Apologia,* 36; cf. *De triplici praeparatione auri et argenti,* in *TC,* 4:371–88, on 378 and 385.

[61] Jean Collesson, *Idea perfecta philosophiae hermeticae,* in *TC,* 6:143–62, on 154.

[62] Newman, *Gehennical Fire,* 118–41.

[63] Suchten, *Secrets,* 63–80; Newman, *Gehennical Fire,* 135–41.

[64] DuClo, *Apologia,* 36; note that while DuClo is explicit that this "animation" is to be understood as a purely *metaphorical* description of the greater "heat" of the prepared mercury, Collesson states that this "animation" occurs by impregnation with the "anima mundi generalis." *Idea perfecta,* 154, 160, and passim.

animated Mercury can then be used for the "radical dissolution" of gold, and the mixture of Mercury and gold (after some further operations) digested into the Philosophers' Stone itself. The Mercury's action on gold is seen as twofold. First, its radical dissolution of gold liberates the "seeds of gold" hidden within the innermost recesses of the metal; second, like common water acting on plant seeds, this Mercury loosens their shells and nourishes them; indeed, DuClo calls it "alimentum seminum metallorum," the aliment of metallic seeds.[65] The germination, nourishment, and cultivation of the golden seed allows for the multiplication of its virtue, as from one grain a plant grows to yield a hundred grains or more, and thus the Philosophers' Stone is capable of transmuting base metals into gold. Common gold is dead, but the animated Mercury animates the gold in turn and makes it grow. This germination and growth process is visible, according to some authors; Collesson writes that the Mercury and its employment are not right "unless the common gold visibly vegetates and the peacock's tail [varied colors] appears."[66]

Incalescent Mercury as Philosophical Mercury

Rather than beginning chronologically with the little-recognized origins of Boyle's quest for the Philosophical Mercury, I shall begin in medias res by recalling one of those long-acknowledged connections of Robert Boyle to transmutational alchemy. I refer to Boyle's peculiar paper in the 21 February 1675/6 issue of *Philosophical Transactions* entitled "Of the Incalescence of Quicksilver with Gold."[67] I will now set this hitherto isolated item into the full context of Boyle's alchemical studies, and into the yet larger context of seventeenth-century chrysopoetic schools. I will demonstrate how this well-known (but imperfectly understood) publication is truly the tip of an alchemical iceberg.

Boyle's curious paper, whose authorship is rather feebly disguised under the reversed initials "B. R.," describes a specially prepared mercury that not only amalgamates unusually easily with gold but also generates sensible heat in so doing. Ordinary mercury neither amalgamates so easily with gold nor generates any heat in the process. While the alchemical overtones of this paper have long been known, our general unfamiliarity with alchemical thought has allowed modern readers to miss its great importance in a chrysopoetic sense. The paper must be read with unusual care, for it is crafted both to reveal and to conceal. Boyle's tone throughout is strongly cautious,

[65] DuClo, *De triplici praeparatione,* 375 and 382. Note, however, that DuClo explicitly rejects a vitalistic interpretation of alchemy; for him the words *seed, growth, animation,* and so forth, are purely analogical. See Principe, "Gaston DuClo."

[66] Collesson, *Idea perfecta,* 149; "nisi ad oculum vegetetur & cauda Pavonis appareat." Cf. 146.

[67] Boyle, "Of the Incalescence of Quicksilver with Gold," *Philosophical Transanctions of the Royal Society of London* 10 (1676): 515–33.

and his intimations of the great importance of this substance are easily lost amid a tangle of qualifications and double negatives. He begins by discussing the possibility that there exists a mercury "more subtle and penetrant than that which is common . . . a more Philosophical Mercury" that bears a special affinity with gold and would grow hot ("incalesce") when mixed with it.[68] He notes that the existence of such a Mercury is asserted primarily by those who write positively of the transmutation of metals, but a greater number of writers "have reckon'd this sort of Mercuries among the *Chimaera's* and *Non-Entia* of the bragging Chymists."[69] The mention of chymical *non-entia* again recalls the contemporaneous antialchemical tracts of Werner Rolfinck and "Utis Udenis," which, as we have seen (chapter 2), assailed belief in metallic Mercuries during the vigorous debate on their existence in the 1670s. Yet Boyle notes that "several prying Alchymists," in spite of their labors, trials, and inquiries, told him that they had never actually seen an incalescent mercury. In this way, Boyle asserts and stresses the rarity of the product of which is he about to claim himself master.

Boyle continues by recounting how he believed that common mercury might be freed from heterogenous, "recrementitious" parts and made to incorporate other materials that might render its effects, though not its external appearance, quite different from normal. Here we clearly find echoes of DuClo, Collesson, and other Mercurialists who prescribe the purgation of common mercury from heterogeneities, and of van Suchten and Philalethes who mention mercury's acquisition of "volatile gold." Boyle claims that at length he succeeded in preparing such a substance, mixed it in the palm of his hand with gold powder, and felt a distinct and sometimes strong heat. For verification's sake, this experiment was performed in the hands of Henry Oldenburg and Lord Brouncker, president of the Royal Society. Boyle then underlines the importance of this product by explicitly connecting it with chrysopoetic authors, claiming that "some years after" he was in possession of this Mercury, he found in alchemical texts "some dark passages, whence I then ghess'd their knowledge of it, or of some other very like it; and in one of them I found, though not all in the very same place, an Allegorical description of it."[70]

The eleventh section of the paper—both the most circumlocutious and the most significant—leaves no doubt as to what Boyle believes his incalescent mercury to be. There he casts the paragraph in terms of what he *will not* determine in regard to incalescent mercuries, and then in that context goes on to make dramatic claims by a modified Ciceronian *praeteritio*. In the first place, he refuses to determine whether or not all Mercuries extracted from metals and minerals will incalesce (implicitly signaling his belief in

[68] Ibid., 517.

[69] Ibid., 518.

[70] Ibid., 521; this reference is probably to Philalethes, *Introitus apertus*, 658–59. Boyle's use of the phrase "not all in the same place" reveals his recognition of the standard alchemical "principle of dispersion," which he himself employed.

the real existence of such Mercuries), but takes the opportunity to note that he "found *Antimonial Mercury* to do" so. In a splendid example of double-talk, he declines to affirm that "every Metalline Mercury . . . is the same as that which the Chrysopaean Writers mean by their *Philosophick Mercury,*" and then at once remarks, "Nay, I would not so much as affirm, that every Mercury, obtained from the perfect Metals themselves, must needs be more noble and fit (as Alchymists speak) for the Philosophick work [making the Philosophers' Stone], than that which may with skill and pains be at length obtained from common Mercury."[71]

Thus Boyle here circumlocutiously implies that his method of purifying and impregnating common mercury provides a material *as fit as any* for making the Stone. He reiterates his claim in the following paragraph by noting that the "Solvent of Gold" mentioned by the "most approved Spagyrists" (i.e., the Philosophical Mercury) is "of kin, and perhaps not much more Noble than one that I had," so that one might be confident that his incalescent mercury "should be of more than ordinary use, both in *Physick* and *Alchymy.*"[72]

Boyle's reticence to speak plainly about this product is striking throughout the paper. While he clearly describes the *effects* of the mercury and details the surest ways of testing other mercuries to see if they will heat with gold, he never mentions the *method* he used to prepare his mercury. Boyle's results are therefore irreproducible and unverifiable. While he mentions the names of two witnesses to the truth of his assertion, Boyle's presentation of some gold and an unexplained mercury derived from unknown sources, followed by his mixing them in the hands of two Royal Society members, is epistemologically identical to the "public transmutations" performed by projectors all over the Continent during the seventeenth century. Contrary to some claims, this paper does not make very much of anything public—Boyle declines to describe the preparation of this mercury and flatly refuses to answer any inquiries about it.[73] In fact, a stipulation for the paper's publication, spelled out by Boyle, requires that no questions may be asked about the paper's content: "divers Queries and perhaps Requests, (relating to this Mercury) . . . I would by all means avoid, for divers Reasons, . . ."[74] This stipulation is echoed in a brief preface by the publisher Henry Oldenburg, where he requests that readers "not fruitlessly endeavour to put a person . . . upon making unseasonable answers to any Verbal or Epistolary Questions about things, wherein some considerations, that he thinks are not to be dispensed with by him, do as yet injoyn him silence."[75] Boyle's recognition of the need to keep the details of the process secret is emphasized by his comment to the reader that since disability prevents him from writing with

[71] Boyle, "Incalescence," 525.

[72] Ibid., 526.

[73] Shapin and Schaffer, *Leviathan,* 72.

[74] Boyle, "Incalescence," 528.

[75] Ibid., 516.

his own hand, "I know, you will not think it fit, I should about such a subject employ that of an *Amanuensis.*"[76] Boyle uses a more traditionally alchemical excuse for his secrecy when he cites his fear of the "political inconveniences which might ensue if [the Mercury] should prove to be of the best kind and fall into ill hands."[77] These "inconveniences" are the disastrous consequences to economical and political stability if the secret of gold making were to fall into the hands of an unscrupulous person, and such fears are a stock feature of alchemical texts:

> . . . this Science must ever secret be,
> The Cause whereof is this as ye may see;
> If one evill man had hereof all his will
> All Christian Pease he might hastilie spill,
> And with his Pride he might pull downe
> Rightfull Kings and Princes of renowne.[78]

Such overt secrecy from Robert Boyle himself, not for either priority or proprietary protection, set amid the pages of the *Philosophical Transactions* should seem odd; it underscores the need for qualification of the usual perceptions of Boyle, the Royal Society, and open communication. The incalescence paper reinforces the pattern of secrecy for chrysopoetic matters seen elsewhere in Boyle's published and private writings.

Boyle strongly reiterates his commitment to secrecy in regard to his incalescent mercury five years later in *Producibleness of Chymical Principles.* There Boyle recounts how he made a mercury whose incalescence with gold was so intense that it was able to dissolve even thick pieces of gold almost instantly. Boyle then boldly asserts that he will never make this mercury again, and adds melodramatically that "for the sake of Mankind, I resolve not to teach the preparation."[79] The implication is that his highly active mercury is the resolution of one of the most closely guarded secrets of the chrysopoeians—a metallic radical dissolvent of gold, the first and most crucial step to the Philosophers' Stone and limitless amounts of gold. Boyle's concern over the misuse of chrysopoeia recurs in a manuscript he compiled outlining the "excuses of Philaletha for concealing the grand Arcanum," where he cites that open knowledge of the Stone would "much disorder the affairs of Mankind, Favour Tyranny, and bring a general Confusion, turning the World topsy-turvy."[80] Boyle thus seems to have come to a conclusion regarding the need for secrecy in alchemical matters identical to that promoted by the adepti.

[76] Ibid., 528.
[77] Ibid., 529.
[78] Thomas Norton, *Ordinall of Alchymie,* in *TCB,* 1–106, on 14.
[79] Boyle, *Producibleness,* 214.
[80] BP, 19:187*v;* published in Newman, *Gehennical Fire,* 254.

Boyle's Early Work with the Incalescent Mercury

Boyle's acquaintance with this incalescent mercury begins long before the 1675/76 description of it in the *Philosophical Transactions.* For there he mentions that his trials afforded him positive proof of the existence of such mercury "about the year 1652."[81] The year 1652 falls within Boyle's earliest phase of natural philosophical investigations. Recent reassessments of Boyle's early life have resulted in a reevaluation of the starting date for his interest in natural philosophy and experimentalism; it is now clear that Boyle's career as an experimentalist did not begin before 1649, and the origin of his intense interest in scientific issues (which characterizes his mature career) dates to the early 1650s.[82] In light of these new findings, the date of 1652 for the discovery of so crucial a substance as the Philosophical Mercury seems quite precocious. Indeed, it would be almost unbelievable had Boyle in fact discovered this Mercury on his own, but he did not; his knowledge of this ingredient essential for the Philosophers' Stone came instead from the foremost chrysopoeian writer of the second half of the seventeenth century—the man behind the mask of the pseudonymous chief of the Mercurialist school—George Starkey, alias Eirenaeus Philalethes.

In the early 1650s Boyle was involved in correspondence with the Hartlib circle—a diverse group of natural philosophers, writers, educators, and entrepreneurs connected by their acquaintance and correspondence with Samuel Hartlib, a German émigré with utopian aspirations.[83] In terms of chymical sophistication, the most important of these workers was the young American George Starkey. Starkey was born in Bermuda of English parents, moved to New England, and was there graduated in the Harvard class of 1646. After several years of chymical experimentation in the colonies, Starkey left for England, arriving there in 1650. He was soon introduced to the Hartlib circle, where he was highly respected for his chymical expertise (at so young an age) and especially for his possession of a number of manuscripts reputed to have been authored by an adept in New England. According to Starkey, this master—bearing the pseudonym Eirenaeus Philalethes—had successfully prepared the Philosophers' Stone, and Starkey had been witness to several transmutations, as well as to the use of the Stone to revive a withered peach tree and to make an old woman grow new teeth. While Starkey eventually died in obscurity, the great adept Eirenaeus Philalethes—

[81] Boyle, "Incalescence," 521.

[82] Hunter, "How Boyle Became a Scientist"; Principe, "Virtuous Romance," 392–95; "Style and Thought of Early Boyle," 256.

[83] Charles Webster, *The Great Instauration: Science, Medicine, and Reform 1626–60* (London: Duckworth, 1975); *Samuel Hartlib and the Advancement of Learning* (Cambridge: Cambridge University Press, 1970); George Turnbull, *Hartlib, Dury and Comenius: Gleanings from Hartlib's Papers* (London: University Press of Liverpool, 1947), Boas, *RBSCC*, 5–30.

a creation of Starkey's "teeming brain" as we now know—gained enormous respect, and Philalethes' works achieved great popularity.[84]

Boyle, who had been affiliated with the Hartlib circle for some years (well before the onset of his own experimental career in 1649), was soon introduced to Starkey, and the two young men began to collaborate on chymical projects. One goal of their collaboration was the production of iatrochemical remedies such as the *ens veneris* (essence of copper), a pharmaceutical of Helmontian origin, which was intended to serve as an economical remedy for rickets. An account of this collaborative project appears in Boyle's *Usefulnesse of Experimental Natural Philosophy* (1663).[85] Starkey and Boyle were also interested in the subject of volatile alkalies, believed to be akin to the much sought after Helmontian alkahest, the universal solvent that could reduce all substances into their constituent principles and eventually to simple water, the first matter of all things according to Helmontian theory. The Boyle-Starkey collaboration was most intense during the early 1650s, although their interaction continued throughout the rest of the decade. Boyle's manuscript "Philosophical Collections" of the early 1650s, for example, contains numerous recipes transcribed from Starkey's notes that are recorded as choice secrets.[86]

Most important to the present study of Boyle's incalescent mercury is a fragmentary letter from Starkey to Boyle that has been dated to April/May 1651. This letter, discovered and published by Newman in 1987, contains Starkey's "key into antimony," a practical process lucidly desccribed that results in the production of "a mercury that dissolves the mettals, gold especially."[87] Newman has also discovered the missing segment of the letter (in a published German translation) where Starkey explicitly remarks that by his method "[common] mercury can be made into a *mercurius Philosophorum.*"[88]

Given the ubiquity of mercury in chymical texts and the vast array of meanings the word can hold even within the Mercurialist school, the relationship between Starkey's mercury and Boyle's is not immediately apparent. Starkey, for example, never mentions that his mercury grows hot with gold. The identity of the mercuries of Starkey and Boyle can, however, be rendered certain. Starkey considers that the alloy of antimony and silver that he prescribes to be amalgamated with common mercury functions to "purge all superfluities from it," as witnessed by the black powder that the amalgam spews out when it is ground. Boyle likewise, writing in the *Philo-*

[84] Newman, *Gehennical Fire,* is the definitive work on Starkey-Philalethes.

[85] Newman, *Gehennical Fire,* 71–72; Clericuzio, "From van Helmont to Boyle," 315–16.

[86] BP, 25:343–46 (1651/52); BP, 8:140–48 (1654/55); Newman, *Gehennical Fire,* 54–78.

[87] William R. Newman, "Newton's *Clavis* as Starkey's *Key,*" *Isis* 78 (1987): 564–74; original MS at BL, 4:99–100*v,* and a partial transcript at BP, 30:499–506.

[88] Newman, *Gehennical Fire,* 58 and n. 26; the German translation of the full letter to Boyle is published in *Dr. Georg Starkeys Chymie* (Nuremburg, 1722), 416–58; the quotation used here is from 448.

sophical Transactions, states that his mercury is "common mercury skilfully freed from its recrementitious and heterogeneous parts."[89] Starkey notes that the work is "laborious"; Boyle too calls the process "troublesome."[90] The 1651 date of Starkey's letter also corresponds excellently with Boyle's reference to 1652 as the date for his first preparation of the incalescent mercury.

These textual signs of the identity of Starkey's philosophical antimonial mercury with Boyle's incalescent mercury can be confirmed by experimental chemistry. Starkey's directions are sufficiently clear that I was able to carry them out in the laboratory. With minor adjustments, I obtained (after an accurately described "laborious" process) a quicksilver that, when mixed with gold calx, did in fact grow sensibly warm, exactly as Boyle describes.

I have remarked on other occasions how chemical knowledge and the replication of historical experiments, rightly and sensitively applied, can assist historical researches.[91] The experimental fact that Starkey's mercury incalesces with gold (a property not described by him) confirms Starkey as the source of Boyle's incalescent mercury. The odd mercury described in Boyle's *Philosophical Transactions* paper is nothing other than the product of Starkey's laboratory transmitted to Boyle, and the centerpiece of the widely respected chrysopoetic writings of Philalethes. Furthermore, it should be pointed out that the fact that some of the phenomena described in alchemical texts—even those dealing with the *arcana maiora*—can be successfully reproduced in the modern chemical laboratory clearly refutes the claims of those who assert that alchemy had no real experimental component. Had the alchemists not actually performed and described real laboratory operations, or had they been mere victims of self-induced delusional states, or described the "irruption" or "projection" of the unconscious on matter (as many followers of Carl G. Jung claim), they obviously could not have accurately described actual, reproducible chemical phenomena.[92]

The entire Philalethean corpus turns upon the Philosophical Mercury described by Starkey in his letter to Boyle. Starkey's clear plaintext description of its preparation in his 1651 letter reappears in the Philalethean *Introitus*

[89] Boyle, "Incalescence," 525.

[90] Ibid., 526.

[91] Principe, "'Chemical Translation,'"; "Apparatus and Reproducibility in Alchemy." The latter paper contains an account of further work on Starkey's Philosophical Mercury.

[92] See Principe, "Apparatus and Reproducibility in Alchemy"; Principe and Newman, "Historiography of Alchemy"; and Luther H. Martin, "A History of the Psychological Interpretation of Alchemy," *Ambix* 22 (1975): 10–20. Required reading on Jung is Richard Noll, *The Jung Cult* (Princeton: Princeton University Press, 1994). On Jung's view of alchemy, see Carl G. Jung, "Die Erloesungsvorstellungen in der Alchemie," in *Eranos-Jahrbuch 1936* (Zurich: Rhein-Verlag, 1937), 13–111; in English, "The Idea of Redemption in Alchemy," in *The Integration of the Personality,* ed. Stanley Dell (New York: Farrar & Rinehart, 1939), 205–80; also, C. G. Jung, *Collected Works,* vol. 9, pt. 2 (London: Routledge, 1959); *Psychology and Alchemy,* vol. 12 of *Collected Works* (London: Routledge, 1953); *Alchemical Studies,* vol. 13 of*Collected Works* (London: Routledge, 1967); *Mysterium Conjunctionis,* vol. 14 of *Collected Works* (London: Routledge, 1963).

apertus (1667) as a fantastic allegorical image involving dragons, rabid dogs, and the doves of Diana.[93] The lengthy *Ripley Reviv'd* (1678) is little more than an extended and repetitious reiteration of the preparation of this sophic mercury, there also with exotic imagery rather than with the straightforward descriptions of the private correspondence with Boyle.[94] These popular alchemical works may well be the works wherein Boyle states he "found . . . some dark passages, whence I ghess'd their knowledge of" the incalescent mercury. Surely it is not accidental that in the *Philosophical Transactions* Boyle mentions "divers Philalethists," the interest they should show for his paper, and their surprise at his ability to prepare their prized secret by more than one method—including one, he writes, that does not employ "Antimony and solid Metals as *Mars* [iron]"; this is a direct reference to two of the ingredients necessary in the method he had learned from Starkey twenty-five years earlier.[95] (So well did Starkey hide his authorship of the Philalethes tracts that even his collaborator Boyle was not aware of it.)[96] It is interesting to note that while Boyle invokes the "Philalethists," he suppresses any mention of Starkey, who had died in the Great Plague of 1665, and attributes his discovery of this prized alchemical product merely to his own trials and "God's blessing."[97]

This connection now made between Boyle's process and George Starkey, his Philalethean corpus, and the whole extended Mercurialist school begins to reveal the real importance of the incalescence paper, which has hitherto been seen as an isolated item. The paper guardedly reveals Boyle's possession of the chief key to the most popular chrysopoeian corpus of his time and to an entire school of chrysopoeian thought. Properly understood in its alchemical context, Boyle's incalescent mercury paper fulfills two functions—first, it advertises Boyle's success in an approach to the Grand Arcanum; second, it requests further enlightenment from the adepti. Boyle makes the latter clear when he explains that he hopes he

> may safely learn . . . what those that are skilful and Judicious enough to deserve to be much considered in such an affair, will think of our Mercury . . . The knowledge of the opinions of the wise and skilful about this case, will be requisite to assist me to take right measures in an affair of this nature. And till I receive this information, I am obliged to silence.[98]

[93] Philalethes, *Introitus apertus,* 657–59.

[94] See Newman's brilliant exegesis of the allegorical introduction to *Ripley Reviv'd,* in *Gehennical Fire,* 118–33.

[95] Boyle, "Incalescence," 529–30.

[96] Not only does Boyle compile a list of arguments supporting and refuting Philalethes' concealment of the Stone (BP, 19:187*v*–88; printed in Newman, *Gehennical Fire,* 254–55), but there is also a record (dated 17 September 1691) of his ownership of "a paper in 4to ty'd with Packthred writ upon Cosmopolita" (presumably Philalethes rather than Sendivogius), which was kept "in the upper shelf of the Press of Mr. Boyle's Bedchamber" (BP, 36:123).

[97] Boyle, "Incalescence," 521.

[98] Ibid., 528–29.

It seems that Boyle was unable to find the means of progressing from the Philosophical Mercury to the Philosophers' Stone. The "opinions of the wise and skilful," that is to say, of the adepti, might inform Boyle of the "right measures" to take "in an affair of this nature." Note that Boyle refuses to answer any questions regarding his Mercury but nonetheless hopes to gain further knowledge from those readers "who deserve to be much considered in such an affair." He does not want the attentions of those less experienced than himself, but rather the assistance of those more experienced. Thus this *Philosophical Transactions* paper should not be read as a *publication* in the sense of making knowledge public but rather as an attempt to contact adepti and obtain instruction from them. It extends and makes more specific Boyle's general desire (expressed elsewhere) to be "willingly and thankfully instructed" by adepti.[99]

In order to fulfill the function of an advertisement, the paper had to be accessible to a great many people in the hope that it might reach adepti. Accordingly, Boyle (with the help of Oldenburg as a translator) addresses the paper to as large an audience as possible by publishing it in both English and Latin in parallel columns—a publishing practice both unprecedented and unrepeated in the *Philosophical Transactions.* This desire for exposure is the likeliest explanation for a curious occurrence six years later when Oldenburg's Latin translation was published in the *Miscellanea curiosa* of the German Academia Naturae Curiosorum under the name of Heinrich Screta.[100]

Heinrich Screta Schotnau a Zavorzitz (1637–1689) was a physician in the Swiss city of Schaffhausen. His publications include works on hearing loss and fevers, and a paper in the *Miscellanea curiosa* on anatomy; he also developed and popularized an opium-based pill but showed no interest in alchemy whatsoever.[101] It seems improbable that Screta's republication of the incalescence paper is a mere plagiarism. First, Screta was already a competent scientific and medical author who did not need to steal a paper. Second, if Screta were to steal a paper, it is unlikely he would choose so odd

[99] Boyle, *Sceptical Chymist,* [xiii]; *Producibleness,* [xiv].

[100] Heinrich Screta, "De mercurio cum auro incalescente," *Miscellanea curiosa sive Ephemeridum Medico-Physicarum Germanicarum Academiae Naturae Curiosiorum* 1682, Decuria II, Annus primus (Nuremberg, 1683), 83–93. The existence of this republication was noted by Partington, *History of Chemistry,* 2:499 n. 8.

[101] Heinrich Screta, *Dissertatio inauguralis physico-medico de laesa auditione* (Heidelburg, 1670; Basel, 1671); *Kurzer Bericht fon der allgemainen anstekenden Lagersucht* (Schaffhausen, 1676; further editions in 1685 and 1716, Latin translation *De febri castrensi maligna,* Schaffhausen, 1686; Basel, 1716); contributions to Franciscus Hadrianides Piens, *Tractatus de febribus* (Geneva, 1689); on Screta's *pilulae anodyne regiae,* see Johann Maurice Hoffmann, *Acta laboratorii chemici Altdorfini* (Nuremberg and Altdorf, 1719), 149–50; "De gallo gallinaceo ova ponente," *Miscellanea curiosa* 1672, Decuria I, Anno 3, 332 ff. On Screta, see Otto Keller, "Der Schaffhauser Apotheker-Arzt Heinrich Screta," *Schweizerische Apotheker-Zeitung* 121 (1983): 194–97; Franz Schwerz, "Führer für das Heinrich-Skreta-Zimmer im Museum der Allerheiligen" (Schaffhausen, 1938); "Der Schaffhauser Arzt Heinrich Skreta," *Mitteilungen der Naturforschenden Gesellschaft Schaffhausen* 12 (1935): 71–142.

a paper on a subject of which he had neither interest nor experience. Third, the "plagiarism" is simply too blatant; Screta reproduces Oldenburg's translation word for word, and the notion that an obscure medical doctor from Schaffhausen would have had the opportunity to involve Henry Oldenburg (who had been dead for five years) and the president of the Royal Society of London as witnesses to his own experiment would have been unbelievable to readers. Finally, there was very little chance indeed of such a plagiarism's going unnoticed by Boyle. Not only was Boyle notoriously sensitive about plagiarism, but the *Miscellanea* was reguarly reviewed in the *Philosophical Transactions*—Oldenburg had been a patron of the German academy, and Boyle received copies of its journal. Extant notes by Henry Miles among the Boyle Papers list seventeenth-century issues of the *Miscellanea curiosa* that are arguably Boyle's own and include the very issue in which Screta's publication was printed.[102] It is hard to believe that Boyle would not have insisted upon some notice of such a plagiarism in either the *Philosophical Transactions* or the *Miscellanea,* or that there would be no indication of his ire in either his papers or his correspondence, or in the correspondence of the Academia Naturae Curiosorum.[103]

I suggest that Screta was acting on Boyle's behalf as a "front" for his chrysopoetic message and request to adepti. It would be consonant with the aims of the paper for Boyle to have arranged for Screta to republish "Incalescence" in a journal that had a wider currency than the *Philosophical Transactions* in Germany, a land known for its pursuit of chrysopoeia, in order to increase his chances of contacting adepti. Screta could have screened Boyle from unwanted attentions and transmitted to him any responses received from those "Judicious enough to deserve to be much considered in such an affair."

Although an eighteenth-century German translation and abridgment pointed out that Boyle's and Screta's papers were identical (while erroneously claiming that Screta was the Latin translator), many commentators remained unaware of this fact.[104] In his *Chemisches Archiv,* Lorenz Crell cites both Boyle's *Philosophical Transactions* paper and Screta's *Miscellanea curiosa* paper without noticing their identity. Indeed, he published a summary of Screta's version (he calls the author *Soreta*) but refused to do so for Boyle's, which he declared to be "incapable of summary, and moreover, this essay is written in the typical alchemical style"![105]

Whether Boyle received the kind of responses from adepti for which he hoped is unknown. One response to the paper is, however, well recognized; I

[102] BP, 36:173, 194.

[103] I thank Dr. Mason Barnett for checking the correspondence of the Academia Naturae Curiosorum preserved in Halle.

[104] *Der Römisch Kaiserlichen Akademie der Naturforscher auserlesene Medicinisch- Chirurgisch- Anatomisch- Chymisch- und Botanische Abhandlungen* (Nuremberg, 1762), pt. 11, 48–52; "Diese Ablandlung ist aus den philosophical. Transactions des Boyle Febr. 21 1675/6, n. 122, fol. 516, genommen, und von D. Screta ins Lateinische gebracht" (52).

[105] Lorenz Crell, *Chemisches Archiv* (Leipzig, 1783), 1/2:71a–72a.

refer to the letter of Isaac Newton to Henry Oldenburg, dated 26 April 1676.[106] This letter has been examined previously in terms of Newton's alchemy and has likewise been cited in reference to Boyle's alchemical interests.[107] Newton recognizes at once not only Boyle's authorship of the paper but also that he is "desirous of the sense of others in this point," and particularly that "of a true Hermetic Philosopher, whose judgement . . . would be more to be regarded in this point than that of all the world beside to the contrary." Newton commends Boyle's secrecy, saying that he "does prudently in being reserved," and that "the great wisdom of the noble author will sway him to high silence." Newton further believes that "the fingers of many will itch" to learn the preparation, and that if it is in fact the true Philosophical Mercury, it could not "be communicated without immense dammage to the world if there be any verity in the Hermetick writers." Yet in spite of these sentiments, Newton declares that he does not believe that there is any "great excellence" in Boyle's product, and refers its incalescence to the particles with which the mercury is impregnated; these may be "grosser" (heavier) than those of the mercury itself, thus capable of giving the gold a "greater shock, & so put into a brisker motion" than simple mercury can, and this greater motion is sensed as heat. Newton compares Boyle's mercury to acids wherein saline particles are intermingled with common water, which enables them to dissolve metals and generate heat in doing so.

Although this letter to Oldenburg is well-known, a subsequent letter to Oldenburg written two weeks later has never been remarked upon. There Newton states simply, "I perceive I went upon a wrong supposition in what I wrote concerning Mr. Boyles Experiment."[108] Newton's exact meaning is unclear; *which* supposition was in error, his postulated mechanism of incalescence or his assumption that the Mercury was of no "great excellence"? More to the point, how did Newton "perceive" that he had gone astray? It may be that Newton and Boyle communicated directly in the interval, and Boyle changed Newton's mind in some way, perhaps by a further demonstration or explanation of its properties. At any rate, the inquisitiveness—not to say acquisitiveness—that Newton displayed to John Locke after Boyle's death in regard to Boyle's alchemical receipts (including that of the incalescent mercury) suggests that Newton's dismissiveness to Oldenburg was feigned. The trail of Boyle's incalescent mercury will lead back to Newton shortly.

Far-Flying Mercury: A Forty-Year Quest

Thus far the study of Boyle and his incalescent/Philosophical Mercury has involved only the years 1651–1652 and 1676, but the great importance

[106] *Correspondence of Isaac Newton,* 2:1–2.

[107] Dobbs, *Foundations,* 194–96; More, *Life,* 215–17. Note, however, that More's understanding of the alchemical context of the incalescent mercury is confused.

[108] Newton to Oldenburg, 11 May 1676, *Correspondence of Isaac Newton,* 2:6.

Boyle attached to this product causes it to surface—if only allusively—throughout his entire scientific career. Following this Ariadne's thread through Boyle's career not only indicates the tenacity of Boyle's esteem for and investigation of the chrysopoetic potential of this mercury but also provides the opportunity of introducing an array of Boyle's chymical colleagues and thus situating him in a network of social interactions.

Boyle's first mention of this special animated Mercury occurs in his earliest purely natural philosophical work, the unpublished essay "Of the Atomicall Philosophy," datable to ca. 1653. A fragment of the text was published by R. S. Westfall in 1956, and the whole has been discussed more recently by Antonio Clericuzio, Robert G. Frank, Michael Hunter, and William Newman.[109] This essay really does not deal directly with atomism as its title implies, but rather contains a brief history of atomism followed by a fairly lengthy recitation of observations and experiments that show the existence of effluvia, which seem to be its main concern; only secondarily are such effluvia used to support the doctrine of a particulate matter theory. At any rate, here Boyle records that he can prepare a "Mercury incomparably more subtle and volatile then is made by any of our knowne chymicall Processes."[110] This mention of its subtility recalls the need for the Philosophical Mercury to be highly penetrating in order to dissolve and open the dense body of gold, and also harkens back to Starkey's explicit comment that his mercury is "twice as volatile" as common mercury. Boyle's mention of his mercury in this context is quite allusive and easily missed except by one who would recognize the significance of an "incomparably more subtle" mercury. This meaning would not be missed by alchemically attuned readers in the Hartlib circle among whom it was presumably circulated. The allusion to this Philosophical Mercury is not actually important to the text but might rather have been included only as a subtle advertisement of the arcana of which the young Boyle was already master.

Other chymists with whom Boyle was in contact during the 1650s sought the same Philosophical Mercury. For example, Hartlib's son-in-law Frederick Clodius showed great interest in Starkey's Philalethean manuscripts and attempted to carry out the process.[111] The collection of processes that John Matson sent Boyle in 1676 contains accounts of Matson's direct acquaintance with Clodius, and at one juncture he refers to Clodius's preparation of a special mercury by "Sterke's way, by ☾ [silver] and regulus of ♁ [antimony]."[112] This manuscript may represent a return to Boyle of Starkey's

[109] BP, 26:162–75; Westfall, "Unpublished Boyle Papers"; Frank, *Harvey and the Oxford Physiologists,* 94–95; Clericuzio, "Redefinition of Boyle's Chemistry," 568–70; Hunter, "How Boyle Became a Scientist," 68–69; Newman, "The Alchemical Sources of Robert Boyle's Corpuscular Philosophy."

[110] BP, 26:166.

[111] On Clodius, see Dobbs, *Foundations,* 74–79; Newman, *Gehennical Fire,* 59–60, 76.

[112] BP, 29:144.

1651 "Key," which Boyle apparently let Clodius copy in 1653 and send to the Amsterdam chymist Johann Morian.[113]

Sir Kenelm Digby (1603–1665), another alchemically minded colleague of Boyle's in his early period, also figures in the tale of incalescent mercury.[114] The relationship between Digby and Boyle has not been fully explored, but it is clear that they communicated regularly. In "Of the Atomicall Philosophy," Boyle approvingly cites "our deservedly famous Countryman Sr Kenelm Digby," and many receipts in Boyle's mid-1650s collections are attributed to Digby.[115] Further, Digby is the only contemporary mentioned by name in the draft of the *Sceptical Chymist,* where he testifies to several transmutational experiments.[116] Finally, Royal Society MS 10, a collection of alchemical treatises, which I have recently shown to have been owned by Boyle, was a gift from Digby; it bears not only his monogram *KD* on the spine but also his autograph marginalia within.[117] In one of these marginal notes Digby refers to a preparation of mercury "which maketh the matter take greate heate when the gold is putt to the mercury."[118] Thus among Boyle's early chymical colleagues, Starkey, Clodius, and Digby, at least, were involved in similar pursuits.

The 1660s also saw references to Boyle's prized Mercury. In 1663, Olaus Borrichius, the Danish chymist, polyhistor, and traveler, visited England and conversed with Boyle at London. He recorded in his diary on 4 August 1663 that Boyle spoke to him of a mercury which amalgamated strikingly quickly with gold and grew hot in doing so.[119] Their conversation that day was filled with the topic of metalline Mercuries, and Boyle's comments signal a belief in their real existence; the reader should note that this discussion occurred within two years of the publication of the *Sceptical Chymist.* Seven months later Borrichius sent Boyle a letter from Paris, thanking him for his hospitality and conversation, and further speculating on alchemical topics. Borrichius's most interesting topic regards the constitution of ordinary mercury. Taking a stance based on Helmontian theories, Borrichius refers the fluidity of quicksilver, its inability to be frozen, and its volatility to its com-

[113] Newman, *Gehennical Fire,* 76.

[114] On Digby and alchemy, see Betty Jo Teeter Dobbs, "Studies in the Natural Philosophy of Sir Kenelm Digby," *Ambix* 18 (1971): 1–25; 20 (1973): 143–63; 21 (1974): 1–28; *Foundations,* 74–79. On Digby in general, see T[homas] L[ongueville], *The Life of Sir Kenelm Digby* (London: Longmans, Green and Co., 1896); E. W. Bligh, *Sir Kenelm Digby and His Venetia* (London: S. Low, Marston and Co., 1932); R. T. Petersson, *Sir Kenelm Digby: The Ornament of England* (Cambridge: Harvard University Press, 1956).

[115] Boyle, "Of the Atomicall Philosophy," BP, 26:162; "Promiscuous Observations begun the 24th of September 1655," BP, 25:153–56.

[116] Boas [Hall], "Early Version," 161–63; *Sceptical Chymist,* 275; Hall notes how Digby drops out of the printed edition and is cited only as "a learn'd Person." Some readers, however, recognized Digby in the *Chymist* anyway; see Becher, *Supplementum in physicam subterraneam,* in *Physica subterranea,* 315.

[117] Principe, "Newly-Discovered Boyle Documents," 58–59.

[118] RSMS, 10:54.

[119] *Olai Borrichii Itinerarium,* 3:65–66.

position from "most minute globules that have an exceedingly solid metallic shell and a center of common water."[120] If these shells are broken open and the water allowed to escape, Borrichius argues, "all the shells will, perhaps, be easily fused together by fire into the noble metal."[121] Ordinary acids cannot break through this metallic covering because their particles are too large; consequently mercury is not "radically" dissolved by the acids and can always be recovered in its original state. The Philosophers' Stone, however, is sufficiently subtle to penetrate the metallic coverings, and the noise that adepti say is heard when the transmuting powder is projected upon mercury is the sound of the common water breaking forth from its confinement within the particles of mercury. Borrichius then suggests that Boyle seal up quicksilver in his air-pump, because "if by means of that machine the intrinsic watery liquid can be drawn out of it by force, I do not doubt but that some significant mutation would follow"; in other words, the air-pump might function as a transmuting engine by drawing out the central water of quicksilver, allowing the residual metallic shells to remain as gold.[122]

During the 1660s Boyle wrote a number of additional essays for his lengthy *Usefulnesse of Experimental Natural Philosophy,* published in 1663 and extended in 1671. Some of these essays were never published and are now lost in their original English versions, surviving only in Latin translations prepared in the 1680s by Thomas Ramsay, a Lithuanian theologian whom Boyle employed as a translator. The extant fragment of one of these unpublished essays deals with the value of chymistry for producing new types of substances. There Boyle again makes an allusion to the preparation of his sophic Mercury, but without mentioning what the product actually is, why it should be important, or for what it is used. Rather, he merely hints at it by claiming "that by melting together stellated martial regulus with silver we obtained a friable mass which we thought would be of great utility in purifying mercury."[123] Boyle makes this declaration in a rather offhand manner without elaborating on why such a "friable mass" might be thought useful in purifying mercury or how it would do so. Rather, his brief allusion leaves it to the alchemically attuned reader to recognize the importance of the purification of common quicksilver by means of an antimonial alloy. This allusion, like the mention in "Atomicall Philosophy," may serve only as a subtle advertisement or may be part of a grander scheme of "dispersion of knowledge" through Boyle's corpus. This passage would not attract any attention from modern readers lacking a thorough knowledge of Boyle's

[120] BL, 1:88–89*v;* Olaus Borrichius to Boyle, 30 March 1664; on 88*v;* "argentum vivum ex paene infinitis minutissimis constare globulis, quibus cortex solidissimus metallicus est, medulla autem aqua vulgaris."

[121] Ibid., 89; "disrumpatur ille cortex, ut aqua exhalet, cortices omnes facile igne in nobile metallum forsan coalituros."

[122] Ibid.; "si suus illi ingenitus liquor aqueus exhauri vi possit istâ machinâ, non dubito, quin insignem mutationem sit subiturum."

[123] BP, 25:411–13; "memini Nos per eliquationem Reguli [martis] Stellati simul cum argento, obtinuisse massan friabilem, quae magno usui in purificando/interpolando [mercuri]o fuisse arbitramur."

chrysopoetic interests and their origins, for without cognizance of the details of the Starkey-Philalethes process this allusion would be missed. The difficulty of recognizing the import of typically brief allusions and asides is a recurring issue in the consideration of Boyle's chrysopoetic pursuits. Without familiarity with the terms, theories, processes, and products of seventeenth-century alchemy such recognition is impossible.

Boyle's laboratory notes likewise provide additional material on his Philosophical Mercury. One receipt among a richly alchemical collection datable to the 1670s occurs in the following coded form:

> Take pure Negerus, Dakilla, imbrionated banasis ana, mix them very well together & drive off all that you can in a Retort with a strong fire of sand . . . [it] dissolves Gold readily, and that with sensible heat.[124]

When decoded, this receipt for incalescent mercury employs antimony ("imbrionated banasis") and copper ("Dakilla") instead of antimony and silver. This apparently simple substitution is significant, for while Starkey prescribed the use of silver in his 1651 letter to Boyle and in most of the Philalethes' tracts, his 1654–55 *Marrow of Alchemy* repudiates the use of silver (specified under the *Deckname* "Diana's doves" throughout the Philalethean corpus) as "tedious labor" and substitutes instead "Venus" or copper.[125] Clearly Boyle had either heard of this substitution directly from Starkey or had gleaned it from his own reading of the *Marrow.* In either case, the change shows Boyle's interest in perfecting the process.

Another laboratory record begins "take negirus animated with Artephius's mineral." Here Boyle uses his common *Deckname* for mercury and refers to antimony as "Artephius's mineral," alluding to the much-respected author Artephius, whose *Liber secretus* begins, "antimony is of the parts of Saturn."[126] An index (intermittently encoded and datable to the 1680s) of some of Boyle's laboratory processes lists several preparations of incalescent mercury, more than one "process with Columbe Diana" or "Diana's Doves," "the most Compendious way to animate Negirus [mercury]" and "to prepare a ☿ [mercury] that is thought to surpass that taught by Iraeneus [Philalethes]."[127]

[124] BP, 25:55. The last words are altered from the original "without sensible heat" by a later correction.

[125] Philalethes, *Marrow of Alchemy,* pt. 2, 15–16.

[126] BP, 25:63. *Artephius his Secret Booke,* trans. Irenaeus Orandus (London, 1624); reprinted in *Nicholas Flamel: His Exposition of the Hieroglyphicall Figures,* ed. Laurinda Dixon (New York: Garland Publishing, 1994), 54–80. The Artephius corpus is very heterogeneous, containing both works from medieval Arabic originals and works accreted much later; the *Liber secretus* falls into the latter category, having been written no earlier than 1600. Boyle mentions Artephius twice in the *Dialogue on Transmutation.* See H. D. Austin, "Artephius-Orpheus," *Speculum* 12 (1937): 251–54; G. Levi della Vida, "Something More about Artefius and the *Clavis sapientiae,*" *Speculum* 13 (1938): 80–85; Halleux, "Le mythe de Nicolas Flamel," 251.

[127] BP, 36:102–11*v;* on such mercuries, see entries 516, 1085, 1090, 1106, 1109, 1378, and 1810.

In the mid-1670s Boyle published his paper in *Philosophical Transactions,* and at the end of the decade, when Boyle was involved with Georges Pierre, the incalescent mercury surfaces again. In one of Boyle's first letters to Pierre he must have revealed the lines of alchemical experimentation along which he was working, for in the earliest of Pierre's extant letters he replies that "I think you have come across something good in animating your common mercury with regulus [of antimony]." Pierre goes on to assert that this is still not "the correct method" and promises a better one. [128] Pierre sent this better method to Boyle, purportedly from Philalette himself (and "written in his own hand"), in his next letter. Although Pierre requests that Boyle destroy the note after reading and memorizing it, Boyle did not do so, for Philalette's note is surely the unsigned document in French preserved elsewhere in the Boyle Letters that gives two methods of animating mercury, the first of which is very close to the process Boyle learned from Starkey.[129] Yet another French manuscript elsewhere in the archive provides a further recipe for the animation of common mercury with regulus of antimony.[130]

In 1680 Boyle revealed more in print about his incalescent mercury in the third essay of the Mercury section in *Producibleness of Chymical Principles.* Part of this essay, "Of the Dissimilitude of Running Mercury," recapitulates "Incalescence," but here Boyle also expands his discussion of the incalescent mercury and claims authorship of the anonymously published *Philosophical Transactions* paper. Again Boyle strongly assails negators of the reality of metallic Mercuries, claiming that "those who looke upon incalescent Mercuries as Chymical *non-entia* . . . are not competent judges of the possibilities of things."[131] Boyle, employing the phrase "animated Mercury" used in DuClo, Collesson, and other Mercurialists, asserts such mercuries to be "matters of fact" and claims that his discourse will be useful for readers who might "vigorously to prosecute in a Spagiricall way, the more noble sort of Mercuriall Experiments." These "more noble" ends are named allusively when Boyle immediately notes that such Mercuries as he has prepared are "fitted to have potent operations, as well upon humane bodies, as the more stubborn ones of Metalls and Minerals"—an allusion to the epithet "medicine of Men and Metals" commonly given to the Philosophers' Stone.[132] Thus Boyle sees his tract as useful to fellow aspiring adepts.

Mercurialist Theories in Producibleness

Significantly, in *Producibleness* Boyle adopts not only the Mercurialist chrysopoeians' language to write about this product but also their theoretical explanations. First, Boyle reproduces exactly the distinction made by

[128] Pierre to Boyle, 13/23 December 1677, BL, 4:106.
[129] Pierre to Boyle, 17/27 January 1678, BL, 4:107; Philalette's note is 6:74–75.
[130] BP, 19:175.
[131] Boyle, *Producibleness,* 213.
[132] Ibid., 224–25.

chrysopoeian Mercurialists between the external and internal cleansing of common mercury to produce Philosophical Mercury, by specifying an "externall Depuration" and then "another that is internall."[133] Second, he employs Mercurialist explanations to speak of the "Impregnation" of common mercury by "Corporeall or Spirituall" means. Recall that van Suchten (followed by Starkey/Philalethes) attributes the activity of animated Mercury to the presence of a "volatile Gold" that the common mercury incorporates during its cleansing by the martial regulus of antimony; Boyle says exactly the same thing in *Producibleness.* He notes that his incalescent mercury was made penetrating and noble because "some subtle parts of another body are so intimately associated and united" with it. More specifically, he asserts that the noble Mercury he prepared showed a specific gravity greater than that of common mercury, even though he had used no "corporeall gold" in its preparation (an admixture of gold, the only substance known to Boyle that was denser than mercury, would be the only obvious way to increase the density of mercury). But Boyle writes that the higher specific gravity "seem'd to argue, that even spirituall or volatile *Gold* (for no visible *Gold* was employ'd)" was able to increase the specific gravity of his mercury.[134] What Boyle means here by "spiritual or volatile gold" is quite unclear without reference to the chrysopoetic treatises of van Suchten.

Van Suchten notes that the presence of "volatile gold" in his animated Mercury can be proven, if one should

> evaporate the *Argent vive* from *Luna* [silver], so remaineth the *Aurum* [gold] volatil, that is *Mercury* of *Mars* [iron], with the *Luna,* and tingeth the *Luna* into the highest colour of *Sol* [gold].[135]

Thus van Suchten asserts that when a drop of animated mercury is evaporated from a silver plate, it leaves behind its volatile gold as a golden spot. Van Suchten notes that this golden color is "not fixed," for it is driven away by a stronger fire. Boyle writes *identically* of his prepared mercury that

> a drope or Globule of which, being evaporated from a thin piece of Silver . . . left upon it a rugged substance . . . of a colour very neare that of *Gold,* from whose nature perhaps it was not very remote.[136]

Elsewhere, Boyle muses "why we may not hope" for great effects from common mercury "well purifyed and impregnated with the Sulphur and finer parts of such bodies as volatile *Gold,* or *Venus,* or *Mars,* or *Antimony.*"[137] Here Boyle not only mentions the volatile gold of the Mercurialists again but uses the term *Sulphur* in its fully alchemical sense, as a part of

[133] Ibid., 204–5.
[134] Ibid., 207–8. 221–22.
[135] Suchten, *Secrets,* 80.
[136] Boyle, *Producibleness,* 210–11; note that the *not* before *very remote* is missing from the text itself but is supplied in the list of errata.
[137] Ibid., 226.

metals and minerals, saying that it is only this part of the substances employed that impregnates the Mercury. The reader should note that the mercury is to become "animated" by the process, that is, endowed with *anima* or soul, a term synonymous with *Sulphur* in alchemical usage. Boyle not only uses alchemical terminology but employs it together with and in the context of Mercurialist explanatory theory. Note also that Boyle here names, in a rather offhanded allusive way, all the materials essential for making the incalescent mercury—copper, iron, and antimony—although without making the connection clear; this is perhaps another example of "dispersion of knowledge."

Boyle also asserts that he can make more than one kind of noble Mercury employing different methods. Boyle's "animation or spirituall impregnation" may occur in various ways, and "by impregnating them with this, or that Minerall, or metall, [one] may much diversify their qualities and operations." This impregnation of mercury in diverse ways to produce diverse mercuries seems clearly drawn from the fourteenth-century Bernard Trevisan (who, remember, appeared as a character in Boyle's *Dialogue on Transmutation*). In his *Response to Thomas of Bologna,* Bernard claims that mercury concocted with various metals is like water boiled with various vegetables or herbs: each vegetable imparts a characteristic virtue to the water. Just so, these mercuries "take up other natures and qualities" dependent upon the metals with which they are treated.[138] This letter of Bernard Trevisan was specifically recommended to Boyle by Starkey in his crucial 1651 letter.[139]

Finally, Boyle admits an observation that supports the dyad theory of metals. Referring to the process of "precipitating mercury *per se*"—that is, keeping mercury near its boiling point in the presence of air for extended periods of time, thus forming the solid red mercuric oxide from the liquid metal—Boyle notes that while he expected great medicinal effects from the red precipitate made from his animated mercuries, he found that they could not be turned into the red precipitate by the usual treatment. "As if that disposition to be calcin'd, (as the Chymists are pleased to speake) or turn'd into powder, required the presence of the recrementitious or more separable part of *Quicksilver,* that a Chymist would perhaps call it[s] *Sulphur,* which was a discovery I could willingly enough have miss'd."[140]

The notion that mercury contained within itself a certain Sulphur whereby it was coagulable into a dry powder (recall that one of Sulphur's chief roles is as a coagulator of Mercury) dates back as far as Geber.[141] Boyle's observation has, however, more proximate precedents. One is Starkey's *Marrow of Alchemy,* where he asserts that metalline Mercuries

[138] Bernard Trevisan, *Responsio ad Thomam de Bononia,* in *BCC,* 2:399–408, on 399; Boyle, *Producibleness,* 224–25.

[139] Newman, "*Key,*" 573.

[140] Boyle, *Producibleness,* 219–21, on 220.

[141] Newman, *The Summa Perfectionis,* 713.

cannot "by the fire (by circulating oft) . . . unto powder dry be turn'd."[142]. Van Helmont likewise attributes the precipitation of mercury to a separable external Sulphur.[143] The same claim is made by Otto Tachenius in his *Hippocrates chymicus*. This work was first published in Latin in 1666, reviewed in the *Philosophical Transactions* in 1669, and published in an English translation by "J. W." in 1677. This "J. W." is John Warr, Sr., and his autograph translation manuscript survives as volume 32 of the Boyle Papers.[144] Tachenius, a chymist and physician settled in Venice, is best known for his propagation and extension of the acid-alkali theory of Francois de la Boë's (Francescus Sylvius, 1614–1672). Boyle assails Tachenius's notion (without naming him) in his "Reflections upon the Theory of Alkali and Acidum" in the collection *Mechanical Origine of Qualities* (1675). Clearly Boyle had read the *Hippocrates*.[145]

In regard to mercury and its precipitation, Tachenius invokes the authority of Geber "and other wise Men" and his own experiments to declare that "Mercury, hath a *Sulphur,* external and separable" which is the cause of its precipitation ("coagulation") into a dry red powder upon prolonged heating. Tachenius further notes that metalline Mercuries "although boiled a whole year . . . can be precipitated *per se* in no degree . . . and therefore must needs want that external *Sulphur.*"[146] Boyle's "willingness to miss" this discovery may stem from its apparent confirmation of either an observation of a man whose theories he had criticized or of the more general notion that Sulphur is the cause of coagulation.

In *Producibleness,* Boyle also seems to respond to an account of an incalescent mercury made by Johann Joachim Becher, himself an adherent of the Mercurialist school. In his *Experimentum chymicum novum* (1671), the first supplement to his *Physica subterranea* (written, as I have mentioned previously, against Rolfinck's attack on metalline Mercuries), Becher mentions, as Boyle would do four years later in *Philosophical Transactions,* the incalescence of a specially prepared mercury with gold. Becher recounts the story of a special mercury that "upon its first contact with gold so heated and boiled up, that there was danger that the glass mortar would be broken, for the hand with which I held it had to be removed on account of the

[142] Philalethes, *Marrow of Alchemy,* pt. 2, 20.

[143] Van Helmont, "Progymnasa meteori," in *Ortus medicinae,* 79.

[144] Otto Tachenius, *Hippocrates chymicus* (Venice, 1666); review in *Philosophical Transactions* 4 (1669): 1019–21; Tachenius, *Hippocrates chymicus . . . Translated into english by J[ohn] W[arr, Sr.]* (London, 1677; reprint, 1696). See also John T. Harwood, *The Early Essays and Ethics of Robert Boyle* (Carbondale: Southern Illinois University Press, 1991), 249, and Hunter, *Guide,* xxxv.

[145] On Tachenius, see Partington, *History of Chemistry,* 2:291–96; on the acid-alkali theory and Boyle's rejection of it, see Marie Boas Hall, "Acid and Alkali in Seventeenth-Century Chemistry," *Archives Internationales d'Histoire des Sciences* 9 (1956): 13–28; and on Sylvius, E. D. Baumann, *Francois de la Boë Sylvius* (Leiden: Brill, 1949).

[146] Tachenius, *Hippocrates* (London, 1677), 99–100, 103.

heat."[147] Later, in the second supplement (1675), he indicates that one can make these incalescent mercuries by "animating" common mercury, and he spends much time describing their properties. Becher states that the property of incalescence is lost quickly after the mercury is prepared.[148] Boyle seems to respond to Becher when, immediately after recounting an experiment displaying vigorous incalescence in a glass mortar (almost mimicking Becher's account), he takes special care to note how the incalescence of his mercury "was not a transient and easily vanishing one," and years later (1691) he published a brief notice on "the durableness of the Faculty of a certain prepar'd *Mercury* to grow Hot with Gold."[149]

Boyle's Late Work on the Mercury and Its Legacy

As great a prize as it was, the Philosophical Mercury is nonetheless only an inlet to the Grand Arcanum; indeed, I have argued that the *Philosophical Transactions* paper is an attempt to gain information on how to proceed. There is evidence of Boyle's attempts to push on from this Philosophical Mercury to the confection of the Philosophers' Stone itself. According to Mercurialists, Philosophical Mercury is to be digested with gold for a long period, eventually producing the Stone. Accordingly, Boyle treats the digestion of gold and his animated mercury at length.[150] A good sign for aspiring Sons of Art (mentioned by Collesson) is that the gold "visibly vegetate and show the peacock's tail." Indeed, Boyle records seeing "very pretty vegetations, and sometimes, which is far more considerable, odd changes of colours" upon digesting gold with his Mercury. In his 1651 letter to Boyle, Starkey also wrote excitedly of how his Mercury causes metals to "grow in the forme of trees," making "gold to puffe up, to swel, to putrefy, to grow with sprigs and branches, to Change Colours dayly."[151] Whether Boyle himself saw any further encouraging signs we cannot now determine directly, for he is silent on that score.

There is a letter to Boyle from one Pellegrini, who had just visited Boyle in London, which describes the very same experimental result in allegorical guise:

> What of our sea? When all hope had almost been abandoned, trees appeared before me, as if to another Columbus, swimming above the waves . . . Ah, if only

[147] Becher, *Supplementum,* in *Physica subterranea,* 317; "Mercurius *ad contactum* primum auri ita *excaluit* & efferbuit, ut periculum fuerit, ne *patella* vitrea *rumperetur:* Manus enim, qua eam tenebam, prae calore amovenda erat."

[148] Ibid., 379–83, 414; on 381.

[149] Boyle, *Producibleness,* 214; cf. "Incalescence," 527–28; *Experimenta et observationes physicae* (London, 1691), 142–43.

[150] Boyle, *Producibleness,* 215–20.

[151] Ibid., 220; Newman, "*Key,*" 573; my own successful replication of these metallic trees is recounted and photographed in Principe, "Apparatus and Reproducibility in Alchemy."

> it be that for which we hope! . . . Venus and Vulcan seem to have joined together so beautifully the children of Leto, and to have returned them as friends to their ancestral sea.[152]

The sea is the Philosophical Mercury, the trees the "vegetating gold" that Boyle, Starkey, Collesson, and others describe. The children of Leto are Apollo and Diana and represent the sun and moon or, in their alchemical equivalents, gold and silver, which in Pelligrini's experiment are "returned to their ancestral sea"—that is, radically dissolved back into their primordial Mercury by the action of the Philosophical Mercury.[153] In conjunction with this typically alchemical allegory, Pellegrini refers to an ingredient as "Tora," which is another of Boyle's own code words (in this case meaning *sulphur*). Apparently Boyle shared his code with Pellegrini (clearly a sign of trust), and Pellegrini must be counted as another member of Boyle's circle of chrysopoetic collaborators.

At the end of Boyle's life the Philosophical Mercury again makes an appearance in an autograph memorandum. This memo, probably written during the last year of Boyle's life, lists the fifteen most important items to be taken care of before his death. There Boyle lists his will, the endowment of the Boyle Lecture for Christianity, the translation of his books, the collection of his papers, and, amid these other items, "the Animated Mercury."[154] This brief document does not make clear exactly what Boyle wanted done with his Philosophical Mercury, but other evidence fills in the picture. Apparently, in the last year of his life, Boyle entrusted the preparation of this precious material to one or more of his friends. One of these friends was the philosopher John Locke, who in 1691 was compiling Boyle's notes for a *General History of the Air* into a publishable form. Locke had a long-standing interest in chymistry, including chrysopoeia. He attended the chymical courses at Oxford led by Peter Sthael in 1663 (where he was "prating and troublesome"), and among the Boyle Papers there exists a receipt for "Mercurius antimonii Lockii."[155] Locke's correspondence with Boyle often refers to experimentation, and Boyle must have had a good opinion of Locke's abilities since he chose him as an examiner of his chymical papers.

Locke's diary entry of Friday 25 September 1691—three months before Boyle's death—contains the following receipt in shorthand amid other items attributed to Boyle.

[152] Pellegrini to Boyle, 11 July 1686, BL, 4:100; "quid de mari [*sic*] nostro? Spe omni feré destituto veluti alteri Columbo apparuerunt arbores supra mare natantes et proximam terram promittentes. Oh utinam quam optamus! . . . videntur adeo pulcré Venus et Vulcanus Latonis filios coniuxisse, et mari patrio amicos reddidisse ipsorum."

[153] The Lune, or silver here, is probably not the element Ag but rather a "philosophical silver."

[154] BP, 36:87.

[155] Guy Meynell, "Locke, Boyle and Peter Stahl," *NRRS* 49 (1995): 185–92; BP, 26:102.

> Take mineral water lb 1, mineral soap ℥ [ounce] ii, wash therewith 7 or 8 times, each time you will find filth come away. Let the water stand and you will see sediment fall. With pure suds wash the fire lady and your water will make ☾ [silver] shine like ☉ [gold].[156]

This rather obscure private note elucidates a published letter from Locke to Boyle. Almost exactly a month after the diary entry, Locke writes to Boyle regarding his difficulties in editing *History of the Air* but closes with the cryptic statement, "give me leave now to tell you, that I have water, and I have vessels, I only want soap to be at work."[157] These incomprehensible remarks about soap and water recall the "first period" of the process from Boyle's papers that Locke sent to Newton during their well-known exchange of 1692, with which I began this volume:

> Recipe ☿ [mercury] lb x, cleanse it well with lb i of flowers of 🜍 [sulphur] in 24 hours. To these lb x, take ℥ [ounce] i of minerel soap, shake it with the ☿ so as it may first embody with it and afterwards by further agitation be spued out of it. This worke may last 24 hours or more. To the same adde ℥ i more of the Soap & worke as before. This doe 7 times. Then before any Durca be added . . .[158]

What do these obscure references to soap mean? Recall that according to the Mercurialists, common mercury is transformed into Philosophical Mercury by an "internal depuration." For many Mercurialists, this cleansing agent was an antimonial alloy. Thus we may now resolve Locke's hitherto unintelligible references to "soap" as an appropriate *Deckname* for that "cleansing" alloy and recognize these receipts as preparations of the incalescent/Philosophical Mercury. Newton says as much to Locke when he notes that the "first period" sent him by Locke bore a marginal note that the mercury produced would "grow hot with gold."[159] Note that the process sent to Newton includes both the external purification (with flowers of sulphur) and the internal (with the antimonial alloy, or "soap"), as mentioned by Boyle in *Producibleness* and by Mercurialists in general.

But a further code remains in Locke's diary: "the fire lady," which corresponds with "Durca" in his letter to Newton. *Durca* is used in Boyle's coded papers as a *Deckname* for *copper.*[160] This identification of the fire lady with copper is confirmed by a copy of the diary process transcribed into one of Locke's journals, which directs, "bathe ♀," using the standard symbol for Venus (i.e., copper). The use of the prepared Philosophical Mercury upon copper suggests that the first period actually produces the Mercury of Venus, "which is always green" according to both Starkey's 1651 letter to

156 This reference was given to me by Prof. Henry Schankula.

157 Locke to Boyle, 21 October 1691, *Works,* 6:543.

158 Locke to Newton, 26 July 1692, *Correspondence of Isaac Newton,* 3:216–17, on 216.

159 Newton to Locke, 2 August 1692, ibid., 3:217.

160 Principe, "Robert Boyle's Alchemical Secrecy," 65–66 n. 19.

Boyle and Alexander van Suchten's *Mysteria gemina antimonii.* Boyle did in fact write of a green mercury in his *Experimenta et observationes physicae* (1691) and mentions such a substance to Georges Pierre.[161] Locke's diary entry also asserts that the final product will make "silver shine like gold," recalling the existence of "volatile Gold" in the mercury claimed by van Suchten and demonstrated by both him and Boyle through evaporation from a silver plate.

While Locke's exchange with Newton in 1692 marks the last appearance of Boyle's incalescent mercury, it also mentions a certain "red earth" somehow related to the process. This red earth is another of the long-acknowledged connections of Boyle to transmutational alchemy. Less than a month after Boyle's death, Newton appended a postscript to a letter to Locke, remarking on how he knows that Boyle "communicated his process about the red earth and mercury" to them both before his death. In midsummer, Locke sent Newton a sample of this red earth he had received from Boyle, receipts copied from Boyle's papers, and some information on the process from his own notes to replace the portion Newton had purportedly lost of what Boyle had told him about the process. Newton's last and lengthiest letter on this subject ties the whole together and provides a glimpse of the entire process. Boyle's process contained three parts. The first part, or "period," yields incalescent mercury; the other two "periods" are not extant, but they presumably result in this "red earth." According to Newton it was on account of this earth that Boyle had the antitransmutation statute repealed in 1689. What this "red earth" might have been is unclear, but it is worth noting the Mercurialists' stipulation that Philosophical Mercury is to be digested to a red powder, which could then be exalted into the Philosophers' Stone. Boyle probably believed that his "red earth" was a rudimentary form of the Stone, needing further fermentation or multiplication (see chapter 3) to bring it to a higher order of operation.

This further operation on the red earth may be the information Boyle requests in his 1682 letter to the mysterious "Monsieur Le Green." Le Green's response notes that "You write to me, Sir, of a multiplication that cannot be learned while dreaming. It can be achieved in several ways, depending upon whether the 41 [Elixir?] has been made by the wet way or mainly by the dry way . . ."[162] The balance of the letter does not seem terribly prescriptive (our present inability to decipher the numbers aggravates the matter) for Le Green is very wary of his letters' falling into the

[161] *Georg Starkeys Chymie,* 448; Suchten, *Secrets,* 88; Boyle, "Strange Reports" in *Experimenta et observationes,* 26–28 (the draft is RSMS, 186:66*v*–67); Pierre to Boyle, 14/24 March 1678, BL, 4:110*v.*

[162] Le Green to Boyle, 19 September 1682 (O.S.), BL, 3:150–55, on 150*r–v;* "vous mescrives Monsieur dune multiplication que lon na prend pas en songeant, elle se fait de plussieurs fachons suivant que 41 a esté construit ou par voye humide ou voye seiche principalement . . ."

wrong hands and promises to write more fully at a future date when communication can be more secure.

While Newton's letters to Locke are dismissive of Boyle's process, Newton certainly wasted no time after Boyle's death in descending upon his papers, and his entire exchange with Locke can hardly be described as anything other than "cagey." But Newton's comments shed further light on Boyle's attitudes. Newton claims that Boyle had not in fact brought the process to completion—that is, presumably, to full projective transmutation. Rather, Boyle seems to have "farmed out" the process among several groups of practitioners; this agrees nicely with what has been revealed in this and the preceding chapter about Boyle's "satellite laboratories"—his employment of Godfrey Hanckwitz in making phosphorus and his placing the German projector under Godfrey's care—and the collaboration with De Saingermain and others. Further, Newton bolsters our conclusions regarding Boyle and secrecy when he reveals that he and Boyle communicated in a somewhat prickly air of mutual distrust. Newton complains that Boyle offered his secrets only under strict conditions, and even after Newton agreed to these, Boyle still concealed crucial steps. Newton writes that in offering some other receipts, Boyle "cumbered them with such circumstances as startled me and made me afraid of any more." But, concludes Newton, "I suspect his reservedness might proceed from mine."[163] Newton's assessment of Boyle to Locke contradicts what he had written some years previously to Fatio de Duillier, where he claims to have declined taking up Boyle's offer of correspondence on (presumably) alchemical topics because he felt Boyle was "too open & too desirous of fame."[164] Of course, Newton apparently *did* communicate on such topics with Boyle and found to his chagrin that Boyle was not in fact open at all, at least not to him.

It is ironic that the tortuous trail of Boyle's Philosophical Mercury should end with Newton. Boyle's first experience with animated mercury came, as I have shown, from Starkey's letter of 1651; amusingly, that very letter of Starkey's contains the text of the *Clavis* of Keynes manuscript 18, which Dobbs considered so central to Newtonian alchemy.[165] Thus Starkey's process came to Newton by two very different routes. Curiously, there is no evidence that Newton recognized the identity of the processes, or that he even realized that his own *Clavis* manuscript provided incalescent mercury. It seems strange that after his communication with Boyle and Boyle's detailed instructions on testing "noble Mercuries" in the *Philosophical Transactions* paper Newton would not have tested the *Clavis* product, but if he ever did so, there is no record of it. One can, however, adduce some circumstantial evidence that in spite of his overtly dismissive attitude toward

[163] Newton to Locke, 2 August 1692, *Correspondence of Isaac Newton,* 218.

[164] Newton to Fatio de Duillier, 10 October 1689, ibid., 3:45; the letter is mutilated at key junctures. For more on the alchemical interactions between Newton and Boyle, see Principe, "The Alchemies of Robert Boyle and Isaac Newton."

[165] Newman, "*Key*"; Dobbs, *Foundations,* esp. 175–86.

Boyle's process, Newton did in fact set about carrying it out straightaway. It was later in the same year that Locke sent him Boyle's process that Newton suffered his well-known mental illness, which, judging from the exceedingly high levels of the toxic substance found by the microanalysis of Newtonian hair samples, may have been brought about by acute mercury poisoning—the kind of poisoning that might very well follow from the multiple distillations and digestions of mercury necessary in Boyle's Mercurialist process.[166] One might question as well whether Boyle's own chronic sickness was caused or aggravated by similar mercury poisoning.

Conclusion

The trail of Boyle's incalescent/Philosophical Mercury stretches over his entire career as a natural philosopher. Contemporary characters as diverse as Starkey, Clodius, Digby, Becher, Borrichius, Locke, and Newton, as well as earlier chrysopoetic authors including Geber, Bernard Trevisan, Artephius, DuClo, and van Suchten, are all involved in one or more parts of this forty-year saga. (Indeed, this preparation of mercury survives into the eighteenth century, where it was pursued by several chymists, including Wilhelm Homberg at the Académie Royale des Sciences. Surely this was one of the most enduring avenues of chrysopoetic study!)[167] The visible signs of Boyle's product in the *Philosophical Transactions* and *Producibleness of Chymical Principles* have been linked to other allusions in numerous other public and private sources. The incalescence paper is not an isolated item but is only a brief public glimpse of a lifelong pursuit of enormous importance to Boyle. This dogged pursuit fixes Boyle squarely within the Mercurialist school of chrysopoeia, the most active school of chrysopoetic alchemy in the seventeenth century. The objection that Boyle was only borrowing illustrative experiments from "the chymists" does not hold here, for Boyle took not only experiments from this school but also its traditional theories, terminology, and explanations, all the time gazing steadfastly toward the goal he shared with these alchemical writers—the production of the Philosophers' Stone. Boyle's animation and impregnation of common quicksilver to provide more "noble and penetrating" Mercuries cannot be artificially separated from the identical process described in identical terms by chrysopoeians; the incalescent/Philosophical Mercury seats Boyle amongst them.

Similarly, Boyle's addiction to secrecy—revealed by his extensive use of ciphers and codes solely in chrysopoeia-related topics, his employment of

[166] L. W. Johnson and M. L. Wolbarsht, "Mercury Poisoning: A Probable Cause of Isaac Newton's Physical and Mental Ills," *NRRS* 34 (1970): 1–9; P. E. Spargo and C. A. Pounds, "Newton's 'Derangement of the Intellect': New Light on an Old Problem," *NRRS* 34 (1970): 11–32.

[167] A paper I delivered at a 1996 Workshop on Corpuscularianism held in St. Andrews, Scotland, treated Homberg's chrysopoeia at length. The papers from this workshop are due to be published as a collected volume.

the classical technique of "dispersion of knowledge," and his remarkable "reservedness" with Newton—shows that he invested traditional alchemy with a special, priviliged status.

All the foregoing material, culled from both printed and archival sources, leaves no doubt as to Boyle's continued and intense involvement and belief in traditional chrysopoetic alchemy. In the succeeding chapter I explore the reasons behind this interest: Was chrysopoeia merely another branch of natural philosophy open to Boyle's investigations, or did it carry a special, further, perhaps greater, importance to him?

CHAPTER VI

Motivations: Truth, Medicine, and Religion

THE LAST three chapters have inventoried and organized some of Boyle's important alchemical—specifically chrysopoetic—endeavors and set them in their proper context. Now it is necessary to explore the driving forces behind Boyle's dogged pursuit of traditional chrysopoetic alchemy. What were, in fact, the advantages he expected to reap from this activity? Undoubtedly, although Boyle pursued some of the same topics for forty years, his motivations changed and developed; thus it would be wrong to exaggerate the importance of a single motivation at the expense of others. It would likewise be inappropriate to expect to find in all cases neat and fully satisfactory explanations for the actions and pursuits of a historical character, for the choice of a career or activities stems both from those external influences readily amenable to historical inquiry and certain internal urgings that cannot, by their very nature, lie within the historian's realm of inquiry. Historical writing is too prone to the fault of overdetermination; as a guard against such unnecessary explanatory exuberance it is judicious to recognize at the outset that even beyond the difficulties of re-creating the web of influences surrounding a historical character, such characters are human beings, and their actions (like our own) are not always traceable to what we would be able to recognize readily as *causes*. Having posited this caveat, one can, perhaps with as much completeness as is reasonably possible, attribute Boyle's great interest in chrysopoeia to three chief heads: natural philosophy, medicine, and theology. The last of these is the most intriguing and provides the most provocative new dimensions to Boyle and his thought.

Service to Natural Philosophy

We already know much about Boyle's interest in chymistry and his application of it to natural philosophy. His apologetic preface to his "Essay on Nitre" goes far toward explaining his interests in chymistry as a tool for probing nature. There he cites his desire "to beget a good understanding betwixt the Chymists and the Mechanical Philosophers, who have hitherto been too little acquainted with one anothers Learning"—a desire that he hopes will "conduce to the Advancement of Natural Philosophy."[1] In a similar vein Boyle remarks on the great value of "associating Chymical Experiments to Philosophical Notions."[2] These sentiments are well docu-

[1] Robert Boyle, "Essay on Nitre" in *Physiological Essays,* preface, [viii–ix].
[2] Boyle, "History of Fluidity and Firmnesse," in *Physiological Essays,* 141.

mented in classical portrayals of Boyle's chymistry, and I will add only some lesser-known examples, especially those where chrysopoeia is specifically addressed. For example, an extended fragment of Boyle's unpublished essay "Usefulness of Chemistry to the Empire of Man" contains a tidy summary of Boyle's early views on the value of chymistry. Boyle mentions this work in the preface to *Producibleness of Chymical Principles,* and he clearly (in terms of its title, literary style, and address to Pyrophilus) intended it to constitute part of his *Usefulnesse of Experimental Philosophy.*[3]

In this essay Boyle clearly reexpresses his defense of chymistry against the Peripatetics who have never cultivated it, and against the general view of it as a "useless if not deceitful" study. He also cites its importance, for

> I believe that the chymists themselves have not been sufficiently equitable to their own art when in their definitions the most learned Sennert and others confine its usefulness to medical preparations and the amelioration of metals [i.e., transmutation]. Indeed, I think that at least two other uses might be added to these twin uses of chymistry: first, its ability to *service* natural philosophy in general, and second, the *promotion* of the practical arts.[4]

In regard to its service to natural philosophy, Boyle is specific here on how "chymists may produce things of great moment and remarkable metamorphoses in natural bodies," and notes that "the many phenomena exhibited in the laboratories of chymists that supply us with arguments (which the Aristotelians could not have thought of even in their dreams) are as well serviceable for illustrating truth as for confuting error."[5] The processes of

[3] The finished English original is lost; only part of Thomas Ramsay's 1680s Latin translation survives (as is the case with the *Dialogue on Transmutation*) at BP, 24:135–49: "Tentamen II. Quod Imperium Hominis promoveri possit per Physicorum in Chymicis peritiam, sive, Utilitas Chymiae ad Philosophiam Naturalem." The essay's composition must date before 1666 because Boyle refers tangentially (fol. 145) to his *menstruum peracutum* and its ability to sublime gold (see chapter 3), and remarks that "we will, God permitting, give you a full account [of the process] in another treatise," indicating that the essay predates the publication of the "full account" published in 1666 in *Origine of Formes and Qualities.* In *Producibleness,* preface, [xiii], Boyle mentions he wrote the essay "many Years since." This fragment and a translation thereof is to be published by Michael Hunter as an adjunct to the forthcoming complete works.

[4] Boyle, "Usefulness of Chymistry to the Empire of Man," BP, 24:139; "opiner ipsosmet Chymicos non satis aequos in propriam Artem fuisse quando, etiam in ipsis definitionibus, Doctissimus Sennertus et alii, utilitatem ejus ad medicas praeparationes et Metallorum meliorationem restrinxerunt. Quippe ad geminos hosce Chymiae usus putem ad minus binos alios adiici posse, velut primo Aptitudinem *subserviendi* Philosophiae naturali in genere consideratae, tum practicarum Artium *promotionem.*" Sentiments extremely similar to these were eventually published in 1674 in "Advertisements about the Experiments and Notes Relating to Chymical Qualities" (esp. 2–4), in *Mechanical Origine.*

[5] BP, 24:141; "non potest mihi non videri probabilis Chymicos posse res magni momenti, insignesq[ue] metamorphoses, in corporibus naturalibus producere . . . multa Phenomena, per Chymicorum laboratoria praebentur/exhibentur, quae nobis suppeditant argumenta (de quibus Aristotelici ne per somnium quidem cogitarunt) aeque ad illustrandam veritatem ac ad errores confutandos inservientia."

the chymists exhibit many potentially instructive phenomena, and their means of operation (particularly using fire, "the most active and vigorous" body) allow for the manipulation of matter in powerful ways. Herein lies one of Boyle's chief contributions toward modern chemistry, namely, the explicit expansion of the realm of chymistry, with special attention directed toward its elevation to a philosophical status by the use of chymical processes and experiments for the elucidation of the general principles of natural philosophy.[6] Boyle advocates the application of chymical *effects* in a newly systematic way toward the explanation of *causes.*

What is important to remember is that since the topics we might now classify as alchemy were then embraced by Boyle's term of *chymistry,* such topics too were included within his notions of the general usefulness of chymistry as an investigative and illustrative tool. Early in his career, Boyle deployed chymical studies toward the elucidation of both natural philosophical and theological problems in "Essay of the Holy Scripture" and "Of the Study of the Booke of Nature."[7] Similarly, in the rewritten 1659 *Seraphic Love,* Boyle cites two well-known alchemical figures—Paracelsus and van Helmont—as "Learn'd expositors" of the Book of Nature, and he compares "Chymical Furnaces" to the dissecting knives of anatomists for their ability to lay open the recesses of nature.[8] Chrysopoeia—the "amelioration of metals" mentioned in the *Usefulnesse* essay—held particular promise for displaying or elucidating natural processes, particularly in terms of the constitution of matter and the nature of change. Boyle was especially concerned with these topics throughout his chymical investigations. His interest in great effects produced by comparatively trifling causes has already been mentioned, and the Philosophers' Stone, by his own explicit mention, was a superlative example of such activity.[9] Boyle invoked the transmuting action of the projected Stone on several occasions to instance phenomena that could illustrate more general principles of the corpuscularian or mechanical philosophy he was advocating.[10]

Boyle's interest in the ability of alchemical arcana to contribute powerfully to his program of "associating Chymical Experiments to Philosophical Notions" undergirds his lifelong interest in the alkahest.[11] This universal solvent lauded by "those Patriarchs of the *Spagyrists, Paracelsus* and *Helmont*" was supposedly capable of dividing all substances into their constituent principles, and, according to the Helmontian principles Boyle subscribed to in his early career, extended digestion in the alkahest would strip even

[6] Clericuzio, "Carneades and the Chemists."

[7] These ca. 1650 essays are discussed in Hunter, "How Boyle Became a Scientist." The material applying chymistry to theology was eventually published in *Some Physico-Theological Considerations about the Possibility of Resurrection* (London, 1675).

[8] Boyle, *Seraphic Love* (London, 1659), 56.

[9] See chapter 4.

[10] See for example, Boyle, *Sceptical Chymist,* 158–59, and "Fluidity and Firmnesse," 230.

[11] Boyle, "Fluidity and Firmnesse," 141.

these principles of their "seminal forms" and leave only insipid water—the universal substratum of matter in Helmontian theory.[12] After performing this feat, the alkahest could then be separated from the divided elements by distillation and thus recovered with no loss of its potency or virtue. Boyle notes that if the true adepti have

> among other rare things some *Alkahestical* or other extraordinarily potent *Menstruum,* or way of penetrating and working upon mixt bodies; they may for ought I know be able to obtaine such substances from them, as may induce me, and perhaps the Chymists too, to entertaine other thoughts about the constitution of compounded bodies . . . than either I or they now have.[13]

Boyle's alchemical authorship appears once again in the context of the alkahest, for his ca. 1680 list of "Tracts Relating to the Hermetical Philosophy" cites a work entitled "Of the liquor Alchahest and other Analizing Menstruums," and he refers to this unpublished essay in the preface to *Producibleness.*[14] This tract, like so much of Boyle's writing, had an extremely long gestation, for an early form of such inquiry is listed on a 1650s inventory as "Of the Attempts of the Chymists, an universall Medecine, the Alkahest & the Elixir."[15] Boyle's early interest in this topic may well stem not only directly from his reading of van Helmont but more immediately from his early association with George Starkey, who was deeply involved in the search for the alkahest.[16] Even though Boyle's tract has not survived in any form, it is clear that his interest in this alchemical *arcanum maius* stemmed from its ability to "dissect bodies" and reveal more about their constitution for the benefit of natural philosophy.

In a preamble to the collection of *particularia* (composed probably in the 1680s), Boyle specifically commends the value of chrysopoeia to natural philosophy. There Boyle notes that at one time he was little concerned to collect *particularia* because they are chiefly valued for their "Lucriferous" nature, a thing in which he had no interest, having "neither wife, children, or ambition to grow rich."[17] Yet upon further consideration that "most particulars worke notable changes on the mettals or other Bodys that afford

[12] Boyle, *Sceptical Chymist,* 76–78 and 344; on Boyle's understanding of the alkahest, see 116–17; Clericuzio, "Redefinition of Boyle's Chemistry"; on the alkahest, see chapter 2. While Boyle's early works frequently mention "seminal forms" (the *Sceptical Chymist* refers to them no fewer than sixteen times), these organizing principles disappear from works written after the mid-1660s, as if Boyle became disenchanted with the notion.

[13] Boyle, *Producibleness,* preface, [xii].

[14] RSMS, 198:143*v.* On the dating of this list, see appendix 1; *Producibleness,* preface, [xii]. A manuscript in the hand of Boyle's amanuensis Robin Bacon that records ten properties or effects of the alkahest survives at BP, 30:489–90 and 491; this may represent some stage in the writing of the alkahest tract.

[15] BP, 36:70.

[16] Starkey was the author of *Liquor Alchahest,* and his notebooks are full of attempts at this marvelous solvent; Starkey's long-term effect on Boyle will be fully treated in the forthcoming book-length study of the Boyle-Starkey collaboration by William Newman and myself.

[17] Text D, appendix 2.

them," he concluded that they "may probably be apply'd to the discovery of unobvious Truths" and "are like much to help a sagacious person in the discovery of the nature of mettals & minerals."[18] These same potential benefits of chrysopoeia were on Boyle's mind years earlier when he noted (1661) that conversation with "true *Adepti*" might instruct him greatly "concerning the Nature and Generation of Metals."[19]

Boyle touches also upon the potential financial value of chrysopoeia in another document, which was to serve as a preamble to what he termed his "Hermetic Legacy," written after he had "grown old." There he endeavors to answer the apparent gossip among some of his friends that in spite of his chymical labors, he still possesses no lucrative (particularly chrysopoetic) chymical processes. Boyle protests that he actually *does* have such processes (in fact the text in question was intended as an epistolary introduction to them), but that he has forborne to publish them because they are intricate, whereas the chymical experiments he did publish were chosen because of their simplicity and facility, and thus their greater ability to illustrate the principles of natural philosophy. As in the case of the preamble to his collection of *particularia,* in the preface to his "Hermetic Legacy" Boyle contrasts his lack of interest in financial gain with his real interests in such chrysopoetic processes. For,

> being a Batchelor, and through Gods Bounty furnish'd with a Competent Estate for a younger Brother, and freed from any ambition to leave my Heirs rich, I had no need to pursue Lucriferous Experiments, To which I so much prefer'd Luciferous Ones, that I had a kind of Ambition (which I now perceive to have been a vanity) of being able to say that I cultivated Chymistry with a Disinterest[ed] mind, neither seeking, nor scarce careing for any other advantages by it, than those of the Improvement of my own knowledg of Nature, the gratifying the Curious and the Industrious, and the Acquist of some useful helps to make Good & Uncommon Medicines.[20]

While Boyle denies having any personal interest in the moneymaking byproduct of such processes, he nonetheless realizes that others would find such effects appealing. Indeed, after mentioning the processes' utility in the service of natural philosophy and truth, Boyle notes that such *particularia* "being skilfully wrought, even in small quantitys may enable a poor and industrious Artist, especially if he be a single man, to get a Livelihood, thô not to grow rich."[21] Boyle even proposes what amounts to an alchemical industry, for "these meaner particulars requiring many hands, Materials & Instruments to carry them on with profit wil set many poor people at work & thereby releive great numbrs enabling them or at least assisting them to

[18] Ibid.
[19] Boyle, *Sceptical Chymist,* preface, [xiii].
[20] Text C, appendix 2.
[21] Text D, appendix 2.

get maintenence for themselves & their distrest familys."[22] Thus chrysopoeia could also be useful in the expression of Boyle's well-known philanthropy.

"Extraordinary and Noble Medicines"

Concerns both philanthropic and personal lie behind another aspect of traditional alchemy that Boyle cites both in these preambles and elsewhere, namely, the production of new and powerful medicines. The completed Stone was reputed to be a universal medicine, and many chrysopoeians noted that its preparation was closely related to that of *aurum potabile,* potable gold, the second most potent medicine in the alchemists' cupboard. Boyle in fact noted that if the Philosophers' Stone "were not an Incomparable Medicine, but were onely capable of transmuting other Metalls into Gold," its discovery would not "much advantage Mankind."[23] Certainly for a man as sickly and as oversolicitous of his health as Boyle, and for a man as genuinely concerned about dispensing charity to the sick as he, the promise of a universal medicine would have been extremely alluring. I have already mentioned how Boyle's reminicences of the *Sceptical Chymist* specifically mentioned his interest in correcting the errors of the "vulgar chymists" in preparing medicines.[24] Even Boyle's famous air-pump experiments seem to have had a medical origin, for in recalling them, Boyle states that "he fell on his Experiments about Air to examine the qualities of it, and what pure Air above the Atmosphere is, which he hoped might direct him in many usefull things for the Regiment of our health."[25]

Scholarly attention has recently begun to investigate the medical dimensions of Boyle's studies. Boyle's Christian philanthropy required his charitable communication of medical receipts that might prove of value, but he found it difficult to practice such charity without either trivializing his status by publishing mere collections of recipes or trespassing into the professional space of accredited physicians. Indeed, one reason we have not always recognized Boyle's strong medical interests and activities is that he suppressed his examinations and critiques of contemporaneous medicine because of fear of censure by the medical establishment of which Boyle was not a member.[26]

Boyle's earliest work in experimental chymistry often involved medicines.

[22] Ibid.

[23] Boyle, *Excellency of Theology, Compar'd with Natural Philosophy* (London, 1674), 134.

[24] See chapter 2.

[25] Recalled in the "Burnet Memorandum" printed in Hunter, *Robert Boyle by Himself,* 29.

[26] Barbara B. Kaplan, *"Divulging Useful Truths in Physick": The Medical Agenda of Robert Boyle* (Baltimore: Johns Hopkins University Press, 1993); Hunter, "Reluctant Philanthropist"; "Boyle versus the Galenists: A Suppressed Critique of Seventeenth-Century Medical Practice and Its Significance," *Medical History* 41 (1997): 322–61.

While collaborating with George Starkey in the early 1650s, Boyle was keenly interested in preparing the Helmontian pharmaceutical *ens veneris,* which was reputed to be an economical remedy for rickets. The 1663 *Usefulnesse* is full of medical matters dating from this early period. Later in life, Boyle's dealings with Georges Pierre often involved the exchange of medical receipts, and Pierre sent him a coded recipe for potable gold. In the preamble to a collection of *particularia,* Boyle cites among his reasons for publishing not only that such processes will "convince diverse virtuosi" of the reality of transmutation, but also that such particularia "may probably be applyed to . . . the Preparations of good medicines."[27] In a related text explaining why he "consented to Tryals made upon differing matters" in his attempts to prepare "the grand Elixir," Boyle notes that one of his "cheif designe[s]" was "to attain good medicines."[28] Again, in a text dealing with a chrysopoetic process involving the "augmentation of gold," Boyle cites as one of his reasons for publishing it that "there is a certain stage, or nick of time, at which part of the Powder being taken away is by a very slight change turn'd into an excellent Medicine." Rather strikingly, Boyle's conflicting concerns over communicating useful medicines and maintaining alchemical secrecy surface here. Boyle writes that he is willing to publish this particular chrysopoetic process because the amount of gold produced is (unlike the case of the Philosophers' Stone) not so great that it might "threaten the welfare of States, if it should fall into unworthy hands."[29] Similarly, Boyle aspired to do useful things in medicine with his incalescent Mercury, hoping that the substitution of this more noble Mercury for common mercury in pharmaceutical preparations would lead to more safe or more efficacious remedies. Furthermore, one of the "Tracts Relating to the Hermetical Philosophy" that Boyle lists ca. 1680 is entitled "Of Chymicall arcana medicinall."[30]

The use of chrysopoetic chymistry in the service of natural philosophy and medicine is hardly surprising. As I have endeavored to point out, these uses merely reunite "alchemical" topics with Boyle's already well recognized "chemical" topics. There is for Boyle, however, one further promise of chrysopoeia that the "less high" parts of chymistry could not offer. Indeed, this use of chrysopoeia marks out what are probably the strangest and most unorthodox of all Boyle's beliefs—that the acquisition of the Philosophers' Stone would facilitate communication with angels and rational spirits. Once such a link between alchemy and religion is made, motives for the study of alchemy become much more compelling for Boyle, and chrysopoetic alchemy expands to fill a crucial place in his thought. We shall see that Boyle's alchemical quest functions as a meeting place of the two activities that drove

[27] Text D, appendix 2.
[28] Text F, appendix 2.
[29] Text E, appendix 2.
[30] RSMS, 198:143*v.* On the dating of this document, see appendix 1; *Producibleness,* preface, [xii].

his life—the advancement of natural philosophy and the advancement of the Christian faith.

Spiritual Alchemy

Before launching into an analysis of the spiritual aspects of Boyle's alchemy, we must first clarify the spiritual content of alchemy as a whole; only in this way can Boyle's beliefs be accurately understood in context. This preliminary exercise is all the more needful because of the prevalent misconceptions regarding alchemy and its relation to religion, the supernatural, and magic. Most older secondary literature classes alchemy alongside magic and the occult sciences, thus implying that these topics bear a necessarily close relationship with one another. Popular misconceptions of alchemy are based largely upon such a grouping. Furthermore, in terms of religion, it has become nearly a stock assumption that alchemy was somehow distinguished from subsequent chemistry by virtue of some (generally ill-defined) spiritual, religious, or esoteric content. While it cannot be denied that alchemical texts are full of religious and metaphysical expressions and references, the common view that alchemy is closely, essentially, and inextricably linked to spiritual or religious topics exaggerates the situation. The importance of such elements in alchemy has been artificially magnified because the historical topic has been viewed through the distorting lenses of the Enlightenment, romanticism, and, especially, the Victorian occult revival.[31]

The earliest alchemical texts of the Latin West do not reveal any radical incorporation of or dependence upon religious, spiritual, or esoteric dimensions. The thirteenth-century *Summa perfectionis* of the pseudo-Geber is a fairly cut-and-dried scholastic treatise, as are its contemporaries. The *Summa* does, however, employ what has been termed an "initiatic style," which invokes God's blessing on the aspiring adept and notes that such a blessing is necessary for success. These petitions have been traced to Islamic sources, particularly to the *Liber de septuaginta,* a production of the "real" Jābir ibn-Ḥayyān, or rather of the ninth- and tenth-century Ismaʿili sect, the Ikhwān aṣ-Ṣafa, or Brethren of Purity.[32] This style, propagated throughout later Western alchemy, was reinforced and codified by the self-aggrandizing (and thus authority-constructing) effects of the creation of an "elect" body of adepti favored by God into whose number aspirants were to be initiated either by a master or directly by the grace of God. Of course, the piety of many alchemical authors and the pervasiveness of religious thinking before the Enlightenment are also powerful contributors to the continuation of this religious tone. To attribute all of it to concerns of authority would be an

[31] The effect of these historical events on the historiography of alchemy is fully recounted in Principe and Newman, "Historiography of Alchemy."

[32] Newman, *Summa Perfectionis,* 90–98. On the Ikhwān aṣ-Ṣafa and Jābir ibn-Ḥayyān, see Kraus, *Jabir ibn Hayyan.*

oversimplification and perhaps unnecessarily cynical or crass; it is likely that most later chrysopoeians *did* believe that knowledge of the *arcana maiora* was divinely privileged.

While the initiatic style became quite standard, the overt implication of spiritual, supernatural, or mystical elements in alchemy remained largely absent until the late fifteenth century. At that time, writers like Pico della Mirandola (1463–1494) in Italy and the abbot Trithemius (1462–1516) and Agrippa von Nettesheim (1480–1535) in Germany linked some alchemical notions with the cabala, Hermetic and Neoplatonic mysticism, and natural magic. Paracelsus (1493–1541) intensified such linkages and devised a world-system populated with a vast number of supernatural beings and elemental spirits and where natural and sympathetic magic played a central role in an organic cosmos. The material *tria prima* of Mercury, Sulphur, and Salt was linked in a web of correspondences that included the Triune Godhead and the threefold nature of man: spirit, soul, and body. The exposition of these complex developments would lead us far from our topic. It suffices to point out that while these developments spawned new and modified schools of alchemical thought, they did not supplant the older schools but rather coexisted with them; in other words, alchemical writings could now be found across a broader spectrum.

An important distinction must be made here, namely, the difference between the use of theology and the supernatural or miraculous as an *instrumentality or mechanism,* on the one hand, and as a *source of metaphor and allusion,* on the other. By the fifteenth century a well-developed set of correspondences or parallels had been set up between alchemical theory and religious truths. The *Rosarium philosophorum,* for example, attests to such development with its use of Christ as an emblem for the Philosophers' Stone—the matter of the Stone undergoes "death" and "resurrection" like Christ.[33] The use of such metaphorical language, however, does not permit us directly to conclude that such authors meant to imply some innate mystical union between events alchemical and religious, or that they intended such metaphors to act as explanations of processes. To draw such conclusions hints at an inability to read metaphor as such.

The observations of chemical processes by alchemists and their attempts to explain or describe them would naturally have led to the construction of such similitudes. The need for alchemical secrecy greatly intensified this tendency as alchemical writers (particularly chrysopoeians) strove to create an intricate set of *Decknamen* within allusive or metaphorical frameworks of correspondences that drew freely from ancient mythology, everyday experience, or religious doctrine and imagery. Many writers believed or *expected*

[33] *Rosarium philosophorum: ein alchemisches Florilegium des Spatmittelalters* (Weinheim: VCH, 1992); see also Betty Jo Teeter Dobbs, *Alchemical Death and Resurrection: The Significance of Alchemy in the Age of Newton* (Washington, DC: Smithsonian Institution Libraries, 1990), although I think the linkages made there and the reliance upon vitalism in alchemy are exaggerated.

that alchemical processes should show similitudes to the greater processes of Creation or Salvation; the famous Hermetic dictum "as above, so below" was often interpreted in just this way. But this view is an expression of minds more accustomed and attuned to the drawing of parallels and the reading of signs and emblems than ours.[34] It does not imply a radical fusion of religious belief or supernatural causation with alchemical theory. In just the same way, the use of similitudes or images drawn from animal or vegetable life to explain or illustrate alchemical processes does not automatically imply belief in a fully animate and vitalistic world, as has sometimes been presumed.

Boyle and Supernatural Arcana

The foregoing introduction to the relationship between chrysopoetic alchemy and religious or spiritual thought is intended to show that Boyle's belief in chrysopoeia and his adoption of various aspects of alchemical theory and explanation do not *automatically* imply a link between Boyle's alchemy and religion. The chrysopoetic authors whom Boyle cites most approvingly—pseudo-Lull, Gaston DuClo, Basil Valentine, Alexander van Suchten—espouse no close link between their Art and matters spiritual outside of the usual initiatic style and metaphorical or allusive descriptions described above. The fact that Boyle did (at least in his later years) closely link alchemy and spiritual agencies is in its own right a remarkable discovery and is not at all an unavoidable consequence of his chrysopoetic pursuits.

The spiritual dimensions of Boyle's alchemy are revealed most clearly in two unpublished documents that have recently come to light. The first of these is the set of autobiographical notes dictated by Boyle to Bishop Gilbert Burnet—the "Burnet Memorandum" cited previously.[35] Readers of the document will be struck not only by how much of the document (nearly half) is devoted to projection accounts and spirit experiences, but also by how smoothly and readily Boyle passes from one topic to the other. Chrysopoeia and the spirit realm are linked in two of the accounts of spirit-apparitions where the persons involved ask the spirits specifically about the preparation of the Philosophers' Stone. Although the questioners were disappointed in their hopes of learning the grand secret from the spirits—in one case the knowledge was delivered but "went so clearly out of Memory that he could never trace it," and in the other, the spirits "lookt very angry" when the question was asked—these accounts show the connection that had been made in some circles, and that Boyle evidently accepted, between al-

[34] On the "emblematic world-view" and the importance of recognizing its distinctness, see William B. Ashworth's trenchant paper "Natural History and the Emblematic World-View," in *Reappraisals of the Scientific Revolution,* ed. David C. Lindberg and Robert S. Westman (Cambridge: Cambridge University Press, 1990), 303–31.

[35] Hunter, *Robert Boyle by Himself,* 26–34.

chemy and the spirit realm.[36] Hunter's study of this document emphasized the moral aspects of such knowledge, namely, the morality of obtaining knowledge by magical or supernatural means.[37] It seems that Boyle's scrupulosity of conscience, which has been cited as important to an accurate understanding of him, also threw up moral impediments to an unbridled pursuit of alchemy that might have made use of spirit communications or magical practices.

A more provocative source for the spiritual dimensions of Boyle's alchemy is a fragment of a remarkable dialogue that may be part of Boyle's "Of the supernatural arcana pretended by some Chymists" listed in his ca. 1680 inventory of alchemical writings.[38] This document was first cited by Hunter, and it is presented in its entirety here as appendix 3.[39] The dialogue takes the form of a discussion among four interlocutors—Arnobius, Cornelius, Timotheus, and our old friend Eleutherius. Timotheus meets the other three coming out of the house of their friend Parisinus and asks them if they have taken notice of the "new sett of books" Parisinus has acquired. Arnobius responds that they did in fact find Parisinus utterly engrossed in reading books about the Philosophers' Stone. Timotheus expresses his amazement that Parisinus, "a man versd in rationall and experimentall Phylosophy . . . should with so much application of minde peruse those dull Authors that deserve to be left in as great an obscurity as they affect in their aenigmaticall writings: and that he should so industriously seek the Philosophers-stone."[40] After all, Parisinus does not want for money and has no need for the gold that the Stone could produce if it were real. But Eleutherius responds that he has certain suspicions that his friend Parisinus has "higher aims then the acquiring of the skill to Transmute inferior mettalls into gold." Arnobius, who has apparently discussed this matter with Parisinus, confirms Eleutherius's suspicion that Parisinus's "ambition reaches higher then gold or anything that gold can purchase." Timotheus is confused by this odd response until Arnobius notes first that the "adept philosophy" can provide medicines worth more than gold, but then further reveals that Parisinus aims yet higher, for their studying friend believes that "the acquisition of the Philosophers-stone may be an inlett into another sort of knowledge and a step to the attainment of some intercourse with good spirits."[41]

This shocking motivation amazes Timotheus, who cannot believe that Parisinus, as a "rationall person and as a good christian," would attempt

[36] Ibid., 30 and 32.

[37] Hunter, "Alchemy, Magic and Moralism."

[38] See chapter 3. It should be pointed out for readers unfamiliar with seventeenth-century usages that *pretend* did not at that time carry the intimation of falsity but rather retained a close relationship to its Latin origin in *praetendere,* to hold forth or present, or the French *prétendre,* to claim, without judgmental value.

[39] Hunter, "Alchemy, Magic and Moralism," 396–98.

[40] Boyle, *Dialogue on Spirits,* appendix 3, 310–16.

[41] Ibid., 312.

such a thing, which is neither "possible nor lawfull," and urges the others to dissuade Parisinus from this course of action. He notes that there is great danger in attempting to contact spirits, for even the Devil can disguise himself as an angel, and so in attempting to contact good spirits, Parisinus might instead, at great peril to his soul, encounter one originating from the "bottomlesse pitt" of Hell. Additionally, St. Paul seems to dissuade men from intercourse with angels, for Christ alone is the true Mediator between God and man. Finally, says Timotheus, the belief that the Philosophers' Stone could somehow attract angels is fitter for some "whimiscall Enthusiast" than for so learned a man as Parisinus.

Eleutherius, playing his role as observer, agrees with Timotheus's expression of disbelief in the possibility of contacting spirits with the Stone—"for what affinity or congruity can there be betwixt the stupid and inanimate Elixir and a rationall and immortal spirit that these happy beings should delight to hover about it . . . ?"[42] Arnobius calmly replies that all these arguments are well-known to Parisinus, and in conversations among Parisinus, Arnobius, and Cornelius, Parisinus has answered them. Eleutherius and Timotheus ask for a recapitulation of what Parisinus's beliefs and arguments are, and Arnobius and Cornelius comply. They first render it probable that men and spirits may in fact communicate directly. They cite scriptural and contemporary accounts of spirits and witches to verify the existence of manifold orders of incorporeal spirits (both good and evil) who have had commerce with men. That point having been granted by the simple authority of Scripture, Arnobius moves on to demonstrate that good spirits, not just evil ones, have been contacted by men, and he cites an instance from the *Demonomanie* of Jean Bodin (1530–1596) as evidence.[43] Eleutherius chimes in with his own examples, recounting the story of a "physican and excellent mathematician" of his acquaintance who had been "much cajold and at last perverted by a famous writer who by many both friends and foes is thought the subtilest Atheist of our times," presumably Thomas Hobbes. This man later became a devout and learned Christian, and Eleutherius, having asked him how it was that he came to be rescued out of atheism, was told that his conversion was effected by a learned natural philosopher who convinced him of his ability to converse with spirits both good and evil.[44] Eleutherius adds another account of a devout prince, then still in power, whose admirable statecraft was due to the counsel and direction given him by amiable spirits.

Unfortunately, the fragment breaks off at this point, and so the sections dealing with the exact relationship of the Philosophers' Stone with angel communication are missing. This defect can be partially remedied by the summary lists of arguments prefixed to the fragment. Under "arguments for the affirmative" Boyle notes that we know very little about the "nature,

[42] Ibid., 313.

[43] Jean Bodin, *De la Demonomanie des Sorciers* (Paris, 1580); it was frequently reprinted.

[44] Boyle, *Dialogue on Spirits*, appendix 3, 316.

communities, laws, Politicks and government of spirits" and that there may well be some which are of a sociable nature with men. Furthermore, in terms of the Philosophers' Stone's effects, "there may be congruities or magnatisms capable of inviting [spirits] which we know nothing of," and "such persons as the Adepti may well be supposd to be under a peculiar conduct and to have particular priviledges" in regard to spirit contact and communication.[45]

The reader may at this point experience some degree of déjà vu, for there is an admirable uniformity between Boyle's method of argument here and in the *Dialogue on Transmutation.* In both cases, the reality of the thing (whether spirit communication or projection) is demonstrated by the use of experiential accounts. In the present case, Boyle has the benefit of scriptural accounts, which by virtue of their source are clearly more reliable than any other sort, and so he spends little time insisting upon them. Further, in both cases, the means by which a seemingly unbelievable phenomenon occurs is left open, and our lack of understanding or inability to explain it is explicitly rejected as a reason for doubting the reality of the phenomenon. This is a recurrent theme in Boyle's thought and presumably the origin of the "imputation of credulity" with which he was taxed in the early eighteenth century and probably (quietly) in his own day as well. Boyle was adamant in his opposition to human arrogance in the face of the immense and incomprehensible possibilities implanted by God within His Creation. In the *Dialogue on Transmutation* Boyle condemns "the haughtiness attendant upon ignorance, so that we think ourselves the standards by which things are to be judged, and by the same reason conclude that whatever things we cannot explain or accomplish are merely hypothetical or mere impossibilities."[46]

Boyle reiterates this stance in many works, especially in later years. Significantly, he employs this axiom not only in natural philosophical concerns but also in theological ones. Boyle was remarkably consistent in his view of knowledge, whether natural philsophical or theological, and would wholeheartedly have endorsed the prince of Denmark's warning that

> There are more things in heaven and earth, Horatio,
> Than are dreamt of in your philosophy.[47]

Would Boyle not have seen dismissals of the marvelous possibilities of God's Creation as dangerously close to, if not integral with, a dismissal of the marvelous and incomprehensible power, goodness, and freedom of God Himself? Indeed, it is unlikely to be mere chance that the two protagonists of the dialogue on spirits—Timotheus and Arnobius—are also chief expositors in Boyle's theologically significant *Things Above Reason* and *Judging of Things said to Transcend Reason,* where the limits of human knowledge are

[45] Ibid., 310–11.
[46] Boyle, *Dialogue on Transmutation,* appendix 1, 253.
[47] *Hamlet,* 1.5.162.

stressed.[48] This pair of works has been shown to address pressing theological controversies of the day, particularly Socinianism, a threat that troubled Boyle from the early 1650s.[49] Socinians espoused the doctrine that Scripture and theology are to be judged by human reason; accordingly, they promoted several pernicious views such as the rejection of the divinity of Christ. Boyle was clear and firm in his dismissal of human reason as a criterion for judging God's activity and the doctrines of orthodox Christianity. His voluntarist theology insisted upon God as the "most free Agent," whose range of possible actions was to be neither circumscribed nor adjudged by human notions.[50] "There is no necessity," writes Boyle, "that intelligibility to a human understanding should be necessary to the truth or existence of a thing."[51] Clearly, this firm theological conviction of Boyle's is integral with his equally firm refusal to limit the possibilities of nature—the visible sign of God's activity—based upon our feeble and incomplete understanding of it.

Origins of Boyle's Belief

The connections that Boyle made between alchemy and the spirit realm are very unusual in terms of mainstream alchemy. The vast majority of chrysopoetic authors, including those contemporary with Boyle, eschew such metaphysical or magical attributes to their Art. How then did Boyle come to adopt the view that alchemy was intimately linked with angelology and the spirit realm?

One source is the general credo that knowledge of the Philosophers' Stone was itself a *donum Dei.* Angels or other spiritual messengers, who had certainly brought revelations to man during biblical times, might continue to be one means of transmission for this divine knowledge to men. Paracelsus was said to have been visited by an angel who revealed secret knowledge to him, and quite a number of alchemical authors claimed to have gained inspiration from spiritual visitors to their dreams.[52] In Boyle's own circle,

[48] Wojcik, *Boyle and the Limits of Reason,* esp. 102–4, 112, 151–88. See also her comparison of Boyle's views with Newton's in "Pursuing Knowledge: Robert Boyle and Isaac Newton," in *Canonical Imperatives: Rethinking the Scientific Revolution,* ed. Margaret J. Osler (Cambridge: Cambridge University Press, 1998). Boyle's two works on reason are paired in the 1681 London edition; Arnobius replaces Sophronius as the key speaker in the second work.

[49] Jan W. Wojcik, "The Theological Context of *Things Above Reason,*" in *RBR,* 139–55. On Boyle's early fears of spreading Socinianism, see Hunter, "How Boyle Became a Scientist," 74–75.

[50] On Boyle's voluntarism in natural philosophy, see Wojcik, *Boyle and the Limits of Reason,* 189–211; on seventeenth-century voluntarism more generally, see Margaret J. Osler, *Divine Will and Mechanical Philosophy* (Cambridge: Cambridge University Press, 1994); "Fortune, Fate, and Divination: Gassendi's Voluntarist Theology and the Baptism of Epicureanism," in *Atoms, Pneuma, and Tranquillity: Epicureans and Stoic Themes in European Thought,* ed. Margaret J. Osler (Cambridge: Cambridge University Press, 1991).

[51] Boyle, *Appendix to the Christian Virtuoso* (first published 1744), *Works,* 6:694.

[52] Much alchemical literature is written in the form of dream revelations; see, for example, John Dastin, *Dastin's Dream,* and William Blomefield, *Bloomfields Blossoms,* in *TCB,* 257–68 and 305–23, respectively.

George Starkey claimed to have been visited in a dream by a *Eugenius,* or good spirit. In a letter to Boyle dated 26 January 1651/52, Starkey tells how this spirit appeared to him while he was strenuously searching for the alkahest. Starkey lay down to rest for a moment in his laboratory and dreamed that a man entered the room, with whom he conversed for a while, and then, when Starkey asked for an explanation of the alkahest's preparation, the man uttered a single line, which although cryptic in itself, Starkey immediately understood fully, for an "ineffable light" entered his mind as he heard the utterance.[53] Boyle, at least in his younger days, viewed such an origin of chymical knowledge with skepticism, for he then "dare[d] not affirm . . . that God discloses to Men the Great Mystery of Chymistry by Good Angels, or by Nocturnal Visions."[54] Yet such revelations by spirits resemble the accounts of less successful spirit communication recounted by Boyle in the "Burnet Memorandum."

The most celebrated instance of angel contact occurs in the accounts of the Elizabethan magus John Dee and his companion Edward Kelley.[55] Although these events transpired in the century before Boyle, they were popularized by Meric Casaubon in 1659 in a thick folio volume of transcripts from Dee's records of his spirit conversations. The surviving fragments of these journals tell of years of communications with spirits achieved through the use of various glasses or gazing stones and an elaborately designed and furnished "Holy Table." Casaubon's aim in publishing this curious matter was not only to show the real existence of spirits but also to illustrate the dangers of attempting to contact them. Much of his concern was to check the growth of enthusiasm, which had burgeoned during the interregnum.[56] Casaubon's warning hinges on the revelation that Dee's supposedly angelic visitors turned out to be demons in disguise—"he mistook false lying Spirits for Angels of Light, and the Divel of Hell for the God of Heaven."[57] The "angelic" messages were not new divine revelations, as advertised, but rather, as Dee learned too late, the wiles of the Devil conniving both to delude Dee and Kelley and to subvert the political and religious order of the world. These accounts are surely a chief background to the concern shown in Boyle's dialogue over the discrimination between good and evil spirits. Boyle's first "argument for the negative" in the dialogue is "that there is great danger of and in mistaking an evill spirit for a good one." This concern he defuses by considering "that tis not difficult for the Adepti to discriminate good spirits from evill ones," thus elevating these chrysopoetic masters to a privileged status.

Chrysopoeia appears fairly regularly in Dee's spirit conversations. The

[53] BL, 5:133.

[54] Boyle, *Usefulnesse,* pt. 1, 112.

[55] Deborah E. Harkness, *Talking with Angels: John Dee and the End of Nature* (Cambridge: Cambridge University Press, 1998); for other references on Dee, see chapter 4, 97 n. 24.

[56] Michael Heyd, *"Be Sober and Reasonable": The Critique of Enthusiasm in the Seventeenth and Early Eighteenth Century* (Leiden: Brill, 1995), esp. 72–92.

[57] Meric Casaubon, *A True and Faithful Relation* (London, 1659), preface, [xxvi].

spirits repeatedly promise him the secret of preparing the Philosophers' Stone. Dee's assistant, Edward Kelley, is reputed to have possessed some of the powder, which he found (according to one story, and perhaps under the guidance of a spirit) in the crypt of a ruined abbey in England.[58] A more widely disseminated story about how Edward Kelley got the Stone is told by Daniel Georg Morhof. Morhof states that Kelley, under warrant of arrest in London for forgery, fled to Wales and there found an ancient book and an ivory box full of red powder in the possession of his Welsh innkeeper. The innkeeper knew nothing of what they were, but told Kelley that he had gotten them from a thief who desecrated a bishop's tomb. Kelley (who as a scrivener could read the ancient writing of the book) bought both book and powder for a pound sterling. The story takes on special importance for this study because although Morhof states simply in print that he got the story from "the mouth of an illustrious man who had it from a kinsman of Kelley's," in his correspondence he reveals that the tale—which thereafter became the standard account of Kelley's acquisition of the powder—was told to him by none other than Boyle himself. Morhof met Boyle in 1670, and Boyle recounted to him then "various histories of metallic transmutations," including the account of Kelley that Boyle had in fact obtained from one of Kelley's kin. Clearly, then, Boyle was interested (as early as 1670) in the tale of Kelley and Dee and had apparently initiated inquiries in order to learn more about their history.[59]

Casaubon takes a very dim view of Dee and Kelley and their angelic and chrysopoetic affairs. He applauds "*Chymistrie* as it is meerly natural, and keeps itself within the compass of sobriety . . . the wonders of God and Nature are as eminently visible in the experiments of that Art as any other natural thing." Yet immediately he emphasizes that "*it is not improbable that divers secrets of it came to the knowledg of man by the Revelation of Spirits*"—a suggestion we have already encountered.[60] Furthermore, while Casaubon expresses his belief in the possibility of the transmutation of metals into gold, he wholly condemns the notion of the Philosophers' Stone as the means of effecting it; "*the Phylosophers Stone,* is certainly a meer cheat, the first author and inventor whereof was no other than the Divel." (The reader should note this sentiment as an example of how a person could support the possibility of metallic transmutation but deny the Philosophers' Stone; clearly this is not how Boyle saw matters.) Casaubon denies utterly

[58] Elias Ashmole tells how Kelley got the powder from a crypt in Glastonbury Abbey, in *TCB,* 481.

[59] Morhof, *Epistola ad Langelottum,* 189–90; "ex ore illustris viri, qui a famulo Kellaei illam habet." Morhof's letter mentioning Boyle is printed in *Commercii epistolici Leibnitiani,* ed. Johann Daniel Gruber (Hanover and Göttingen, 1745), 1353–54; "Historiam eius notabilem, quam narro, Boylius a famulo eius habet." See also R.E.W. Maddison, "Studies in the Life of Robert Boyle, F.R.S.: Part IV, Robert Boyle and Some of His Foreign Visitors," *NNRS* 11 (1954): 38–53, on 43–44.

[60] Casaubon, *True and Faithful Relation,* preface, [xxxix]. Casaubon also asserts that the natural magic of Trithemius and Paracelsus came directly from the devil, [xxxviii].

that "any good Angels did ever meddle in a practice commonly attended with so much imposture, impiety, and cousenage" as the Stone. Here his concerns about enthusiasm reemerge, when he condemns together those who await, expect, or seek out spirit guidance in the confection of the Stone with those who daily "hunt after *Revelations,* and Prophecies, and unlawful Curiosities."[61]

These accounts of angel communication and even of the revelation of alchemical secrets by spiritual messengers are all, however, still quite distant from Boyle's belief that the Stone itself could attract angels. Yet a barely recognized school of supernatural alchemy seems to have developed in England in the early seventeenth century. (Again, this evidences the diversity of schools within alchemy.) A general tenet of this school is that there exist "supernatural Stones" of far greater power than the physical, transmutatory Philosophers' Stone, which was, according to their opinion, the lowest order of Stone, being only physical and natural rather than magical and supernatural.[62]

A noteworthy exponent of this school's view is Elias Ashmole, the avid collector of alchemical, astrological, and magical manuscripts, and early member of the Royal Society. In the preface to the *Theatrum chemicum britannicum,* Ashmole briefly defends the value and veracity of the alchemical art, and after praising its power, he notes that the Philosophers' Stone, or "Minerall Stone"—so sought after as the alchemical *summum bonum*—is actually only the beginning of the adepts' knowledge. Ashmole's embrace of supernatural alchemy is rather curious in terms of its juxtaposition with the manuscripts he edited for the *Theatrum,* for those texts show no hint of supernatural involvement or of any arcana beyond the Philosophers' Stone.

Ashmole cites the existence of the "*Vegitable, Magicall,* and *Angelicall Stones.*" The first of these allows for the perfect knowledge of "the *Nature* of *Man, Beasts, Foules, Fishes,* together with all kinds of *Trees, Plants, Flowers,* &c," and by its operation the adept can cause plants to bud, flower, and fruit at any season of the year. The "Magicall, or Prospective Stone" seems to be akin to a gazing Stone, for "it fairely presents to your view even the *whole World,* wherein to *behold, heare,* or *see* your *Desire.*" It also gives the owner the ability to understand the language of animals. Finally, the greatest of all these Stones is the Angelical, the possession of which caused Hermes to "give over the use of all other *Stones.*" According to Ashmole, "*Moses* and *Solomon,* (together with *Hermes* were the only

[61] Ibid., preface, [xxxx]. A less charitable reading of Casaubon's intentions comes from John Webster, the physican and author of the mineralogical text *Metallographia.* Webster claims that Casaubon's previous censure of enthusiasm as pure delusion was so severe that it opened him to the charge of atheism, and in order to repair his reputation he published this volume on Dee's communications to advertise his own belief in the existence of spirits. Webster notes that this repair of Casaubon's reputation was made at the expense of Dee's; *The Displaying of Supposed Witchcraft* (London, 1677), 8.

[62] I am presently collecting materials for a paper on the topic of the "supernatural school," which will appear in due course.

three, that) excelled in the *Knowledge* thereof, and who therewith wrought *Wonders.*" This supreme Stone "affords the *Apparition* of *Angells,* and gives a power of conversing with them, by *Dreams* and *Revelations:* nor dare any *Evill Spirit* approach the *Place* where it *lodgeth.*" It seems, however, that this last Stone would have been rather difficult to get hold of, for "it can neither be *seene, felt,* or *weighed,* but *Tasted* only."[63]

Although Ashmole's views seem quite idiosyncratic, he is actually drawing upon tenets current in the aforementioned English school of supernatural alchemy.[64] Ashmole does not reveal his sources directly but rather only ascribes his recognition of these higher arcana to a book entitled *De occulta philosophia,* attributed to St. Dunstan, and claims that it "will chiefly back" what he declares on this topic.[65] Nonetheless, much of what Ashmole writes is actually taken almost verbatim from the "Epitome of the Treasure of Health," a curious work by one "Edwardus Generosus," dated 1562, which itself references the work of St. Dunstan. What this "Book of St. Dunstan" is remains unclear.[66] While the "Epitome" was never published, it must have circulated fairly widely, for copies exist in several libraries and among the papers of Ashmole and Newton.[67]

Boyle may have encountered Ashmole's claims in the printed text of his *Theatrum* or, equally likely, received them orally from personal contact with Ashmole. A view similar to Ashmole's appears in a short British Library manuscript (dated 19 October 1660) that recounts the effects of supernatural stones of awesome spiritual power.[68] This manuscript tells of how the *Lapis Angelicus* not only can purge the soul but also "is an Oraculous

[63] Ashmole, *TCB,* Prolegomena, [vi–ix].

[64] The singularity of Ashmole's beliefs (and their possible origin in oral relations from Sir William Backhouse, Ashmole's alchemical "father") are noted in C. H. Josten, *Elias Ashmole (1617–1692),* 5 vols. (Oxford: Clarendon Press, 1966), 1:76–92, esp. 86.

[65] Ashmole, *TCB,* Prolegomena, [vi].

[66] I have been unable to track down such a text. Works spuriously attributed to St. Dunstan are published in *Philosophia maturata, or An Exact Piece of Philosophy,* ed. Lancelot Colson (London, 1668), and a fifteenth-century Latin manuscript of this book was in Dee's library and is now Corpus Christi College Oxford MS 128; *Renaissance Man: The Reconstructed Libraries of European Scholars 1450–1700. Series 1: The Books and Manuscripts of John Dee, 1527–1608,* vol. 2, *John Dee's Manuscripts from Corpus Christi College Oxford* (Marlborough: Adam Matthew Publications, 1992), 29. The text is, however, cobbled together from George Ripley, and contains no trace of supernaturalism, and so cannot be the source of the references made by Ashmole or "Edwardus Generosus."

The evil spirit whom Dee believed to be the archangel Raphael promised him "the secret knowledge and understanding of the Philosophers' Stone, and of the Book of St. Dunstan's" (Casaubon, *True and Faithful Relation,* pt. 2, 34). This book of St. Dunstan might be identified with the ancient tome Kelley bought along with the Stone, which would unite Morhof's story with Ashmole's story of Glastonbury Abbey, where St. Dunstan's shrine was located. Clearly much work needs to be done on this school and its sources.

[67] I have located these copies: Bodleian Library Ashmole MS 1417:57–82*v;* Cambridge Library Keynes MS 22; British Library Sloane MS 2502:70–81*v;* University of Glasgow, Ferguson MS 199:19–72.

[68] British Library MS Sloane 648:99–100.

Method or Instrument for the seeing, opening, and discovery of spirits in generall & of such invisible Powers & Substances as the Eye can noe way see." Yet greater, however, is the *Lapis Evangelicus,* which is "yet more stupendous, more immediately Divine." This supreme Stone works upon the human soul, transmuting it into one of perfect charity and godliness, "giveing a heart larger than the Sand of the Sea, yea larger than that of Solomon." Significantly, this manuscript is written in a hand often found in documents circulated in the Hartlib circle. This unidentified scribe seems to have been employed by Hartlib to copy documents for distribution. Several items in this hand are preserved among the Boyle Papers, including, among other things, a number of "newsletters" from abroad dated 1661 and a typically Hartlibian text describing the regulations for a hypothetical "college."[69] Manuscripts in this hand are also present in the Hartlib Papers and scattered through several volumes of the Sloane MSS at the British Library.[70] The document was thus at least circulated among the Hartlib circle, if not produced there, and it is quite possible that Boyle himself received a copy. What is certain is that if such notions as these were current among some of the Hartlibians, Boyle is likely to have come into contact with them.

This notion of these "higher" supernatural Stones, however, still does not exactly match Boyle's attribution of spirit-attracting power to the otherwise, as Boyle himself notes, "purelly corporeall" Philosophers' Stone. Yet while most chrysopoeians continued not to attribute spiritual virtues to the Stone, there does seem to have been (not surprisingly) some crossover in England from the supernatural school into more mainstream chrysopoeia. One contemporaneous tract that does in fact attribute such powers to the Stone is an anonymous work entitled *De Manna benedicto.* This work seems to have been fairly widely circulated (Newton, for example, owned a manuscript copy) and to have been written before ca. 1650, as Ashmole quotes from it (without naming it) in his prolegomena to the *Theatrum,* attributing it only to an "incomparable *Authour.*"[71] The author of *Manna* says nothing about the preparation of the Stone but deals instead exclusively with its "lesser known" uses. Some of these uses are purely physical, such as the mercurification of the metals, while others are magical, such as the production of apparitions of the heavenly bodies and of the Creation. At the conclusion of

[69] BL, 1:53–54 and 55–56; BP, 35:205–8.

[70] For example, in a largely Hartlib-related volume, British Library MS Sloane 427:21–28*v,* a work dated 1660 and entitled "De lapide philosophico"; on 65–66*v,* a copy of a letter from Benjamin Worsley to Hartlib containing comments on Boyle's "Discourse on Colors"; and on 84, comments on the declination of the compass, dated 1660.

[71] Newton's copy is in Keynes MS 33. Other copies are Ferguson MS 9:14–24, 199:72–79; Bodleian Library Ashmole MS 1440; and in British Library Sloane MSS 2194, 2222, and 2585. The work is published as *Tractatus de lapide, Manna benedicto,* in *Aurifontina chymica,* ed. John Fredrick Houpreght (London, 1680), 107–43; although the title is Latin, the work is in English. Ashmole's citation is *TCB,* Prolegomena, [vi]: "Gold . . . '*is the cheifest intent of the Alchemists, so it was scarce any intent of the ancient Philosophers*' and the lowest use the Adepti made of this *Materia*"; it occurs on 113 of the printed edition.

his text, however, he glancingly refers to a topic that he "forbear[s] to set down," namely, "to command and converse with Spirits."[72]

These "lesser known" uses of the Stone are undoubtedly among those which Wilhelm von Schröder explicitly cites as being known to Boyle. While discussing the various virtues of the Stone, Schröder writes that "my good friend in England, Mr. Robert Boyle, told me that some hundred virtues which the Philosophers' Stone possesses had been communicated to him, and that he esteemed such knowledge as highly as that of the secret of the preparation of the Stone itself."[73] Certainly communication with angels may have been among these "hundred virtues" communicated to Boyle. The angelic power of the Philosophers' Stone is more explicitly mentioned in another anonymous mid-seventeenth-century English text.[74] Here all the magical properties that Ashmole attributes to his three "higher" Stones seem to be attached to the single Philosophers' Stone. Like others in this school, the anonymous author claims that chrysopoeia is the "basest fruit, beleeve it, that growes upon that tree," for "in the true secret of the Magie, there are many other rare virtues, excellencies & heavenly gifts" hidden in the Philosophers' Stone. Among these "rare virtues," he mentions "the perfect knowledge of all Nature & of Heavenly wisdome by way of a mirror" and, most germane to our present inquiry, the "Communication with good Spirits & Angels."[75]

There thus seems to have been a seventeenth-century English school of supernatural alchemy from which Boyle could have derived his belief in the spirit-attracting power of the Stone. Newton himself may hint at this school when he writes to Oldenburg regarding Boyle's need to preserve "high silence" about the Philosophical Mercury, "there being other things beside the transmutation of metals (if those great pretenders bragg not) which none but [the adepti] understand."[76] It may well prove impossible to identify a clear textual origin for Boyle's unusual belief, for it is quite likely that he

[72] *Manna,* 142.

[73] Wilhelm von Schröder, *Unterricht vom Goldmachen,* in *Deutsches Theatrum Chemicum,* ed. Friedrich Roth-Scholtz (Nuremberg, 1727; reprint, Hildesheim: Georg Olms Verlag, 1976), 1:219–88, on 279. "Mein guter Freund in Engelland / Herr Robert *Boyle,* berichtet mich / ihm wären etlich hundert Tugenden / die der *Lapis philosophorum* vermöchte / *communici*ret worden / welche Wissenschaft er so hoch hielte / als das *arcanum* der *praeparation* des *Lapidis* selbsten."

[74] British Library Add. MS 4459:36–37*v.* A date for the text may be surmised from the list of "approved" authors cited on 37*v.* The latest of these is Jean Collesson's *L'idee parfaicte,* which was published at Paris in 1630, thus providing a sure terminus a quo. Furthermore, the author seems clearly to fall in with the Mercuralist school, and cites authorities generally connected with that school, yet makes no mention of the chief of that group, Eirenaeus Philalethes, thus suggesting that this composition predates the publication of the Philalethean corpus, which commenced in 1655. This manuscript is now located among miscellaneous manuscripts that belonged to Thomas Birch, and is adjacent to many written by Henry Oldenburg, allowing the intriguing but unverifiable possibility that it came from Oldenburg's or Boyle's papers.

[75] British Library Add. MS 4459:36*r–v.*

[76] Newton to Oldenburg, 26 April 1676, *Correspondence of Isaac Newton,* 2:1–2.

received oral accounts of spiritual effects of the Stone from colleagues, visitors, or contacts. I have shown in general that Boyle's beliefs are in fact supported by a contemporaneous English school rather on the fringe of classical chrysopoetic thought; what is really more significant than the precise origins of these beliefs is their value to Boyle's overall program both natural philosophical and theological. The notion of a spiritually active Philosophers' Stone was extremely enticing for Boyle on several fronts.

Alchemy against Atheism

Robert Boyle was above all other things a Christian. His solid and unwavering devotion to biblical Christianity constitutes the backdrop for all his other actions and pursuits. Boyle's interest in the defense and propagation of Christianity runs as a strong, uninterrupted current throughout his life. The writing of moral and devotional treatises characterized his early career; his most reprinted book, *Seraphic Love* or *Motives and Incentives to the Love of God,* dates from that period. Concerns over orthodox religion and piety have recently been seen as powerful driving forces behind Boyle's "conversion" to natural philosophy in 1649.[77] Indeed, his first forays into natural philosophy (before his ca. 1655 remove to Oxford) yielded observations of natural phenomena not for unriddling the mysteries of physical nature but rather for Christian apologetic and devotional use, as is revealed in his early (ca. 1650) essays "The Study of the Booke of Nature" and "Essay of the Holy Scripture." Even more recognizably "scientific" works, such as *Usefulnesse of Experimental Natural Philosophy* (which draws some of its origins from the 1650s), still show a strong concern for propounding the moral and devotional value of natural philosophy.[78] Later in life Boyle financed numerous religious publications, as well as the translation and printing of the New Testament in Irish, Malay, Turkish, and Lithuanian. He funded numerous foreign missions and at the end of his life endowed the annual Boyle Lectures for the defense and propagation of Christianity.[79] Boyle was also an active participant in important theological disputes. The contexts for his theological works are just now coming to be appreciated, as is the depth of Boyle's long-standing concern over threats to orthodox Christianity from such sources as the Socinians.[80] But on a scale larger than the defense of orthodox biblical Christianity from heretical incursions lay a somewhat more crucial opposition to the greatest danger of all to religion, namely, atheism.

Atheism in the seventeenth century has received considerable attention

[77] Hunter, "How Boyle Became a Scientist."

[78] Principe, "Virtuous Romance," 392–94.

[79] R.E.W. Maddison, "Robert Boyle and the Irish Bible," *Bulletin of the John Rylands Library* 41 (1958): 81–101; Maddison, *Life,* 111–12, 141–42, 205–12.

[80] Wojcik, "Theological Context."

during the past twenty years.[81] Much interest has focused on the elusiveness of the phenomenon. Identifying more than a scant handful of atheists by name is quite difficult, and even those who have traditionally stood out as atheists—Thomas Hobbes, for example—generally exhibit at most a sort of a cryptoatheism; this has in Hobbes's case provoked scholarly debate over whether or not he was a "real" atheist, and how his writings, particularly *Leviathan,* are to be read. Regardless of the outcome of the debate on Hobbes's real beliefs, he was in fact seen by many in the late seventeenth century as *the* atheist.[82] Furthermore, in seventeenth-century English usage at least, *atheism* covers a multitude of sins. It was not only the absolute denial of God that counted as atheism, but also a host of less egregious dissensions from lesser theological points that were perceived as "tending toward" a denial of part or all of the orthodox understanding of God's nature.[83] Nonetheless, in spite of modern scholarly debates on the subject, it cannot be denied that "atheism" was in fact seen as a significant threat by many late-seventeenth-century thinkers, including Boyle.

In regard to Boyle's personal concerns, Richard Bentley, the deliverer of the first Boyle Lecture, wrote that Boyle had placed atheists "in the first place as the most dangerous enemies" of Christianity.[84] Clear statements of Boyle's concern about atheism appear widely disseminated in his published works, and in more concentrated form in *Reconcileableness of Reason and Religion* and the *Christian Virtuoso. Reconcileableness* begins with a strongly worded preface noting how religion is "now more furiously assaulted and studiously undermin'd than ever, not only by the vicious Lives of Men, but by their licentious Discourses."[85] In *Christian Virtuoso* Boyle similarly laments the "great and deplorable Growth of Irreligion" and assails the "Wits and Raillers" for their contagious atheistic beliefs.[86] These published advertisements of his concern are augmented by hundreds of pages of notes and drafts against atheism preserved in the Boyle Papers. Indeed, in *Christian Virtuoso,* a marginal note refers the reader to "another

[81] David Berman, *A History of Atheism in Britain: From Hobbes to Russell* (London: Croom Helm, 1988), 1–69; Michael Hunter and David Wootton, eds., *Atheism from the Reformation to the Enlightenment* (Oxford: Clarendon Press, 1992); Michael Hunter, "Science and Heterodoxy: An Early Modern Problem Reconsidered," in *Reappraisals of the Scientific Revolution,* ed. David C. Lindberg and Robert S. Westman (Cambridge: Cambridge University Press, 1990), 437–60.

[82] Samuel I. Mintz, *The Hunting of Leviathan* (Cambridge: Cambridge University Press, 1962), Berman, *Atheism,* 48–67. Among contemporaneous responses to Hobbes, see Thomas Tenison, *The Creed of Hobbes Examined* (London, 1670); Seth Ward, *In Thomae Hobbesii philosophiam exercitatio epistolica* (Oxford, 1656); Ralph Cudworth, *True Intellectual System of the World* (London, 1678).

[83] Hunter, "Science and Heterodoxy," 440–43.

[84] Richard Bentley to Bernard, 1692, printed in *Museum criticum* (Cambridge, 1826), 2:557–58.

[85] Boyle, *Some Considerations about the Reconcileableness of Reason and Religion* (London, 1675), i.

[86] Boyle, *Christian Virtuoso,* preface, [i], 10.

Paper," but an unpublished one, entitled "About Some Causes of Atheism"; fragments of this tract, along with others, survive.[87] These items were composed over a long period of time; some documents are extended discussions stretching over more than a score of folios and thus attest to the level of Boyle's concern. His failure to publish these items is undoubtedly more attributable to his aversion to engaging in public debate than to any hesitancy of conviction.

Demonstrating Spirits

Some seventeenth-century schemes for the refutation of atheism are well-known. One popular tactic was to demonstrate the existence of spiritual beings, which would show the presence of a spirit realm and, by extension, that of God Himself. The description of the titular persona in *The Character of a Town-Gallant* (1675) provides us with an excellent illustration of how the denial of spirits went hand in hand with seventeenth-century atheism. This work also reveals that the outlines and prevalence of the "coffeehouse atheism" of "Wits and Raillers" that Boyle apparently so feared will be difficult to gauge by a scholarly reading of the texts of the "chief atheists" alone.

> His Religion . . . is pretendedly *Hobbian:* And he Swears the *Leviathan* may Supplly all the lost Leaves of *Solomon,* yet he never saw it in his life, and for ought he knows it may be a *Treatise* about catching of *Sprats,* or new Regulating the *Greenland* Fishing Trade. However the Rattle of it at *Coffee-houses,* has taught him to Laugh at *Spirits,* and maintain that there are no *Angels* but those in *Petticoats.*[88]

One well-known project to verify spirit activity is Joseph Glanvill's collection of reports of witchcraft, *Saducismus triumphatus* (1681). Boyle favored and supported Glanvill's mission, and as we have seen, he mentioned alchemy and Seyler's Viennese transmutations in connection with it (chapter 4). "Any one relation of a supernatural phenomenon being fully proved and duly verified," Boyle wrote to Glanvill in 1677, "suffices to evince the thing contended for; and consequently, to invalidate some of the atheists plausiblest arguments."[89] In his early career Boyle sponsored Peter du Moulin's translation of François Perrault's *Devill of Mascon,* an account of the

[87] Ibid., 49; notes for the tract occur at BP, 2:143–45. See also J. J. MacIntosh, "Robert Boyle on Epicurean Atheism and Atomism," in *Atoms, Pneuma, and Tranquillity: Epicurean and Stoic Themes in European Thought,* ed. Margaret J. Osler (Cambridge University Press, 1991), 197–219.

[88] *The Character of Town-Gallant* (London, 1675), 7.

[89] Boyle to Glanvill, 18 September 1677, *Works,* 6:57–59, on 58. On Glanvill's attempts to verify the existence of spirits, see Jackson I. Cope, *Joseph Glanville: Anglican Apologist* (St. Louis: Washington University Studies, 1956), 87–103, and further useful reading on witchcraft and atheism is Michael Hunter, "The Witchcraft Controversy and the Nature of Free-Thought in Restoration England: John Wagstaffe's *The Question of Witchcraft Debated* (1669)," in *Science and the Shape of Orthodoxy* (Rochester: Boydell Press, 1995), 286–307.

carryings-on of a spirit in a household in France. Boyle's correspondence with du Moulin, published as prefatory material to the translation, was the first appearance of Boyle's name in print—evidence of his early concern over the verification of spirit activities.[90] Additionally, at the same time Boyle was corresponding with Glanvill about his project, Boyle brought up the subject of witches in his correspondence with Georges Pierre. He apparently told Pierre about a Scottish witch, and Pierre responded with an account of a group of accused witches brought before the Rouen Parlement.[91]

In terms of verifying the existence of spirits, a spirit-invoking Philosophers' Stone would be most valuable. In the dialogue on spirits Boyle explicitly notes how the invocation of spirits by one "skill[ed] in naturall phylosophy" had in fact already rescued a disciple of "the subtilest Atheist of our times"—no doubt the Monster of Malmsbury—and converted him into "not only a profest Christian, but a very knowing and devout one."[92] The problem with accounts of spirit visitations and witch activity is that they are notoriously difficult to verify, and Boyle notes to Glanvill that he considers most accounts unreliable. But if the Philosophers' Stone had the capability of attracting spirits and facilitating communication with them, then it might be used actually to *demonstrate* the existence of spirits at any given time. Such power promised to be an irresistible weapon against the atheistic deniers of a spirit realm.

Boyle in fact refers (in the early 1670s) to the allure of proving the existence of spirits by invoking them. In *Excellency of Theology,* he writes that

> The Souls of inquisitive men are commonly so curious, to learn the Nature and Condition of *Spirits,* as that the over-greedy desire to discover so much as *That there are other Spiritual Substances besides the Souls of Men,* has prevail'd with too many to try forbidden ways of attaining satisfaction; and many have chosen rather to venture the putting themselves within the power of Daemons, than remain ignorant whether or no there are any such Beings: As I have learned by the private acknowledgements made me of such unhappy (but not unsuccessful) Attempts.[93]

The danger Boyle cites here echoes Timotheus's concern in the *Dialogue on Spirits* regarding the "fatall mistake of a divill for an Angel of light." The

[90] Perrault, *The Devill of Mascon.*

[91] Pierre to Boyle, 18/28 February 1678, response to Boyle's letter of 16 February 1678 (N.S.?), BL, 4:127–28*v.*

[92] Boyle, *Dialogue on Spirits,* appendix 3, 315–16.

[93] Boyle, *Excellency of Theology,* 2–3. Although Boyle explicitly states that the text was written in 1665, the quoted passage has all the marks of a later interpolation and was probably added as the manuscript was being readied for the press. The concluding parenthetical "(But this onely upon the By)" that binds this passage on spirits back into the flow of the surrounding and unrelated discourse is one of the typical markers of an interpolation as I identified them in my study of the revisions to *Seraphic Love;* the consistent use of such markers in many Boylean compositions renders them valuable for historians. See Principe, "Style and Thought of Early Boyle," 255.

mention of "forbidden ways" likewise recalls Timotheus's objections but also savors of the specifically moral issues explored by Hunter, which showed Boyle tormented by qualms regarding activities involving spirits that might be religiously unlawful. The "Burnet Memorandum" reveals Boyle's tortured position of being powerfully drawn to supernatural or magical practices as sources of knowledge about the natural or supernatural world, yet fearful of the possible spiritual perils involved.[94] Thus the question presents itself: If Boyle had succeeded in obtaining the Philosophers' Stone, would he have actually felt it permissible to employ it to summon spirits? In the *Dialogue on Spirits,* Arnobius and Cornelius agree that Parisinus (who represents Boyle) has sufficient arguments to render such transactions not only safe but morally allowable, particularly, it seems, for the adepti. The spirit dialogue fragment therein seems, then, to show some development beyond the fears embodied in the "Burnet Memorandum," suggesting that Boyle had advanced toward removing at least some of his moral scruples. Of course it is difficult to ascertain Boyle's exact views when he uses the dialogue format; yet since Parisinus seems so clearly to represent Boyle here, and Timotheus's objections on the basis of religion and morality are refuted so perfunctorily, it appears that later in life Boyle might well have been more willing to engage in the spirit-related phenomena that he had previously censured as "forbidden ways" of attaining knowledge.[95]

Incorporeal-Corporeal Interactions

Related topics of interest to Boyle are also implicated in his supernatural alchemy. Boyle was very interested in questions of the interaction between the corporeal and the incorporeal. This question undergirds several important theological issues including how an incorporeal God can cause change in His material creation, and how the immaterial human soul is joined to the material body and is able to cause it to move. The physical part of the problem involves how an incorporeal body could push mechanically against a corporeal one and bring it into motion, for the incorporeal would not share the property of impenetrability that allows one corporeal body to contact and push against another.[96] How spirits could communicate with

[94] Hunter, *Robert Boyle by Himself,* lxxiv.

[95] This change may have been the result of his contacting a spiritual adviser other than Burnet. Burnet, as his memorandum shows (Hunter, *Robert Boyle by Himself,* 31–32), was the one who advised Boyle that the offer made to him to manifest spirits was "to be forborne." In a similar case, Boyle accounted a subsequent refusal to look into a magical glass a "great victory over himself"; he must at the same time have been somewhat nettled at the loss of so great an opportunity. Another adviser may have helped him remove or at least lighten his moral scruples, as Arnobius's remarks imply he had done.

[96] Walter Charleton, for example, notes how "it is Canonical, that no Immaterial can Operate upon a Material, *Physically;* the inexplicable activity of the Rational Soul upon the body by the mediation of the spirits, and that of Angelical Essences excepted." *Physiologia Epicuro-Gassendo-Charltoniana* (London, 1654), 236. See also Osler, *Divine Will,* 70–72.

human beings forms a related issue, because it is unclear how their incorporeal bodies could set the intervening corporeal air into motion so that a voice could be heard. This issue constitutes "the difficulty in point of Physick or Phylosophy [which] rises from the distance between the humane nature and that of an unbodyed spirit" alluded to in the dialogue.[97]

These same issues were points of attack by atheists, who either denied the existence of the incorporeal or denied that such a substance could possibly affect corporeal matter. In an unpublished tract on atheism Boyle asserts his belief that "if these Men were satisfied in their scruples about the very notion and Existence of a Spirit or incorporeal Substance Intelligent and Powerfull, and that an Immaterial Substance can worke upon matter, I should thinke them more than halfe way advanced towards the acknowledgement of the Being of God."[98] Boyle touches upon the issue of the action of the incorporeal on the corporeal in many of his later writings.[99] This issue surfaces repeatedly in *Reconcileableness,* and it is notable that in one place Boyle links the problem specifically to Hobbes's "fundamental Maxim" regarding body and motion, which, as Boyle argues in a six-page critique, must terminate in either self-contradiction or atheism.[100] Similarly, in *Christian Virtuoso,* Boyle expresses his inability to comprehend "by what merely Mechanical Ty, or Band, an Immaterial Substance [the human soul] can be so durably (perhaps for 80 or 100 Years) joyn'd and united with a Corporeal." Likewise, he can no "better conceive, how a mere Body can produce Pain, Pleasure, &c by its own mere Action, or rather Endeavor to act, on an Immaterial Spirit."[101] In *Excellency of Theology,* he calls the union of immaterial soul with material body one of the profoundest mysteries, no less remarkable or incomprehensible than the Incarnation itself.[102] For Boyle this incomprehensibility is not a source of disbelief but rather one of awe for the Creator. In *Things Above Reason,* Boyle explicitly names the union of "an immaterial Spirit to a humane Body" as belonging to the class of the "Inexplicable" things above reason.[103] Finally, in a manuscript frag-

[97] Boyle, *Dialogue on Spirits,* appendix 3, 315.

[98] BP, 2: 2–37, on 2–3. The text is in Robin Bacon's hand and thus dates after the early 1670s.

[99] Wojcik, *Boyle and the Limits of Reason,* 67–75.

[100] Boyle, *Reconcileableness,* 12–13, 37–43, and 51–52; cf. *Animadversions to Mr. Hobbes's Problemata de vacuo,* preface, where the theological consequences of Hobbes's physics are given preeminence over any other objections; cf. *Excellency of Theology,* 23. Shapin and Schaffer, *Leviathan,* 201–7, mention the theological problems with Hobbes's physics and its support of his perceived atheism, but these are left quite subsidiary to the sociological issues given preeminence throughout their volume. I would suggest that when Boyle's devotion to Christianity is recognized as *primary,* the "atheism" of Hobbes's physics is itself quite sufficient to explain most of the "nonexperimental" issues behind the Boyle-Hobbes controversy.

[101] Boyle, *Christian Virtuoso,* 35.

[102] Boyle, *Excellency of Theology,* 147–49.

[103] Boyle, *Things Above Reason,* 77–78; Wojcik, *Boyle and the Limits of Reason,* 102–4.

ment echoing many of the concerns voiced in the *Dialogue on Spirits,* Boyle notes that a chief objection to accounts of spirit invocations is in fact the problem of corporeal-incorporeal interactions: "how utterly Improbable it is, that druggs, which are but materiall things . . . should have the power to engage either evil or good Angells, which are granted to be Immateriall beings."[104] But though "improbable" these actions are not "impossible," for here again Boyle invokes our ignorance of the "Nature, Customes & government of the Intelligent Creatures of the spiritual world" and refuses to let our ignorance govern our beliefs.

Boyle's notion of supernatural alchemical activity could help in this critical issue. If the Philosophers' Stone, which "is as truely corporeall as pouder of post [sandstone] or of brick," can in fact "attract or invite incorporeall and intelligent beings," then it provides a precedent for analogous, and more theologically crucial, interactions between corporeal and incorporeal bodies.[105] The "congruities or magnatisms . . . which we know nothing of" that allow for this interaction between spirits and the Stone could instance (though not explain) similarly unknown, inexplicable, or simply unknowable interactions between the body and the soul. Such an instance would be quite sufficient for Boyle—who, as I have stressed, did not require that an *explanation* be part of the demonstration of a thing's reality—for it would provide the clinching part of his standard methodology for proving "unobvious truths": first showing them to be not-impossible, and then exhibiting an instance of their reality either directly or by analogy.

In a related vein, Boyle may have believed (or perhaps hoped) that the spirits linked to the Philosophers' Stone might be the *cause* of transmutation. Boyle's interest in the possibility of spirits acting on matter is revealed first in his already-cited letter to Glanvill. There, writing of the actions of spirits invoked by witchcraft, he notes that "if it appear that there are intelligent agents that are able to increase; whereas men can but determine the motions of matter, the discovery of it may advantageously enlarge our knowledge." Boyle intriguingly notes that spirit activities may "help to enlarge the too narrow conceptions men are wont to have of the amplitude and variety of the works of God."[106] Such a suspicion on Boyle's part would further explain the choice of Bishop Burnet as a witness of projection. While the bishop would have served as a creditable witness of the event, one of Boyle's more chymically inclined colleagues would seem to have been a better choice if the goal were only verification of the physical reality of the transmutation. If, however, Boyle expected or feared some

[104] BP, 2:100–101, fair copy at 102–3; a seemingly related fragment appears at 104–5. These fragments are written to be part of a dialogue and are endorsed "NN speaks." The striking similarity of their argument to that of the dialogue on spirits shows how the two were at least conceptually linked, if not actually intended as parts of the same work.

[105] Boyle, *Dialogue on Spirits,* appendix 3, 313.

[106] Boyle to Glanville, 18 September 1677, *Works,* 6:57–59, on 58.

spirit activity at the time of transmutation, then the choice of Burnet as a spiritual guardian or expert seems more understandable.[107]

Boyle even takes up the notion of angels acting on matter in his *Excellency and Grounds of the Mechanical Hypothesis* (1674). He does not rule out the possibility that some physical changes may be directly produced by incorporeal agencies, using the actuation of the body by the soul as an example.[108] But he asserts that the mechanical hypothesis is not degraded in its applicability by such a possibility because "the chief thing, that Inquisitive Naturalists should look after in the explicating of difficult *Phenomena,* is not so much what the Agent is or does, as, what changes are made in the *Patient,* to bring it to exhibit the *Phenomena* that are propos'd; and by what means, and after what manner, those changes are effected."[109] That is to say, angels as agents may be real, but such explanations are too distant for the natural philosopher's use; mechanism must be invoked as the immediate—though perhaps not ultimate—cause of changes. Boyle's message is that the agents may be either corporeal, in which case their actions are directly reducible to matter and motion, or incorporeal, in which case the means of their effect on matter is incomprehensible (and nonmechanical), but the visible effect must arise immediately from a mechanical change in the patient. "If an Angel himself should work a real change in the nature of a Body, 'tis scarce conceivable to us Men, how he could do it without the assistance of Local Motion."[110] Thus Boyle asserts that the mechanical philosophy has *room* for incorporeal agents, but that it cannot deal with them directly; they are outside its domain but not negated.

Alchemy as a Middle-Term

When we consider the involvement of rational, incorporeal spirits, Boyle's brand of chrysopoeia stands out as a potentially unifying principle between his chief spheres of activity—natural philosophy and theology. Boyle, at least in his early career, saw no tension between the two and moved effortlessly between the Book of Nature and Book of God, readily applying experiences of the laboratory to excite devotion for the Creator, or to illuminate or render more comprehensible certain difficulties in points of exegesis or doctrine. But as Boyle grew older, he seems to have more fully recognized the potential conflicts between the mechanical worldview he promoted and the religion to which he was devoted.

The discomfort and suspicion between mechanistic natural philosophy and religion arose from several issues. In the first place, while mechanical conceptions of the world could explain changes and daily phenomena, they

[107] Hunter, "Alchemy, Magic and Moralism," 405–6.

[108] Boyle, *Excellency of the Mechanical Hypothesis,* 40.

[109] Ibid., 19–20.

[110] Ibid., 22.

had the unwelcome effect of distancing God from the continued operation of His Creation. A mechanical universe, functioning regularly by the operation of physical laws, no longer requires divine intervention to keep it running. God becomes, in various memorable phrases, an absentee landlord, a watchmaker who wound up His Creation at the start and then went away, or a God of the gaps—generously filling in wherever more mundane explanations are wanting. This was widely viewed as either unacceptable for the Divinity or leading directly to deism, if not to atheism. Mechanical explanations dispensed with the operation of spiritual agencies, and the atomism of mechanical systems was still tainted with classical atheism. Thus mechanism tended toward materialism, which led again, in turn, toward atheism. Finally, a mechanical universe is deterministic—a problem unconvincingly answered in antiquity by the notorious Epicurean "swerve," and which remained theologically vexing in a century still very troubled by arguments over free will both human and divine.

Boyle and other seventeenth-century thinkers used several strategies to disengage mechanism from such unfortunate associations and consequences. Gassendi, for example, in Margaret Osler's memorable phrase, "baptized Epicureanism" by making the motions of atoms the direct effect of God's will. Later thinkers like the Cambridge Platonists and Isaac Newton attempted greater emendations to mechanical models, even by tracing the descent of mechanistic atomism from sources less objectionable than Lucretius and Epicurus—in the choicest formulation from divine revelation to Moses.[111] Yet there was no escaping the fact that the mechanical hypothesis which Boyle so successfully championed *did* distance God from His Creation. Even Boyle's favored argument from design had limited effectiveness, for although his metaphor of the world as a great watch implied the existence of a watchmaker, it ensured neither the continued need for the watchmaker nor His interest in the continued running of His handiwork. This leaves one with deism, which the seventeenth-century considered little different from atheism.[112]

The desire to protect religion from the overapplication of mechanism

[111] Osler, *Divine Will;* "Fortune, Fate, and Divination"; "The Intellectual Sources of Boyle's Philosophy of Nature: Gassendi's Voluntarism and Boyle's Physico-Theological Project," in *Philosophy, Science, and Religion in England 1640–1700,* ed. Richard Kroll, Richard Ashcroft, and Perez Zagorin (Cambridge: Cambridge University Press, 1992), 178–98. The connection to Moses was made through the supposed Phoenician Moschus who was at least a Semite if not Moses himself; hence Boyle sometimes calls corpuscularianism the "Phoenician hypothesis." See "Essay on Nitre," preface [viii–ix] in *Physiological Essays,* preface, and *Excellency of Theology,* 165; see also J. E. McGuire and P. M. Rattansi, "Newton and the Pipes of Pan," *NRRS* 21 (1961): 108–43, esp. 114–15, 130, and Danton B. Sailor, "Newton's Debt to Cudworth." *Journal of the History of Ideas* 49 (1988): 511–18.

[112] Boyle was certainly not satified with deism; besides his condemnations of it in *Notion of Nature* and elsewhere, the only published piece of Boyle's verse is an explicit refutation of deism. See R. A. Hunter and Ida Macalpine, "Robert Boyle—Poet," *Journal of the History of Medicine* 12 (1957): 390–92; "Robert Boyle's Poem—An Addendum," ibid., 27 (1972): 85–88.

manifests itself in many of Boyle's later publications. The *Excellency of Theology as Compar'd with Natural Philosophy* seems a corrective to Boyle's own apparent greater devotion to the physical world. Perhaps more important still is his *Christian Virtuoso,* a work of *primary* importance to the understanding of Boyle and his view of his own time. The preface sufficiently spells out the unfortunate state of mutual suspicion between zealots on both sides of the issue—some dismissive natural philosophers and some "ill-informed" religious thinkers—who believed "that religion and Philosophy were incompatible." The result of this standoff was "that the Libertines thought a *Virtuoso* ought not to be a *Christian;* and the Others, That he could not be a true One."[113] The *Christian Virtuoso* is his attempt to defuse and to mediate this confrontation by illustrating the good services that natural philosophy and theology can offer one another; of course, Boyle himself as Christian Virtuoso provides the exemplar of such cooperation.

But Boyle was caught in a difficult dilemma. On the one hand, his attribution of all physical properties and physical change to those "two grand and most Catholick Principles of Bodies, Matter and Motion," had succeeded admirably in providing rational and comprehensible explanations of natural phenomena. Clearly he was deeply committed to mechanism as an explanatory tool. On the other hand, the overemphasis on these very same principles of matter and motion were at the heart of the dismissal by "Epicureans, libertines, and atheists" of the incorporeal, supernatural realm upon whose existence religion depends. The horns of the dilemma appear in juxtaposition in *Reconcileableness,* where in some places Boyle appeals to his own mechanical hypothesis and its chief principles of matter and motion and elsewhere complains that the "new Libertines" dismiss all authorities that do not "explicate things by Matter and Local Motion."[114]

Boyle's eagerness to defend religion did not, however, cause him to compromise his dedication to mechanical explanations. The clearest example is Boyle's dispute with the Cambridge Platonist Henry More, for although Boyle occasionally used notions like "Plastic Power" that were integral to More's thought, he struck back against More's debilitation of mechanism.[115] More, although devoted early in his life to the mechanistic thought of Descartes, soon saw several operational flaws in it (which he communicated to Descartes by letter) and eventually recognized its materialist dangers.[116] Thus More posited the existence of a spiritual agency he termed the Spirit of Nature or *Principium hylarchicum,* defined as

[113] Boyle, *Christian Virtuoso,* preface, [ii].

[114] Boyle, *Reconcileableness,* v.

[115] R. A. Greene, "Henry More and Robert Boyle on the Spirit of Nature," *Journal of the History of Ideas* 23 (1962): 451–74; John Henry, "Henry More versus Robert Boyle: The Spirit of Nature and the Spirit of Providence," in *Henry More (1614–1687) Tercentenary Studies,* ed. Sarah Hutton (Dordrecht: Kluwer Academic Publishers, 1990), 55–76. Boyle's use of "plastic powers" and "seminal virtues" occurs predominantly in early works.

[116] Alan Gabbey, "Henry More and the Limits of Mechanism," in *Henry More Tercentenary Studies,* 19–35, esp. 19–24.

> A substance incorporeal, but without Sense and Animadversion, pervading the whole Matter of the Universe, and exercising a Plastical power therein according to the sundry predispositions and occasions in the parts it works upon, raising such *Phaenomena* in the World, by directing the parts of the Matter and their Motion, as cannot be resolved into mere Mechanical powers.[117]

More believed that proof of such an active incorporeal substance, besides instancing an incorporeal entity, would be "a very useful Dogma for assuring the souls personal subsistence after death." Boyle should have welcomed any such support of orthodoxy. Yet when More used Boyle's own pneumatic experiments to argue for the activity of the Spirit of Nature, he ran afoul of Boyle. Indeed, when Boyle's criticism, eventually expressed publicly in 1672, reached More's ears, More was genuinely surprised, for he had expected Boyle to welcome his "demonstration" of the existence of an immaterial spirit against materialists.[118] Although we must not overlook the deep philosophical and epistemological differences between Boyle and More—Boyle's voluntarism against More's intellectualism[119]—their dispute was largely over the limits of mechanism. While More had correctly divined Boyle's fears over materialism and its expulsion of spirits, angels, and eventually God from the world, he had underestimated Boyle's commitment to mechanical explanations and thus incorrectly judged where Boyle set the limits of mechanism.

For Boyle, More's Spirit of the World was merely unnecessary—"our Hydrostaticks do not need it"—and so Ockham's razor was sufficient to excise it. Further, as "Truth ought to be pleaded for only by Truth," it was dangerous to defend so crucial a doctrine as the immortality of the human soul upon such a "precarious Principle" as More's Spirit of the World. While Boyle was clear about "heartily wishing [More] much success, of proving the existence of an incorporeal substance," he was not about to debilitate the explanatory power of the mechanical hypothesis to grasp after straws in a matter of such enormous importance.[120] The kinds of spirits Boyle might contact through alchemy, moreover, were real entities whose existence was demonstrable by the testimony of Scripture; More's Spirit was a tenuous construct without evidence.

Boyle was forced to fight on two fronts at once. On one side he had to fight against the "atheists" who overextended mechanism and thus diminished or obliterated the role of nonmechanical incorporeal substances, and on the other side against overzealous asserters of the activity of spirits who would diminish the role of mechanism. In the former category fell authors like Hobbes with a mechanical physics but a corporeal God that Boyle rec-

[117] Henry More, *The Immortality of the Soul* (London, 1662), 193.

[118] Boyle, *A Hydrostatical Discourse occasion'd by some Objections of Dr. Henry More* (London, 1672).

[119] Henry, "More versus Boyle."

[120] Boyle, *Hydrostatical*, 142.

ognized as no God at all. In the latter category fell Henry More, whose Spirit of Nature, working as a general factotum, cast away the advances in explanatory power made possible by the mechanical philosophy. While Boyle asserted the power of the mechanical hypothesis to More, he stressed its limits before the libertines and atheists, not only noting the cases where the mechanical hypothesis failed to give an adequate account of the phenomena, but also asserting the importance of studying incorporeal substances themselves.[121] Likewise, the *Excellency of the Grounds of the Mechanical Hypothesis,* particularly in light of its pairing in publication with *Excellency of Theology as Compar'd with Natural Philosophy,* should be read as an attempt to map out appropriate domains for Boyle's two chief interests and their attendant explanatory principles. For Boyle, the difficulty lay in ascertaining a defensible argument regarding where the limits of mechanism lay; there had to be some boundary between the realm of nature where mechanism was sufficient and the realm where it was not. Accordingly, we find Boyle particularly interested in topics that might straddle this divide: miracles, witchcraft, and, above all, his peculiar brand of spirit-tinged alchemy.[122]

Chrysopoetic alchemy in Boyle's version constituted an interface between the natural, mechanical realm and the supernatural, miraculous realm. As such, it could act as a mediator between his two potentially conflicting commitments to mechanical natural philosophy and Christian theology. Chrysopoetic alchemy centered on a physical object prepared by art—the Philosophers' Stone—which as such could be touched, handled, weighed, and subjected to any number of fully physical "philosophical tryalls." But, according to Boyle's belief in its "higher functions," it acted as well upon the suprasensual realm, attracting angels and spirits by what must be non-mechanical means, and could thus instance both the existence of spirits and the possibility of interactions between the corporeal and the incorporeal. The Philosophers' Stone offered the promise of a clean and defensible boundary between a mechanical natural philosophy and Christian theology.

Conclusion

The motivations for the study of chrysopoeia outlined here touch upon all of the topics most dear to Robert Boyle. Boyle's renowned interest in natural philosophy was served by the insights that projection, or other noble arcana such as the alkahest, could yield. A study of projection and the immense physical changes that process entails could not fail to provide illuminating new phenomena to aid him in puzzling out the constitution of matter and

[121] Boyle, *Christian Virtuoso,* preface, [vii–ix]; see also BP, 2:61–62, and *Reconcileableness.*

[122] On Boyle and miracles, see R. L. Colie, "Spinoza in England 1665–1730," *Proceedings of the American Philosophical Society* 107 (1963): 183–219; J. J. MacIntosh, "Locke and Boyle on Miracles and God's Existence," in *RBR,* 193–214.

the mechanisms of its transformations. The promise of such knowledge would powerfully excite the curiosity of a probing natural philosopher like Boyle. Boyle's philanthropy, too, could be furthered by chrysopoeia; the Philosophers' Stone was esteemed not only the medicine of metals but the medicine of man as well. Many potent medicines, administrable both to the sickly Boyle himself and to his fellow men, were supposedly preparable from the Stone in its final and intermediate stages. In terms of financial health, the making of noble metals, especially by *particularia* whose laborious preparation and low potency prevented them from being dangerous to the "welfare of States," could actually provide an alchemical cottage industry to support cadres of able workers. Finally, Boyle's increasing concern over the defense of Christianity from burgeoning atheism could gain perhaps its most powerful weapon from traditional chrysopoeia. Boyle's belief in the ability of the Philosophers' Stone to attract intelligent spirits meant that the Elixir was a potential ocular demonstration against atheism. The possessor of the Stone could use it to summon angels and other spirits, thus demonstrating their existence, and its very ability to do so instanced a motive interaction between the corporeal and the incorporeal, thus providing a precedent to support crucial theological truths. How could Boyle fail to be enormously moved by the allure of such great rewards? Indeed, he was so moved; his dogged pursuit of traditional chrysopoetic alchemy from the time of his earliest forays into experimental natural philosophy to his deathbed is neither curious nor surprising in view of the matchless prizes he expected to reap from it.

EPILOGUE

A NEW BOYLE AND A NEW ALCHEMY

THERE CAN NOW be no uncertainty about Robert Boyle's avid pursuit of traditional chrysopoetic alchemy. The level of his activity varied over the course of his life, never vanishing and sometimes, as in the latter half of the 1670s, reaching a peak of great intensity. The long-acknowledged but rather disconnected hints of Boyle's chrysopoetic pursuits are now revealed as a few brief glimpses of the continuous fabric of Boyle's lifelong alchemical quest. Having reached this point, one might reflect on how natural Boyle's interest should have seemed all along. Had we overcome our prejudices against alchemy and recognized that it was not divorced from chemistry in the totality of seventeenth-century chymistry, we would have been surprised if Boyle, with his known addiction to chymistry, had *not* shown an interest in chrysopoeia.

In the endeavor to portray Boyle's alchemical quest accurately, it was first necessary to be more rigorous about alchemy itself. We needed to return to the primary sources and to analyze their language and their contents in order to construct more rigorous and more historically accurate identifications and definitions of alchemy. We needed to slough off the layers of accretion with which subsequent generations obscured historical alchemy, and upon which the prevailing view of alchemy (even among historians) is still too frequently based.[1] We must be extremely careful in taking for granted the categories and categorizations that have come down to us from earlier historiography.

I have attempted to crystallize this fresh perspective into terminological conventions that, I hope, will enable our descriptions and discussions to have greater precision and to evade ambiguity and unintentional connotations. I have also attempted to sketch out a skeleton for some of the structures within alchemy that future studies (of which many are needed, and many are now in progress) can flesh out. There is a vast amount of primary source material in alchemy waiting to be given the benefit of an unbiased, historically sensitive analysis and to be incorporated into the historical context both intellectual and social. I would like to hope that the era of ahistorical, hodgepodge "picture books" of alchemy is over, or at least that historians of science will more diligently winnow publications on alchemy for use in their early modern studies.[2]

[1] This point will be further explored in Principe and Newman, "Historiography of Alchemy."

[2] What I refer to here are the books generically "on alchemy" that consist of heterogeneous collections of alchemical emblems and quotations wrested from their physical, intellectual, and cultural contexts.

It seems to me imperative that historians be specific in the future about the branches of alchemy they wish to study or to invoke, and be sensitive to the competing schools of thought within each branch. It is such an approach that allowed for the important reevaluation of the *Sceptical Chymist* presented in chapter 2. Without a sensitivity to the diversity found within chymistry, that book's real targets could not have been isolated. This more nuanced and precise approach will provide two benefits. First, it will show with much greater accuracy the dynamism within "alchemy." Far from being static and resistant to development or influences from other fields of endeavor, at least some branches of alchemy will be revealed as much more akin to the canonical "modern sciences." Second, an appreciation of the internal structure of alchemy will help us escape the clouded sight imposed by the smeary lenses of essentialism. These dual benefits will show the inadequacy of the common practice of juxtaposing a monolithic "alchemy" as a foil to early modern science. Now that the alchemical interests of so many figures of the early modern period are being revealed, we must take into consideration that the alchemy of Isaac Newton is not the alchemy of Robert Boyle, and neither of those is identical with the alchemy of John Dee, or of Michael Sendivogius, or of Elias Ashmole, or of J. J. Becher. Alchemy is not a generic spice powder seasoning the pot of seventeenth-century thought. It has many flavors and nuances, and its deployments and hoped-for rewards vary widely. It is critical that we be discerning about these (often extreme) differences if we wish to maximize the accuracy and fruitfulness of our historical enterprise.

In Boyle's case, I have examined in particular his chrysopoeia and his participation in what I have (tentatively) identified as a uniquely English school of supernatural alchemy. His experimental pursuit of the Philosophers' Stone involved many years, many substances, and many people. For much of Boyle's life, however, this quest centered specifically on the Mercurialist school of chrysopoeia, which school believed the Stone was to be begun with a special "philosophical" treatment of common mercury. I have not delved at all into the Helmontian notions that informed Boyle's early thought, even though theories regarding seminal principles, the archeus, and so forth, which might be broadly classed as alchemical, are important to Boyle, particularly in his early years. In these particulars, I refer the reader to the important work of Antonio Clericuzio.[3]

The reader will note that although I followed Boyle's quest of the Philosophical Mercury from 1651 to 1691, his brief pursuit of the "wet way" revealed in the *menstruum peracutum* experiments of the 1650s, and reevaluated the *Sceptical Chymist* of 1661, much of this study deals with Boyle's more mature career after about 1670. There are several reasons for this focus. First, I was at pains to show that chrysopoeia was not to be classed as

[3] Clericuzio, "From van Helmont to Boyle"; "Redefinition of Boyle's Chemistry"; "Boyle and the English Helmontians."

a part of Boyle's juvenilia, but rather that the intensity of his quest continued unabated until his death. Second, I have relied heavily upon archival material, and the archive itself is weighted, by the survival of later material, toward the second half of Boyle's life. Third, and most important, Boyle's early alchemy warrants a study in its own right. To have treated Boyle's early alchemy fully here would have doubled the length of this study or necessitated a universally more cursory examination of Boyle

When Boyle first turned toward natural philosophy around 1649,[4] many of his earliest experiences were oriented toward chrysopoeia and chemiatria. Even before that date he made an unsuccessful attempt to begin chymical studies and even then mentioned an interest in the Stone.[5] The 1650s, the years between Boyle's turn to laboratory experimentalism and the publication of his first books, are clearly of great importance in the development of his scientific interests, methodology, and persona. If in fact chymistry constituted a major part of his formative experimental career, and traditional alchemy constituted an important part of his chymistry, then we must ask what the developmental influences of traditional alchemy were upon Boyle's style and practice of natural philosophy. Was Boyle's laboratory practice, which has been long seen as paradigmatic of early modern science, shaped by formative experiences in traditional alchemy?

This intriguing question forms a theme of a joint study that William Newman and I are presently conducting. As I mentioned in chapter 5, Boyle collaborated during his formative years with several seekers for the Stone—Kenelm Digby, Benjamin Worsley, Frederick Clodius, and, perhaps most important, the young American émigré George Starkey.[6] Starkey's pursuit of the Stone and the alkahest is recorded not only in the famous Philalethes tracts but also in laboratory notebooks, several of which, miraculously, have survived. These notebooks, some of which span the period of active collaboration between Starkey and Boyle, reveal for the first time the day-to-day operations, thought, and methodology of a practicing chrysopoeian alchemist.[7] These documents will constitute a crucial part of our forthcoming study.

The Boyle revealed in the present study manifests several characteristics that have not always appeared in previous accounts. First, he appears here as an *evolving* Boyle.[8] It would be unreasonable to expect that Boyle did not alter or amend his views over the course of his adult life. Unfortunately, much of the standard work on Boyle, until quite recently, has concentrated

[4] Hunter, "How Boyle Became a Scientist."

[5] Boyle to Katharine, Lady Ranelagh, March 1647, *Works,* 1:xxxvi. See also Principe, "Boyle's Alchemical Pursuits," 93 and 99.

[6] See chapter 5.

[7] See Newman, *Gehennical Fire,* 270–74, esp. MSS 3, 4, 12, 15; and Principe, "Newly-Discovered Boyle Documents," 59.

[8] An important exception that relies upon Boyle's evolution is Hunter, "How Boyle Became a Scientist."

on his early natural philosophical work—material largely written at Oxford and published in the 1660s. This temporal window reinforced the notion of an orderly, "modern scientific" Boyle on account of its emphasis on the mechanical philosophy, scientific methodology, and the context of the early Royal Society. In practice it has sometimes proven difficult to move Boyle beyond the air-pump and its environs.[9] But this is not the whole Boyle; it is not even, in a sense, the mature Boyle. His later thought became more complex, more involved with the ramifications of natural philosophical issues, and more concerned with ongoing disputes in science and religion; accordingly, his tracts of the 1670s and 1680s become increasingly diverse and increasingly difficult to digest and to reduce. The more intricate (and the more intriguing) Boyle is the more mature Boyle, one who developed beyond some of his convictions of the 1660s for which he has long been most celebrated.

In terms of chymistry, the continuation of the analysis of the *Sceptical Chymist* to include the 1680 appendix illustrates how some of Boyle's views developed. Boyle's authorship of alchemical tracts like the *Dialogue on Transmutation* in the 1670s and 1680s and his enhanced chrysopoetic endeavors at this time both in and out of the laboratory reveal his *increasing* interest in traditional alchemical topics. Perhaps ironically, Boyle's traditionally most studied period is also his least alchemical. Boyle's certainty about the reality of the Stone and other chief alchemical arcana solidified outside of the 1660s. This development has been tied in large measure to his crucial witnessing of chrysopoetic projections in the late 1670s, events that by his own admission removed his doubts over a thing he was "long distrustfull of." More provocatively, it was in the last two decades of his life that Boyle expanded his alchemical interests to include unusual spiritual operations of the Philosophers' Stone, as outlined in his *Dialogue on Spirits*. Exactly when this development occurred remains unknown, but I have argued that this extension of alchemy into spiritual territory parallels his increasing concerns later in life with the threat of atheism and his increasing recognition of the limits of the mechanical philosophy.

Equally crucial to the new Boyle is a sense of his chymical context. Boyle's work was not carried out in isolation. In terms of intellectual context, this study shows Boyle's debts to and involvement in ongoing disputes over the reality of the Philosophers' Stone and metalline Mercuries, as well as his active participation in one of the most important avenues of research in chrysopoeia. Boyle's social chymical context is perhaps more surprising. This study has revealed a truly *international* Boyle. Boyle never left England after 1648, but his wide net of correspondents and visitors kept him constantly connected to developments across Europe. While Boyle's hospitality

[9] It is significant, however, to note that *all* of the essays in *RBR* take the reader in new directions, and, of course, Hunter, "Conscience of Robert Boyle," which reveals a Boyle far more complex and troubled than cutouts from the 1660s provide.

to foreign visitors has been noted previously, the ramifications of his close ties to Continental affairs have not always appeared sufficiently prominently in studies of him.[10] All too often the history of early modern science has suffered from a rather parochial view, concentrating upon national contexts, centers, or institutions, with preeminence, at least in the past, often going to Britain. While the developments in seventeenth-century Britain are unquestionably of great importance, we cannot be too careful about unwarrantably insulating those developments from Continental influences. We have seen how much of Boyle's correspondence on chrysopoeia was with Continental sources, especially in France. Boyle's fluency in French, Italian, and Latin (and his maintenance of operators like Hanckwitz and Slare who spoke German) no doubt facilitated these communications, but such polyglot talents were by no means as unusual among Boyle's colleagues as they are among modern Anglophones. The example of Boyle clearly demonstrates the close connections that then bridged the Channel, and warns against our circumscribing too narrowly the social circles surrounding figures of interest.

A closely related feature brought forth by this study is the *collaborative* Boyle. While the importance of operators and servants in Boyle's enterprise has been interestingly pursued elsewhere,[11] this study of Boyle's chrysopoeia reveals a Boyle supporting and supported by the direct collaborative work of many other natural philosophers and laboratories. How many laboratories were in fact financed or directed by Boyle? We will probably never know precisely. I have mentioned here the lab-cum-lodgings of Boyle's operator Hanckwitz and the additional workshop, paid for by Boyle, for the use of the would-be projector Hanckwitz oversaw. There is also the anonymous German whose chrysopoetic operations were being replicated by De Saingermain in his lab at Paris, with Boyle acting as the connector between these two centers. Additionally, there is the "company" of chrysopoetic hopefuls in London that Newton mentioned to Locke as the source of Boyle's mysterious "red earth." Boyle was sought out for alchemical advice and sought it in turn from others. The forthcoming study of Boyle's work with Starkey will further enlarge our understanding of Boyle as a collaborator. The forthcoming edition of Boyle's complete correspondence will also help scholars to ascertain whether this picture of internationalism and collaboration holds as well in other aspects of Boyle's enterprise as it does in his chrysopoeia.[12]

[10] R.E.W. Maddison, "Studies in the Life of Robert Boyle, F.R.S.: Part I, Robert Boyle and Some of His Foreign Visitors," *NRRS* 9 (1951): 1–35; "Studies in the Life of Robert Boyle, F.R.S.: Part IV, Robert Boyle and Some of His Foreign Visitors," *NRRS* 11 (1954): 38–53. See also Boyle's influence on the Continent revealed in Clelia Pighetti, *L'Influsso scientifico di Robert Boyle nel tardo '600 italiano* (Milan: Franco Angeli, 1988).

[11] Shapin, *Social History of Truth*, 361–69.

[12] *The Correspondence of Robert Boyle*, ed. Michael Hunter and Antonio Clericuzio (London: Pickering-Chatto, forthcoming 1999).

While the immediate goal of this study was to restore the traditional alchemical dimension to Boyle's work and thought, that restitution necessarily makes some wider changes in our overall perspectives on Boyle and the contemporaneous development of science, particularly chemistry. Primarily, the reader may wonder how the revelation of Boyle as a chrysopoeian affects long-standing views of him as an important figure in the development of modern chemistry. Is there now a conflict that must be resolved?

First, having stressed that the topics which in modern times have been segregated under the differing rubrics of alchemy and chemistry were in the seventeenth century equally encompassed under chymistry, I believe that much of the apparent conflict between Boyle as a seeker of the Stone and Boyle as an architect of early modern chemistry vanishes as an artifact of a historiographic category mistake. There is no conflict; Boyle was simply a seventeenth-century chymist—some of his chymical activities have potential counterparts in modern chemistry, while others continue the traditions of chrysopoeia and spagyria. Both activities fell unproblematically within the domain of the seventeenth-century chymist.

Second, and much more important, the new Boyle and the new alchemy of this study themselves represent the resolution of a conflict. Boyle's enduring place as a key figure in the history of chemistry has been predicated largely upon his elaboration of a mechanical corpuscularianism with special reference to chemistry. Chymical experiments helped Boyle develop his corpuscularianism, and his corpuscularianism, in turn, made new sense of chymical phenomena. In the standard historiography, Boyle's position as a mechanist, combined with the belief that the *Sceptical Chymist* represented a rejection of alchemy, accorded very well with more general accounts of the development of chemistry, specifically the replacement of a vitalistic, irrational alchemy with a physical, mechanical chemistry.[13] As a result, as I showed in chapter 1, there appeared to be a critical disjunction in the history of chemistry at the time of Boyle. This study bridges that gap from both directions.

On the one side, recent studies by Newman have shown the corpuscularian and mechanical traditions alive within transmutational alchemy. Indeed, it now seems quite likely that one source of Boyle's own corpuscularianism is to be found in the writings of chrysopoeians and iatrochemists like the pseudo-Geber and Daniel Sennert. Both van Helmont and Philalethes embraced a mechanistic, corpuscularian understanding of transmutation.[14] Thus chrysopoeian alchemy turns out not to be so far re-

[13] The notion of mechanism's marking a disjunction from a vitalistic alchemy was influentially propounded by Metzger, *Doctrines*. Boas, *RBSCC*, strongly reinforces Boyle's position in this disjuncture.

[14] Newman, "The Alchemical Sources of Robert Boyle's Corpuscular Philosophy"; "Boyle's Debt to Corpuscular Alchemy"; "The Corpuscular Theory of J. B. van Helmont"; "The Corpuscular Transmutational Theory of Eirenaeus Philalethes."

moved from the mechanical conceptions of a Boyle or a Lemery after all. On the other side, we now see that Boyle himself in no way rejected transmutational alchemy but rather pursued it avidly and appropriated several of its theoretical principles. Thus Boyle turns out to be not so far removed from the goals and practices of traditional alchemy. Boyle was not as "modern" as we thought, nor alchemy as "ancient." What we are witnessing, then, is a rapprochement between what have previously been seen as two separate and irreconcilable halves of the history of chemistry.

In further witness of this rapprochement, we see that Boyle was eager to deploy the alkahest and chrysopoetic transmutations—especially via the Philosophers' Stone—toward the elucidation or support of the mechanical system of natural philosophy he advocated. He did not see his belief in the transmuting powers of the Stone as fatal to, or even detracting from, his commitment to a mechanical view of physical phenomena. Boyle, from his favorable vantage point, does not seem to have seen a disjunction between the hidden arcana of the adepts and his own mechanical system; Boyle's chrysopoeia argues strongly for continuity in the history of chemistry.

The new Boyle and the new alchemy display the action of evolution, not revolution, binding together the history of seventeenth-century chymistry with a greater, and more probable, degree of continuity. Boyle was critically involved in important chymical issues of his day, not necessarily ones that we in retrospect view as crucial for the development of our own chemistry, or even of the chemistry of the eighteenth century. Indeed, some of the issues that concerned Boyle had been wholly lost sight of because of a neglect of his context. The debate over the reality of the Philosophers' Stone and of metalline Mercuries persisted throughout the seventeenth century (and well into the eighteenth), and by 1680 Boyle clearly took a position in favor of both. The issue of the licitness of alchemy and its possible links to morally perilous activities likewise erupted intermittently from the thirteenth to the eighteenth century; again, Boyle was involved, and, after what appears to have been a tortured period of doubt and uneasiness, he seems to have decided in alchemy's favor.

This study of Boyle's chrysopoeia has also revealed some items that affect more than his relationship to chemistry. Boyle's long-standing commitment to secrecy—replete with coded papers and letters, published dispersions and allusive references, and a reticence to reveal that shocked even Newton—forces us to reevaluate the way in which Boyle viewed the status of knowledge. Somewhat surprisingly, he seems to have invested alchemy with a privileged status that required its special treatment; the secrets of the adepti were not for everyone. Perhaps most difficult for some readers to accept is Boyle's supernatural alchemy. A Boyle who sought the Philosophers' Stone to facilitate communication with angels is perhaps as far as one can get from the portrayals of Boyle as a rationalist empiricist that held sway relatively unchallenged from the mid-eighteenth century until quite recently.

Yet this surprising dimension, so clearly portrayed by archival materials, reveals two crucial parts of Boyle's intellectual makeup. First, it returns Boyle's religion to center stage. No real understanding of Boyle is possible without an understanding and acknowledgment of his sincere religious disposition and concerns. Sociological, political, philosophical, and intellectual analyses of Boyle miss the mark whenever they fail to give preeminence to his devotion to orthodox biblical Christianity. Second, the nexus that alchemy provides between Boyle's commitments allows it to highlight one of his most important features as a thinker, namely, the uniformity and consistency with which he approached the world. Boyle's views of natural philosophy and of theology form a seamless garment. Just as Boyle had no patience with those who, like the Socinians, wanted to shackle God's actions to what they could understand by reason, so he had no patience for those who dismissed the reality of natural phenomena because of their inability to understand them. Christ is fully God and fully man whether or not we can explain or comprehend His dual nature; the Philosophers' Stone is real, whether or not we can explain or understand its effects. God is a wholly Free Agent in Boyle's voluntarist theology; He could act as He wished in both theological and natural realms. For Boyle, the Book of Nature and Book of Scripture are more than images or mere rhetoric; they are the real productions of a single Author who is supremely self-consistent. Boyle as a reader of both demanded a reciprocal consistency in his understanding and approach. Christ and the Apostles worked miracles regardless of whether or not we comprehend them, for the eyewitness Evangelists testify to the fact; the Philosophers' Stone transmutes lead into gold in an instant, whether or not we have seen it, for eyewitnesses testify to it. The "imputation of credulity" with which Boyle was taxed was, in effect, a consequence of his epistemological consistency—the impugning of witnesses of strange natural phenomena implicitly impugns witnesses of miracles.[15]

The surprising revelations and revisions brought forth by this study of Boyle's chrysopoeia underscore how much remains to be done in terms of historical understanding even in a period as well-traversed as the seventeenth century and even as regards a character as well-recognized as Robert Boyle. If even Robert Boyle is susceptible to the kind of radical reevaluation that this study represents, then how much must remain to be done in the history of early modern science? In these future historical studies it would be worthwhile to recall the phrase that provides the Royal Society's motto, one that Boyle surely played a role in choosing: *Nullius addictus iurare in verba magistri* (dedicated to swear allegiance to the words of no master). Boyle himself ("I admit no man's opinions in the whole lump") was quite probably the best embodiment of this sentiment, and the sense of the motto is good advice for historians as well. If anything is clear from the histo-

[15] Boyle's uniform view of knowledge and the centrality of his Christianity argued for here are strongly corroborated in Wojcik, *Boyle and the Limits of Reason*. Similarly, his refusal to limit the powers of God in Nature resulted in the "diffidence" displayed in Sargent, *Diffident Naturalist*.

riographic portions of this study, it is that the subjugation of historical figures and events to grand ideological schemes invariably blemishes our understanding. Various exponents of "programmes," from Enlightenment deists to current postmodernists, have tried to co-opt Boyle for their own schemes, but it is questionable whether they have produced solid results or only raised phantasms that distract from the pursuit of more valuable studies and that vanish in the light of more substantive research. Much the same thing may be said of the tortured historiographic fortunes of alchemy. It is by far the wiser course to follow historical characters and developments than to try to lead them about by the nose. Undoubtedly, projects now underway by the noted scholars cited throughout this book will continue to add to the accuracy and detail with which we view Boyle and his epoch. Whatever may befall arguments regarding the speed and nature of change in the period, the seventeenth century will remain a crucial era for the development of science and modern culture, and we are only beginning to understand it.

APPENDIX 1

ROBERT BOYLE'S DIALOGUE ON THE TRANSMUTATION AND MELIORATION OF METALS[1]

Introduction

Boyle's *Dialogue on Transmutation* now exists only in fragments, yet the importance of the text warrants its publication in full despite its imperfect state, for although I have devoted large parts of two chapters to the *Dialogue*, I have not exhausted its usefulness to historians. Reconstructing a text from fragments can never be a precise task, but recent studies of Boyle's method of composition clarify the relationship of these fragments to the completed work. The most detailed exposition of Boyle's mature compositional technique is provided by Michael Hunter and Edward B. Davis in their analysis of *Notion of Nature*.[2] On the basis of their study, together with other recent studies that reach very similar conclusions, it is clear that Boyle's works took on a polished form only upon arriving at the printer. In his mature career, Boyle tended to write on single sheets of paper rather than in bound or coherent bundles, and treatises had only a loose, amorphous form until the last moment. The period of composition during which these snatches of text were penned and stored often stretched over many years or even decades. Hunter and Davis argue that Boyle's idiosyncratic (and apparently chaotic) methodology gave him a great deal of flexibility in his deployment of materials for writing treatises, even though it often led to the confusion and mismanagement of documents almost routinely alluded to in Boyle's apologetic prefaces.[3]

As a result we do not find lengthy unpublished compositions in the Boyle archive. Instead, we find a huge number of scraps—the single sheets or brief sections—that Boyle would himself weld together into a tract immediately before publication.[4] Boyle's usual method of composition from conception

[1] The title of the work is actually uncertain. Boyle's working title usually mentions both the transmutation and the generation of metals. But since his last mention of the work (1691) replaces *generation* with *melioration* and there is no discussion of metallic generation in the extant dialogue, I have settled on the title as given here.

[2] Hunter and Davis, "The Making of Robert Boyle's *Free Enquiry*," 209–25.

[3] Ibid., esp. 216–19. See also Principe, "Style and Thought of Early Boyle"; John T. Harwood, "Science Writing and Writing Science: Boyle and Rhetorical Theory," in *RBR*, 37–56; and Boas Hall, "Early Version."

[4] There are a few exceptions: *Swearer Silenc'd* existed in complete form in manuscript until its posthumous publication in 1695. This work, however, dates from 1647, before Boyle settled upon the methods typical of his mature career. Most of the extended sections of unpublished treatises surviving in the archive are due to be published with the new *Works* edition under the editorship of Hunter and Davis.

to printed text may be summarized as a slow and lengthy period of writing, accretion, and evolution rather suddenly punctuated by publication of a final manuscript pieced together from scattered sheets. Thus we cannot expect to reconstruct a smooth text for the *Dialogue* in the absence of a manuscript readied by Boyle for the press or for private circulation; the welding together and polishing of the scattered sheets would have been done at the final stage by Boyle himself according to his mental plan for the completed work.

There is one further complication. The following fragments of the *Dialogue* do *not* all belong together in a single draft; they are of different dates, and some would have been partly or wholly rejected from the completed tract. Rather, the following fragments are parts of the changing *texts* of the *Dialogue* as it evolved. Some pieces are clearly missing altogether, and some pieces present here were probably not used in the final version. One clear example of the latter case is fragment 7, the first part of which is clearly redundant with the prologue; Zosimus's speech on the differences between alchemical and natural gold occurs in an only slightly reworded form in Heliodorus's opening comments—the two could not have coexisted in a final draft. Part or all of fragment 7 was probably rejected from the final draft because of the borrowings from it; in fact, fragment 7 is the only draft fragment of the *Dialogue* that lacks the endorsement "T[ranscri]b'd." The patchwork text presented here represents a blurred image of the dialogue's composition and evolution over many years.

Dating and Evolution of the Dialogue

It is likely that the writing of the *Dialogue on Transmutation* occurred in two distinct periods. The first, dating from the second half of the 1670s, provided most or all of the prologue and part one (fragments 2–10) as well as part three (fragment 17). The second period of composition, dating from the early 1680s, deleted some characters and added others (fragment 1), extended the plot to include a discussion of alchemical literary style and the character of the adepti (part two, fragments 11–16), and probably substituted a new ending.

The first version of the dialogue is outlined in the "Heads of the Dialogue" preserved in fragments 2a–b. These "Heads" accurately summarize the arguments of fragments 2c–5 and broadly allow for the subsequent testimonial accounts of fragments 6–10. They also specify a conclusion in which Pyrophilus enters and comments on the proceedings—seemingly an accurate description of fragments 17a–e, published as *An Historical Account of a Degradation of Gold by an Anti-Elixir* in 1678. Fragment 3 provides a likely terminus a quo for this initial phase of composition. This manuscript contains corrections in the hand of Thomas Smith, who began working for Boyle ca. 1674.[5] Since this is the first draft of a crucial part of the dialogue where the main lines of the argument are set down, it is un-

[5] Hunter, *Guide*, xxxii.

likely to have been written as an interpolation to a preexisting text but probably represents an early phase of composition.[6] This date aligns well with Boyle's renewed interest in chrysopoeia as witnessed by the publication of the incalescent mercury paper in February 1676.

The Publication of Anti-Elixir

The publication of *Anti-Elixir* in 1678 provides a terminus ad quem, which is the firmest date available for the dialogue's composition. The *Anti-Elixir* tract as published, with its well-defined but unexplained setting and its reference to earlier but unpublished arguments, must postdate the substantial, if not entire, execution of the dialogue in its first form. But its publication raises some interesting questions: Why did Boyle sever the conclusion of the work to publish it anonymously as an obviously amputated seventeen-page tract?

It was once suggested that *Anti-Elixir* was a response to Newton's 1676 criticism of Boyle's incalescent mercury paper, but now that Newton's alchemy is well-known, it is clear that Boyle did not need to defend alchemy to Newton.[7] Besides, Newton and Boyle conversed privately, and Newton's criticism was itself private and so hardly required a printed response. It is more likely that the publication of *Anti-Elixir* is a result of the experiences of Boyle's *annus mirabilis alchemicus* of 1677–78.

I suggested previously that Boyle received the anti-Elixir from Georges Pierre during the latter's visit to London in the summer of 1677. This supposition requires a close timing between Boyle's receipt of the anti-Elixir and his publication of the tract about it, and in fact, the surviving drafts of *Anti-Elixir* (fragments 17b–d) suggest that it was written hurriedly only shortly before publication. These drafts are all in the same hand and on the same paper; this consistency among the dispersed fragments (unusual for a Boylean composition) suggests that they were part of the same entire draft and were written within a very short time of one another; otherwise we should expect a diversity of hands and possibly of papers.[8] Fragments 17c and 17d display much revision, identifying them as first drafts, yet the amended text is nearly identical with the published version. This shows that the fair copy from these drafts was published without subsequent revision, implying that little time intervened before publication.

I conjecture that Boyle deployed this impressive experiment as crucial evidence for the concluding episode of his *Dialogue,* but shortly thereafter,

[6] This is not to say, however, that no material gathered at an earlier date was incorporated into the *Dialogue;* fragment 6, containing testimony from Olaus Borrichius, is based upon information probably obtained in the 1660s.

[7] Fulton, *Bibliography,* 92–93.

[8] The hand of fragments 17b and 17c had been identified as Boyle's own (Hunter, *Guide*); however, after discussions between Prof. Hunter and me, the identification of the hand as Boyle's has been rejected, and it will now be known as "Hand C," displacing the very rare "old" Hand C mentioned in the *Guide.* This change will be fully described in a future edition of Hunter's *Guide.*

in early 1678, Boyle witnessed the yet more impressive use of the real Elixir to transmute lead into gold. The impact of this further experience fell close alongside the accounts he was receiving of Seyler's transmutations, and in the context of the increased antialchemical mood of the 1670s to which part of *Producibleness of Chymical Principles* was addressed. These events taken together may have provoked Boyle to act more publicly in his defense of chrysopoeia, even at the cost of wresting a fragment from its context.

Whatever the provocation for publication, *Anti-Elixir* was evidently published in a very small run. Already by the early 1740s, Henry Miles encountered difficulties in locating a copy from which to print the text for Birch's *Works,* and at present only four copies are known.[9] This scarcity may result from a restricted circulation, as was apparently the case with the now unknown 1688 first edition of *Medicinal Experiments*, which was privately printed.[10]

The bibliography of the rare *Anti-Elixir* is confused. Fulton catalogs editions of 1678 and 1739 and an abridgment published in 1782 in connection with James Price's claim of transmutation.[11] Fulton could not trace a supposed 1689 edition or a German translation mentioned by Ferguson. [12] The German edition actually exists in the Ferguson collection and is an anonymous 1783 translation of James Price's 1782 abridgment.[13] Schmieder cites a further German translation, namely, in the *Göttingisches Magazin* of 1783, which is also a translation of Price's 1782 abridgment.[14] The 1689 edition, however, remains a mystery. Gmelin (1797) and Schmieder (1832) cite this edition, and Ferguson presumably copied their information.[15] While it is easiest to dismiss this supposed 1689 edition as a bibliographic error, Schmieder implies that he had a copy which he compared with the 1678 edition.[16] While no firm evidence for a 1689 edition exists, one might speculate that a reissue might have been undertaken by Boyle in connection with his efforts to repeal the Act against Multipliers in 1689. In light of the scarcity of the 1678 edition, it is not unlikely that all copies of a private or restricted printing of 1689 would have perished before the present day.

The draft fragment 17b reveals Boyle as the author of the "publisher's note," thus opening the possibility that other publisher's notes are also by

[9] British Library Add. MS 4314; Fulton, *Bibliography,* 93.

[10] Hunter, "Reluctant Philanthropist," 252; see RSMS, 186:119 ff.

[11] Fulton, *Bibliography,* 92–94; on James Price, see *DNB.*

[12] *BC,* 1:122; Fulton, *Bibliography,* 93 n. 2.

[13] *Versuche mit Quecksilber, Silber und Gold . . . von James Price . . . Nebst einem Auszuge aus Boyles Erzählung von einer Degradation des Goldes* (Dessau, 1783); Ao-b.46 in the Ferguson Collection, University of Glasgow. I am grateful to Mr. David Weston, principal assistant librarian of the special collections for this information.

[14] Schmieder, *Geschichte der Alchemie,* 456. The German edition is "Vom Gold machen des Dr. Price (ein Auszug des Hrn. Prof. Gmelin aus des Doctors Schrift)," *Göttingisches Magazin der Wissenschaften und Litteratur,* 1783 (reprint, Osnabrück: Otto Zeller Verlag, 1977), 3:410–52.

[15] Gmelin, *Geschichte der Chemie,* 105; Schmeider, *Geschichte der Alchemie,* 456; Schmieder erroneously dates the reissue to 1737.

[16] Schmeider, *Geschichte der Alchemie,* 459. He found the two identical.

Boyle. This fragment is a draft (not the first, judging by the paucity of corrections) of the notice explaining the publication and importance of the tract, and includes even a third-person reference to Boyle's affirmation of the truth of the relation. This discovery sheds additional light on Boyle's concern over publication and the proper representation of himself and his writings.[17] Finally, fragment 17e is an unpublished Latin translation of *Anti-Elixir* (without the publisher's note). It shows scattered differences from the published text; these variants are recorded in the footnotes of this edition.

Second Phase of Composition

The second phase of the *Dialogue*'s composition began ca. 1680 with the transcription of the "Heads of the Dialogue" into the notebook now known as RSMS 198.[18] As the notebooks functioned partly as staging areas for Boyle's writings, this transcription seems to mark the start of renewed activity on the *Dialogue*.[19] The adjacent inventory (in the same hand) of Boyle's "Tracts relating to the Hermeticall Philosophy" reinforces the impression of the writer's taking stock of affairs before beginning a new phase. A contemporaneous notebook, RSMS 199, contains most of the remaining fragments presented here.

Fragment 11 is a crucial hinge to the plot; here the "main Controversy betwixt the Chrysopoeans & their Antagonists" has been suspended owing to the arrival of an "illustrious Stranger." This interlude is not mentioned in the original "Heads," even though both the setting and the characters clearly argue that it is part of the same dialogue. Here a few society members begin to discuss the adepts' practice of secrecy even in regard to medical matters. The other RSMS 199 fragments do not follow this introduction very well, for they deal instead with the character, knowledge, and writing style of the adepti. As such these fragments seem akin to Boyle's planned or partly written works "Of the obscure & enigmaticall stile of Chymical Philosophers" and "Of the difficulty of understanding the books [of] Hermeticke Philosophers."[20] It is therefore possible that fragments 12–16 do not belong to the *Dialogue on Transmutation* at all but to one of these other works. Alternatively, the similarity of setting and factions in these fragments to the earlier *Dialogue* argues that Boyle may have spliced the ideas behind these otherwise unknown works into the *Dialogue* during its second phase of composition. However, the reader will notice that the import and arguments of these fragments occur (in abbreviated and softened form) in

[17] See Harwood, "Science Writing."

[18] RSMS 198 is dated by Hunter, *Guide,* xxix.

[19] Both the freedom from corrections and the several instances of the typical transcription error of randomly dropped letters argues that fragment 2a is a copy of an earlier text. Fragment 2b is still later, as the numerals added in the margin of 2a (possibly in Boyle's hand) are smoothly incorporated into 2b.

[20] Boyle had considered such a work for many years; an essay entitled "Of the Writings of the Chymists" is listed in a catalog dating from the mid-1650s (BP, 36:70).

Erastus's opening speech in fragment 2c. Thus a third possibility is that fragments 12–16 were eventually rejected, and their assault on the adepts relocated to Erastus's opening statement in the completed and now lost final version of the *Dialogue* of which fragment 2c is a partial translation.

Alterations in the cast of characters further evidence the evolution of the *Dialogue*. Fragment 1, a "dramatis personae," is presumably an attempt to organize the revision by enumerating the characters and their parties. The three parties headed by Zosimus, Erastus, and Eleutherius recorded here exactly reflect the plot described in the prologue (fragment 2c); however, all the characters from *Anti-Elixir* (fragment 17) are missing. Their absence implies either that Boyle substituted a completely new ending, or that they had been renamed. Since Pyrophilus originally supplied the clinching argument, his absence is problematic, but his replacement may well be the new "illustrious Stranger" of fragment 11, for there must have been some important reason for his introduction. Extrapolating from the usual identity of "strangers" in alchemical literature, I suggest that in the new (and presently lost) conclusion this stranger may have revealed himself as a traveling adept and performed a projection before the assembly or contributed a clinching testimonial account similar to Pyrophilus's.

Was the Dialogue *Completed?*

It is likely that the *Dialogue* was finished at least into a form suitable for private circulation. One piece of evidence is offered by fragment 2c, a Latin translation of the prologue in the hand of Thomas Ramsay, a Lithuanian theologian who worked as a translator for Boyle from ca. 1685 to 1691.[21] Many of Ramsay's translations survive in the Boyle Papers, some preserving the text of otherwise lost English originals.[22] It is rather unlikely that Boyle would have sent off an unfinished work to be translated, so the existence of this translation suggests that the *Dialogue* was finished by the latter 1680s.

A second piece of evidence is supplied by the inventories that Boyle made

[21] The earliest record of Ramsay's work for Boyle is a receipt dated 1686, and after that time (excepting a trip back to Lithuania in 1689–1690) he worked for Boyle until 1691, as witnessed by four surviving letters—BP, 41:123; BL, 5:1–8. Ramsay was in England as early as 1682, when he published *Climax panegyrica vitae Jacobi Eboraci et Albaniae ducis* in London. He continued his bloated style (equally evident in his letters to Boyle) in two other publications—*Apostrophe ad Idaliam matrem de redivivo . . . Caesare Carolo II* (1685), and *Elogium faelicis fati Britanni sub auspiciis Gulielmi III* (1690). All the copies of these works that I have inspected bear corrections in the hand of the author, and the British Library's copy of *Climax* is autographed. Hunter (*Guide,* xxxvi) suggests that Ramsay began working for Boyle prior to 1677 on the basis of a translation at BP, 25:19–50; this translation bears corrections in the hand of Henry Oldenburg, who died in 1677. Prof. Hunter and I have since agreed that this manuscript is not in Ramsay's hand.

[22] For example, the richly alchemical segment possibly belonging to an extended *Usefulnesse* (BP 24: 399–417) and "On the Mechanical Production of Light" mentioned in a catalog drawn up by Boyle on 3 July 1691 (BP, 36:72). Ramsay's translation *Commentarii experimentales de mechanica productione lucis* survives at BP, 20:35–88. These texts are scheduled to be published in the *Works* now in preparation by Hunter and Davis.

of his writings. A list dated 10 July 1684 records "About the Generation and Transmutation of Metalls, a red cover in folio mark'd A" among the contents of "Box B."[23] Its enclosure in "a red cover" implies that the work had progressed beyond the composition stage, which would have left it in scattered sheets or notebook drafts. Although appearance on such a list does not automatically imply completion, many other items on this inventory are candidly described as "loose papers," "collections," "notes," or "imperfect," whereas the dialogue is not. The *Dialogue* is recorded again on 30 December 1690, and the last notice of "Conferences about Melioration & Transmutation of Mettals" occurs in a list entitled "Mr. Boyle's Philosophical Writings not yet printed, set down July the 3d 1691."[24]

Boyle's *Paralipomena* provides a third piece of evidence. The term *paralipomena* was classically applied to the third and fourth Books of Kings, which, considered apocryphal, were appended to the canonical text as supplements. The word in the seventeenth century had just this connotation of a supplement or appendix to a previous work.[25] Boyle used the word to describe writings that formed an appendix to his completed works—in one case, a bundle of sheets that were mislaid and not properly spliced into the press copy of a treatise; in another, the topics that arose as second thoughts to an already-written treatise.[26] Boyle kept a special collection entitled *Paralipomena,* described as "a Bundle of loos[e] Papers, which I had laid together, as they came to hand with a prospect that they might serve as *Paralipomena* or Supplements to Entire Tracts, whether publishd or not."[27] Boyle catalogs twenty-two chapters to his *Paralipomena,* and the eighth comprises "Particulars referable to the Dialogue on the Melioration and Transmutation of Metals," implying that the *Dialogue* was now an "Entire Tract." [28]

This ancillary material may be the "Severall Papers concerning Transmutation of Metalls, Quarto red cover strings sad red," listed in the 10 July 1684 inventory.[29] These same materials probably also constitute part of the

[23] BP, 36:119–20 and 121–22.

[24] BP, 36:69; this document bears the partial date "Decem. the 30" but the year can be confidently added owing to the presence of the first draft of this list in RSMS, 186:33*v*–34*v,* a notebook that also contains (fol. 4) the errata to the published *Medicina hydrostatica,* which appeared in 1690. BP, 36:72, 35:187; and RSMS, 186:86*v*–88.

[25] Posthumous *Paralipomena* were published, for example, under the name of the iatrochemist Daniel Sennert (Wittemberg, 1642), and Henry More published a defense of his writings as *Paralipomena prophetica* (London, 1685). Boyle refers to Sennert's *Paralipomena* in *Usefulnesse,* pt. 2, 85.

[26] RSMS, 189:119*v*–21; 198:114. *Excellency of the Mechanical Hypothesis* is described as "a kind of *Paralipomena* to his Dialogue [about the Requisites of a good Hypothesis]," publisher's advertisement, [i].

[27] Copy of a letter to the secretary of the Royal Society, BP, 35:196 (copy at 199; rejected drafts at BP, 38:34 and 36, and 14).

[28] Boyle was highly solicitious about the ordering of the *Paralipomena,* as no fewer than eleven copies of its catalog survive, and the advertisements attached to several copies make it seem that he wished the collection published. BP, 35:56, 171, 190, 191, and 213*v;* 36:4, 5, 36, 63, 92, 112; the most complete advertisement occurs at BP, 36:3.

[29] BP, 36:59–60.

"two 4to Past-bord Covers relating to the Dialogue about Mettals and some other things disputed of by Chymists" recorded in a list of collected papers and incomplete works made on 21 January 1691.[30] Note that these papers which I tentatively identify as parts of the *Paralipomena* are in *quarto,* while the *Dialogue on Transmutation* is explicitly described as in "a red cover in *folio*"—the two being clearly distinguished by their *separate listing* in the 10 July 1684 catalog. The existence of clearly subsidiary notes implies that the *Dialogue on Transmutation* had attained a state of sufficient completeness that further pertinent information could not be readily incorporated into it, as would still have been possible were it never more than only partially coherent sheets.

Fate of the Dialogue

Boyle began the *Dialogue on Transmutation* in the mid-1670s, published part as *Anti-Elixir* in 1678, revised and extended it in the early 1680s, and completed it in the mid- to late 1680s. But what of its fate? What became of the red folio marked A in box B? Boyle's own catalogs carry us to 3 July 1691, six months before his death; the next catalogs available are those of Henry Miles, dating from the early 1740s.[31] While Miles does list "de generat. & transmut. Metall. Folio"—which is Ramsay's translation (fragment 2c)—he makes no mention of the English dialogue.[32] Its absence suggests that the *Dialogue* was lost prior to the 1740s. In any event, the list of unpublished Boyle texts compiled in the 1744 *Works* lists "Origin of Minerals and Metals—Conferences about the Transmutation and Melioration of Metals" as lost.[33]

The *Dialogue*'s fate is unclear. It may have remained with any of the many people who handled Boyle's papers before the 1740s. John Locke, Edmund Dickinson, or Daniel Coxe, all of whom had strong alchemical interests and who were designated inspectors of Boyle's chemical papers, may have kept it. Peter Shaw also had access to part of Boyle's papers. It is also possible that it was neither accidentally lost nor taken by an interested person but

[30] The fair copy at BP, 36:121 and the first draft at RSMS, 186:36*v*–37, 38–39 bear the partial date 21 January, but the year can be inferred from the position of the draft in RSMS, 186; see n. 24.

[31] BP, 36:141–43, 163–64; attention was first called to this and the other Miles lists by Hunter, *Guide,* xx and xxvi.

[32] BP, 36:157–60*v* and 149. The final page of this list is apparently missing, as one of Miles's memoranda (149*v*) refers to a bundle of papers "markt Ma," while the list cites only bundles A–Z and Aa–La. One of Miles's labels (La) survives pasted down at BP, 26:100. Ramsay's translation is listed on 158, as well as on a scrap pasted down at BP, 26:84. These lists and the losses from the Boyle archive will be fully treated in Hunter and Principe, "The Lost Papers of Robert Boyle."

[33] *Works* 1:ccxxxvi–ccxxxviii. Hunter has wisely provided a caveat for those using this list; however, the foregoing examination of the MSS justifies its use in this case. See Hunter, *Guide,* xv.

rather was destroyed by an "apologist" who saw a work in favor of transmutation as a taint to Boyle's memory.[34]

Boyle may have circulated some copies among friends, and if so, one may possibly have survived somewhere, perhaps separated from the correct attribution. The surprising recent recoveries of the 1648 presentation copy of *Seraphic Love,* the original (and previously unknown) text of *Theodora,* and even of Boyle's student notebook provide some hope that further items may yet come to light.[35] At present, however, the *Dialogue on the Transmutation and Melioration of Metals* can be accessed only in the reconstructed form presented here.

Editorial Remarks and Conventions

Transcription and translation are two tasks more open to criticism than to compliment. In this edited text I have tried to balance the sometimes conflicting forces of truthfulness to the originals and service to the readers. Much of the success of a transcription lies in its ability to be read fluently without awkward hesitations that distract from the author's message; the editor should vanish silently into the background and allow the author to speak in as unencumbered a manner as possible. It seems pointless to publish a text so burdened with editorial conventions and symbols that the final form is at least as visually inaccessible to reader as the originals. In order to provide a readable text, I use only three simple and obvious editorial marks—the caret (ˆ) to mark authorial insertions, the square brackets ([ ]) to enclose editorial additions and original pagination, and a dashed line (----------) to show a small blank space left in the manuscript for the later addition of further information.

Out of respect to the author, an edited text ought to present the latest revised version available, eschewing both the antiquarian desire to preserve and present every scribble and jot of every draft, and the ahistorical temptation to conjecture unwarrantably about what the author *would have* done. I have therefore purposefully placed all deletions and interlineations in endnotes, where they are accessible to those interested in the development of the text, but out of the way of other readers. These are keyed to superscript *numbers.* Only editorial annotations are retained as footnotes, and these are keyed to superscript *letters.* Readers may ignore the numbered notes entirely unless they are seeking alternative readings. Scholars interested in the documents themselves can view them in the microfilm version available through University Publications of America.

I have preserved both the spelling and the capitalization of the originals. Only when the spelling is sufficiently odd to provoke a puzzled pause in the

[34] Hunter, "Dilemma of Biography," 129–33.

[35] Principe, "Style and Thought of Early Boyle," and "Newly-Discovered Boyle Documents"; Hunter, "A New Boyle Find."

reader have I felt it necessary to insert a judicious and clarifying letter, enclosed within square brackets to indicate its origin. The use of *sic* is restricted to clarifying situations where an original solecism might otherwise be thought a transcription error. The thorn (as in y^e) has been silently replaced by *th,* as would have been done at the printer's in the seventeenth century. Similarly, the usual superscripted short forms of *which, with, when,* and so forth have been silently expanded. Paleographic marks, such as the titulus, showing the omission of one or more letters (as in com̃on), have been silently expanded.

Since about half of the fragments are first dictation drafts, there is little punctuation to articulate Boyle's lengthy sentences. Thus to improve readability I have exercised editorial license to insert appropriate punctuation where it was needed to channel the flow of the prose. Like other editorial additions, these marks are enclosed in square brackets, which I trust the reader will not find difficult to read over without hesitation.[36]

[36] Useful comments on the editing of texts may be found in Hunter, "How to Edit a Seventeenth-Century Manuscript: Principles and Practice," *Seventeenth Century* 10 (1995): 277–310; and *Règles et recommendations pour les editions critiques* (Paris: Les Belles Lettres, 1972).

Robert Boyle's *Dialogue on the Transmutation and Melioration of Metals*

In the listing below, the following abbreviations are used: (d)—draft; (f)—fair; and (t)—translation.

Prologue

1—RSMS, 199:147	Hugh Greg
2a—RSMS, 198:145–44*v*	Hand A[a] (d)
2b—BP, 25:305	Robin Bacon (f)
2c—BP, 25:185–215	Thomas Ramsay (t)

First Part

3—BP, 38:160*r–v*	Robin Bacon and Thomas Smith (d)
4—BP, 25:287–88	Robin Bacon (f)
5—BP, 25:291–93	Robin Bacon (f)
6—BP, 25:299	Robin Bacon (f)
7—BP, 10:10*r–v*	Hand Z
8—BP, 25:303	Robin Bacon (d)
9—BP, 25:419–21	Robin Bacon (d)
10a—BP, 8:160–63	Hand B (d)
10b—BP, 8:164	Hand B (f)

Second Part

11—RSMS, 199:121	Hugh Greg (d)
12—RSMS, 199:25–27	(?) (d)
13—RSMS, 199:27*v*–9	Hugh Greg (d)
14—RSMS, 199:46–51	Hand Z (d)
15—RSMS, 199:52–55	Hand Z (d)
16—RSMS, 187:6*v*–8*v*	Robin Bacon (d)

Conclusion

17a—*An Historical Account of a Degradation of Gold* (London, 1678).

17b—BP, 8:171	Hand C (d)
17c—BP, 38:71–72	Hand C (d)
17d—BP, 10:159–60*v*	Hand C (d)
17e—BP, 31:237–74	Robin Bacon (t)

[a] I have here employed the alphabetical labels for hands of unknown amanuenses devised by Michael Hunter in his *Guide*. There are, however, two diversions from the published scheme. The hand labeled here as C was originally confused with Boyle's autograph (introduction to this appendix, n. 16) and is not the C of Hunter's list; the "old C" was sufficiently rare to be now left unnamed. The hand labeled here as Z is a previously unlabeled hand. These designations have been made in concert with Prof. Hunter, who will be producing a revised edition of his valuable *Guide* in due course.

Text of the *Dialogue*

Prologue

FRAGMENT 1 [Characters of the Dialogue; RSMS 199:197 [b]]

Friends to Chymistry nam'd in the Dialogue

Zosimus
Philoponus
Pyrocles
Arnoldus[1]
Aemilius
Bernardus
Guido
Philetas

Adversaries to Chymistry nam'd in the Dialogue

Erastus
Guibertus
Themistius
Julius
Crates
Cleanthes

Persons engag'd in neither party

Eleutherius
Sextus
Eugenius
Heliodorus

[b] The "friends" Arnoldus, Aemilius, Bernardus, and Guido, the "adversaries" Guibertus, Themistius, Julius, and Crates, and the "unengag'd" Sextus do not appear in any fragments.

FRAGMENT 2c [Thomas Ramsay's Latin translation of fragments 2a–b and of a lost English original; BP, 25:185–215]

Summa capita
DIALOGI de Generatione & Transmutatione Metallorum.

Occasio Colloqui, et quaedam[2] Discursus praeliminares.

Capita summa, contra possibilitatem Transmutationis, quorum praecipua sunt haec; alterum[3] quod viae probandi ^illam^ sint minus competentes: alterum quod veritas eius incredibilis sit reddita aeque solidarum objectionum vi, ac ob insuperabiles difficultates.

Primum generale caput haec particularia continet; quod primo, Authoritatis probandae Transmutationi adhibitae, partim sint obscurae, partim suspectae, aut alio aliquo modo insufficientes, secundo quod philosophica[4] fundamenta quae illi probandae substruuntur, vel dubia sint, vel praecaria, vel falsa qualia sunt Metallorum omnium ad aurum tendentia.

Tincturam sive animam auri transferri (*sive transplantari*).

Auri reliquorumque Metallorum, quô queant propagari, Semen. Huic adnectenda est Authorum inter se dissensio[5] (quoad theoriam et praxin) et modorum quibus Transmutationem adstruere nituitur obscuritas.

Tandem[6] quod Pretensi Chymici Philosophici, ne intelligibilem quidem generationes metallorum Rationes reddiderint.

Ad secundum generale caput, referri potest, durabilis & obstinato metallorum natura.

Quâ cum operationes perfici debeat celeritas.

Exigua agentes ad patiens proportio; ut et diversae, transmutationem postulantes qualitates.

[186] Ab altera parte Defendens respondet; Affirmativa probari posse, quantum ad rei possibilitatem, objectiones difficultatesque ab Adversario propositas dilvenda.

Ut et, quod eius realitas demonstrari queat, per particulares historias, aliàque testimonia.

Nec non quod fulciri queat, rerum in natura et actu analogiâ.[7]

Posthaec Societas Pyrophilum adventare edocta, aliquandin subsistit, spatium ei, omnia penitius, quae transacta fuere investigandi, ut et quae ad illa reponere dignaretur, concessura.

FRAGMENT 2a/b [RSMS, 198:145–44*v*, earlier draft;[d] BP, 25:305]

The Heads of the Dialogue about the Transmutation and Generation of Mettals.

The occasion of the conference and some preluminary discourse.

The Heads of the Arguments against the possibility of transmutation whereof the two chief are these: *first* that the wayes of probation are incompetent, and *secondly* that the truth of it, is made incredible by weighty Objections & scarce difficulties superable.

I. The first generall head comprises these particular considerations;

(I)[52] first that the Authorities alleag'd to prove Transmutation are either dark or suspicious or some other way insufficient.

(II) Next that the Philosophical grounds pretended for it, are either dubious (or precarious) or false, such as are,

(1) The Tendency of all Mettals to Gold.

(2) The tincture or Anima of Gold to be transfer'd.

(3) The Semen of Gold and other Mettals whereby they may be propagated.

(4) To which is added, The Disagreement of the Writers (both as to Theory and practice) and the unintelligiblenes of the wayes, which they propose, to explicate Transmutation.

(III) Lastly that the pretended Chymical Philosophers have not given so much as an intelligible account of the Generation of Mettals.

II.[53] To the second general head, may be refer'd

(I) the durable and obstinate nature of Mettals.

(II) The celerity wherewith the Operation must be wrought.

(III) The small proportion of the Agent to the Patient. And the several qualities that must be changed.

I. On the other side the Respondent answers, that the affirmative may be made out, as to the *possibility* of the thing by removing the Objections & difficulties propos'd by the Adversary.

II. And the *reality* of the thing may be prov'd, by particular Histories and other testimonies.

III. And the whole may be countenanced by things Analogicall in nature or Art.

After which Pyrophilus being related, to come in, the company pauses a while, to give him time to inform himself, of what had passed, and to consider what he would say thereupon.

[d] Margin: "T[ranscri]b'd."

[187, Ramsay's Latin continues]

De generatione & transmutatione Metallorum Dialogus
Ad ----------[c]

Nobilissime Domine,

Quanquam per omnia nequeam Tuam curiositatem explere quâ scire desideras, quae in hodierno nostro congressu (ubi expetebaris pariter ac desiderabaris magnopere) transacta fuere, ut pote sub initium absens, tamen sat mature accessi, ad audiendum, quicquid hujus diei Colloquium, in se continebat momenti, et fere, totum illud quod ex condicto temporis supererat, poposcit. Id quod mihi velim, Discursus est quidam, cuius occasionem et praecipua capita, exequar sequentibus.

Postquam a Societate, aliquid temporis, rebus minoris momenti insumptum fuisset, Philetas arreptâ, ab artificialibus quibusdam, immediate ante exhibitis raritatibus occasione, auri frustrum in medium protulit, quod ab illo a quo nactus fuerat, argentum olim extitisse asseveratum [189] est, atque deinceps per Projectionem transmutatum.

Vixdum autem Praeses cum paucis quibusdam proximè sibi accumbentibus, a contemplando illo[8] destiterat, cum consurgens Erastus, propiusque ad Sellam admotus, Si Philetas, inquit, certam nobis viam indicare posset, quâ aurum per Projectionem factum, ab altero discernere valeremus, impetrarem forte a me, ut munus Eius, Raritatis titulo insigniri paterer, multoque magis,[9] (quia clarum artis specimen) aestimandae quam quidem nunc, quando illud, quà praestantissimum quod a natura producitur metallum intueor. Deficiente vero tali Lydio lapide, ad quem auri (ut ajunt) Philosophici a natu^ra^li, queat expendi differentia; quisquis sustineat, frustrum auri, (quod per projectionem factum declaravit alter) obtrudere velut argumentum, pro possibilitate, transmutandi alia metalla in aurum, nae ille, hos quibuscum transigit, nimium credulos supponere censebitur, cum apud nullos praeterea, possit tam elumbis, locum habere argutatis. Quodsi eum qui retulit, probitate non minus ac peritia conspicuum deprehendero, et qui ^id^ de certa scientia noverit, quod loquitur fateor facile me verbis[10] fidem adhibiturum, etiam auro Ipsius nunquam viso; quemadmodum adhibuissem Philetae, si se oculatum hujusce Transmutationis testem asseverasset. Quodsi vero vel fidem eius vel peritiam vocem in dubium, aspectus producti ab Ipso auri, ne γρὺ quidem ex mea diffidentia minuerit, cum Impostori nil aeque facile sit, quam frustrum aliquod auri, simulare elaboratum per artem, quod interea probum naturale ac ordinarium esse noverit, et per consequens (ut pote genuinum) aptum ad sustinendas omnes consuetas probandi rationes, in[11] investigando hoc metallo adhiberi solitas. Verumenimvero, nihil mirum est, ad unum [191] figmentum in arce collocandum, alterius opus esse satellitio: remque tàm, si minus impossibilem at certe in-

[c] The space left for the name of the addressee remains blank.

[Editor's translation of fragment 2c]

A Dialogue concerning the Generation and Transmutation of Metals
to ----------

Most noble Sir,

Although I may well be unable to satisfy fully your curiosity wherewith you desire to know what took place in our meeting today (where you were as greatly besought as desired) since I was absent at the beginning; still I did arrive soon enough thereafter to hear whatever today's discussion contained of importance, and nearly everything that it desired which exceeded the set time. What I would wish for myself is a particular discourse whose occasion and main heads I shall go through in the following.

After the Society had spent some time in lesser matters, Philetas, taking the occasion from certain rarities of artifice that had just then been exhibited, brought forth a bit of gold that was asserted by the person from whom it had been obtained to have once been silver, and thereafter transmuted into gold by projection.

Scarcely had the president and those few seated next to him finished looking at the piece of gold, when Erastus arose, moved closer to the Chair, and said: "If Philetas can indicate to us a sure way whereby we might have the power to distinguish between gold made by projection and other gold, I would perhaps give in and suffer that his gift be distinguished with the title of a rarity, and it would then be much more esteemed (since it would be evidence of the Art) than it is now, when I look upon it only as a most excellent metal produced by nature. Indeed, since we lack such a touchstone whereby the difference between Philosophical gold (as they call it) and natural gold could be examined, whosoever maintains that a piece of gold (which someone else declared to have been made by projection) has the force of an argument on behalf of the possibility of transmuting other metals into gold will indeed be thought to suppose those with whom he is dealing to be too credulous, since with no other kind of person could anything so weak take the place of an argument. Now if I found that he who gives the report were distinguished by no little probity and experience, and knew what he said with sound knowledge, I confess that I would then easily credit his words, even without ever seeing his gold; likewise I would credit Philetas if he had asserted himself to be an eyewitness to this transmutation. Yet were I to call into question either his fidelity or his skillfulness, the sight of the gold he brought forward would not diminish one iota of my diffidence, since it would be just as easy for an impostor to pretend that any bit of gold was made by art, because at the same time he would know that the gold is good, natural, and ordinary, and consequently (being genuine) fit to withstand all the usual means of testing generally employed to try that metal. But indeed, it is not surprising that for one fiction to hold sway, another must be its guard: and so a thing so impossible and surely unbelievable (as is the

credibilem, (prout est lapis Philosophicus,) non aliis posse, quam debilibus atque precariis defendi argumentis.

Quidam ex Societate, melius erga Chymiam affecti, tametsi priorem dictorum *Simplicii* [Margin: NB. quia paulo ante dicebatur Erastus] partem non improbarent, posteriorem tamen tulerunt graviter, satisque alacriter contra ipsum invehi coeperunt, quod Lapidem Philosophorum, ipsamque metallorum transmutationem, tanquam chimaericam, imo etiam qua impossibilem traduxisset. Prudens vero et moderatissimus Heliodorus, occursurus his animi fervoribus, quos mordax Simplicii Oratio excitare potuisset, pro solita sibi, minimeque affectata gravitate, ad societatem conversus; Quando quidem (inquit) laudabilis apud Nobilissimam hanc societatem mos obtinuit, ut magis extimularent homines, ad quicquid ipsis curiosum visum est adferendum, quam a tali officiositate deterrerent, meâ sententiâ, benevolâ mente, excipiendum est, oblatum a Phileta frustrum auri, quod per Projectionem elaboratum fuisse, probabiliter edoctus est. Quanquam enim libenter concedam, primam discursus Simpliciani partem, satis congruam rationi prae se ferre objectionem, tamen istud concedo, solummodo in quantum tali innititur suppositioni, aut ad minus eam comprehendit,[12] cujus veritatem, jure quis queat vocare in dubium. Equidem negari nequit, argumentationem eius, sat firmam futuram, quodsi nulla possit inter factitium et naturale aurum [193] deprehendi differentia: perpendendum tamen ab altera parte est, non ess[e] ἀδύνατον, quominus aurum Philosophicum, omnibus illis qualitatibus sit praeditum, quas consensus vulgi, ad aliquam metalli portionem auri nomine investiendam, requisitas et sufficientes judicavit aut quominus capax sit, aurifabrilis aequè ac monetariae officinae tractationem sustinere.[13] Neque illud ipsum impedit, ne praeterea habeat aliquam proprietatem (aut fortasse una plures) per quam queat ab auro ordinario discriminari velut v. g. per mollitiem vel ductilitatem, specificam item gravitatem, aut per aliquam medicam Virtutem, item per tincturae exuberantiam, quâ queat figere exiguam aliquam plumbi particulam, suae cupellationi adhibitam: brevibus, poterit aut in his aut in aliis qualitatibus, ad quarum singulas periensendas, minus mihi artis suppetit, ab auro vulgari differre: prout quamvis diversis in locis, natura nobis suppeditet Vitriolum, non nos tamen latet, Arte quoque illud posse confici; ac licet utrumque sit verum, factitium tamen praeter facultates[14] Vitriolo essentiales, poterit quasdam habere adventitias[15]; prout Vitriolum artificialiter ex cupro confectum, adhibito idoneo acido menstruo, magis potest (expertus loquor) reddi fusile,[16] magisque prae altero per liquorem dissolubile; idque verisimiliter, ob insignem proportionem metallorum, quae intra se,[17] multe in specie graviora comprehendit, quam vero Vitriolum naturale, terrestrem partem metallicae permixtam, communiter habere [195] solitum. Quantum ad me igitur, ^(pergit Heliodorus)^ gratâ mente, Philetae Nobilissimi munus, suspiciendum censeo, atque in Cimelio nostro reponendum, ut semper in promptu sit, sicubi explorandi illud, talis sese obtulerit occasio, quae ad detegenda eius particularia (dummodo quaedam habeat) maxime oppor-

Philosophical Stone) cannot be defended save with such weak and begging arguments."

Certain others of the Society who were better inclined toward chymistry, though they might not condemn the first part of the sayings of Simplicius [Ramsay's marginal note: note well, for just a moment ago he was called Erastus[e]], still they bore the latter part badly, and so they began to inveigh against him vehemently for asserting that the Philosophers' Stone and the transmutation of metals itself was merely chimerical, and indeed an impossibility. Then the prudent and most temperate Heliodorus, seeking to avert this raging ardor of the spirit that the biting speech of Simplicius had been able to stir up, turned to the Society with his accustomed and wholly unaffected seriousness and said: "Since this Most Noble Society has indeed maintained a laudable custom that they should spur men on toward whatever curious sight be brought before them rather than deter them from such obligation, in my opinion, the piece of gold brought forward by Philetas (which he has quite probably been told was made by projection) should be accepted with a benevolent mind. For although I may freely concede that the first part of Simplicius's discourse presents an objection acceptable enough to reason, still this I concede only insofar as it rests upon a supposition (or at least contains such) whose truth no one could justly call into question. Indeed, it cannot be denied that his argument shall be quite sound enough if no difference can in fact be found between factitious and natural gold. Yet it is to be considered, on the other hand, that it is not an impossibility that philosophical gold is possessed of all those qualities which the generality of common people judge requisite and sufficient for any portion of metal to be invested with the name of gold, or that the gold be able to withstand both a goldsmith's and a mint-master's examen; nor can it be denied that it may have some further property (or perhaps more than one) whereby it may be distinguished from ordinary gold, as, for example, by softness or ductility, or by specific gravity, or by some medicinal virtue, or by an excess of tincture whereby it may fix some small part of the lead added to it in cupellation.[f] In short, it may differ from common gold either in these or in other qualities the individual testing of which seems to me no great art. For in the same way, in a great many places, nature affords us vitriol, and it is not hidden from us that vitriol can also be made by art.[g] While both vitriols are indeed true, yet the factitious can have certain particularities beyond those properties which are essential to vitriol. For example, the vitriol made artificially from copper by the addition of a suitable acid menstruum is more able to be made fusible (I speak knowingly) and is more soluble in liquors than the other; and likewise, on account of the greater proportion of metals that it contains, the artificial vitriol has a much greater specific gravity than

[e] The interchange between the names Erastus and Simplicius persists throughout the fragments.

[f] On this phenomenon, see 94–95.

[g] *Vitriol* is a generic term for either iron or copper sulphate.

tuna videatur. Atque hoc est Viri Nobilissimi, quod mihi de anteriore discursus Simplicii parte restabat dicendum. Quantum enim ad posteriorem, in qua nobis ob oculos posuit, quod Illis lapidem Philosophicum chimaerae nomine insignire placeat, satis superque percipio, diversos Vestrum, in hanc rem animadvertendi flagrare desiderio. Porro quia quaestio, et eximia admodum et non ultimi momenti est, hoc praesertim in saeculo quô ars Spagyrica non excolitur solum verum etiam indies perficitur: et quia haec controversia de Transmutatione metallorum Viliorum in nobiliora, praecipue per Projectionem, non fuit (quoad novi)[18], ab Authoribus, modernae Philosophiae principiis imbutis ^ventilata^; meô judiciô, quandoquidem jam quaestionem hanc movimus, operae pretium erit, ut eam aequa lance trutinemus.[19] Quia vero vix satis temporis ad exantlandam tantam difficultatem nobis suppetit cum Venia vestra Viri Nobilissimi, insinuare Vobis duo mihi liceat, aeque ad regularem eius tractationem, ac citiorem absolutionem spectantia. Horum primum est; Ut ad evitandas incompositas, & supervacaneas tautologias, quaestio ab utraque parte, ventiletur, per singulos, quibus alii qui in eadem sunt sententia, tanquam auxiliares copiae intermisceri poterunt, si quo casu explicandum habeant [197] quod ab Illis praetermissum videbatur. Nollem tamen vel Eleutherio, vel aliis eiusdem cum eo indolis praescribere,[20] quo minus tertiam quasi Sectam conficiant, si forte parum conveniens[21] videbitur, hac occasione (ut solent in multis aliis) constanter alterutri *sese* parti addicere,[22] sed vel susccenturiare[23], vel respondere ad argumenta quae ipsis pro- vel improbabuntur, a quacunque tandem parte fuerint proposita. Exposito primo consilio pergo ad alterum. Hìc autem vellem quemlibet vestrum meminisse, Societatem hanc non Scholasticorum sed Phisycorum esse, proinde parum decorum in Disceptatoribus futurum, tempus et horas, in Vulgaribus argumentis, subnixis vel Aristotelis vel aliorum, minus in metallinis[24] versatorum experimentis authoritate, illoroumve [*sic*] a quibus formarum substantialium supponitur existentia, sive tandem aliis vulgaribus Scholarum doctrinis, quas saepius, Societas haec Nobilissima, aut obscuritatis, aut precariatus, aut etiam falsibilitatis coarguit, conterere: spero igitur ad talia se accincturos[25] argumenta, quae Ingenuos[26] & Philosophos deceant; verbo usuros probationibus ab ipsius rei quae sub considerationem venit, natura petitis.

Maxima pars Auditorum, perspectâ, factae a Praeside admonitionis aequitate non minus ac oportunitate, confestim quorum intererat, loca movere coeperunt. Lapidistae utique (si sic loqui liceat) qui tale quid, ut est Lapis Philosophicus dari contendunt, ad unam sese Camerae partem contulerunt,

does the natural vitriol that generally contains an earthy part mingled with the metalline. Therefore, for my part (proceeds Heliodorus) I am of the opinion that the offering of most noble Philetas be taken up with a gracious mind and placed in our Repository so that it may be ever available, if anywhere such an occasion for examining it that may seem most opportune for uncovering its peculiarities (if it have any) may offer itself. And this, most noble gentlemen, is all that remained for me to say about the first part of Simplicius's discourse. Now, as for the latter part, in which he laid before our eyes that he is pleased to mark the Philosophical Stone with the name of a chimera, I perceived quite clearly enough that some of you burn with the desire of addressing this matter. Furthermore, because the question is altogether worthy and of no trifling importance, especially in this age when the Spagyric Art is not only cultivated but completed daily, and also because this controversy over the transmutation of the baser metals into the more noble ones, especially by projection, has not been (as far as I know) aired by authors versed in the principles of the modern philosophy, then in my judgment, seeing that we have moved now to this question, it shall be worth the effort to weigh it in a fair balance. But because there is hardly enough time to exhaust such a difficult problem, with your indulgence, most noble gentlemen, permit me to suggest two things to you, both for the orderly treatment of the subject and for the quicker discharge of it. The first of them is this: In order to avoid disorderly and empty repetitions, let the question be aired on each side by one person, on whose behalf others of the same opinion can be sent in like relief troops if by some chance there are things to be explained that seem to have been passed over by those who went before. Though I do not wish to compel either Eleutherius or others of the same disposition into making a sort of a third sect, yet if perchance it seems as little appropriate on this occasion (as has been so on many others) for them to devote themselves steadily to one or the other party, then let them either act as substitutes or respond to the arguments that they accept or reject from whichever party the arguments were proposed. Having thus expressed my first piece of advice I shall proceed to the other. Here I should like each of you to remember that this Society is not one of Scholastics but of Physicists; thus it will be unfitting for the disputators to spend time and hours in vulgar arguments supported by the authority of Aristotle or others little versed in metalline experiments, or of those who suppose the existence of substantial forms, or finally with other vulgar doctrines of the Schools that this most noble Society has very often refuted as obscure, begging, or false. I hope therefore that the parties will restrict themselves to such arguments as are fitting for Virtuosi and Philosophers; in a word, that they will use proofs drawn from the nature of the thing under consideration."

Having perceived the justness and advantage of the advice given by the President, most of the Assembly immediately began to change places as was necessary: The Lapidists (if one may so call them), who contend that the existence of the Philosophical Stone is to be granted, collected themselves in

relictô Antilapidistis, alterius partis beneficio, ut et aliquo commodo inter utrosque interstitio, ubi Eleutherius aliique magis ad neutralitatem proclives residerent. Heliodorus immediate post factum schisma[27] priorem Sectam [199] unanimiter in Zosimum oculos conjecisse, quo ipso, Eum suum Prolocutorem visa est designare, eorumque adversarios, de evocando ad causam suam in arce collocandam[28] Erasto, consensisse percipiens, Virtuosorum horum novissimum, a cujus partibus majore numero certabatur,[29] eam ipsam ob rationem ad certamen evocat, quod, quantum memoriâ assequi possum, sequenti stylo aggressus est.

Nisi Viri Nobilissimi, altiùs perpenderem, quantopere naturâ proclives simus ad credendum facile, quod avidè exoptamus, ut et, quod insignis hujusce Quintessentiae[30] jactatores, duo, summis omnium fere votis expetita, Opes videlicet & sanitatem polliceantur, mirarer sanè tot Viros, non ultimô alias ingeniô, tam facile ad fidem aeque ac spem rei nullatenus probabilis, qualis est, quem lapidem Philosophicum dicere placuit, pellici potuisse. Neque enim tale quid dari concedere possumus, nisi simul supposuerimus, granum pulveris nescio qualis, par esse (et quidem in momento) quod centies, millies, imo forte decies millies, excendentem se pondere fluidam, albam, crudam, volatilemque hydrargiri massam, in solidum, malleabile, recoctum ac flavum aurum convertat. Contra hujus tam improbabilis, et nisi fallar impossibilis rei existentiam, possem plurimorum Doctorum Virorum authoritatem, omnium fere aetatum & gentium adducere, quorum plerique non sine applausu caeterorum, conceptis Verbis, inanem hujus Alchumisticae praetensionis fastum condemnarunt. Attamen (pergit Erastus) non utar hac lege: tot enim Philosophorum, Scholasticorum, Jurisconsultorumve authoritatibus, obruere adversarios, haud habeo necessum, ubi ratio plura quam sufficit mihi suppeditat argumenta: brevitatis igitur [201] a laudabili more nostro nobis insinuatae memor, plurima praetermittam, quae quantumlibet satis urgeantur confidenter, et fortasse etiam a viris literatis, tamen non satis firmo subnixa talo existimo, contentus pauca quaedam proponere, quae ad hujus Auditorii ingenium quam proxime accedere videbuntur. Methodi vero gratia ad duo summa capita, illa reducere constitui.

Primum est; Modos probandi, Lapidistis usitatos esse minus competentes.[31] Alterum; Operationes Lapidi Philosophico attributas, solidarum obiectionum vi, non minus ac ob insuperabiles difficultates, factas esse incredibiles.

Primum, horum generalium capitum, multas in se particulares considerationes includit, quas breviter prout sese menti obtulerint attingam.

one part of the room, leaving the use of the other part to the anti-Lapidists, so that with some convenience there was also left a space between the two parties where Eleutherius and the others who were more inclined to neutrality might take their places. Immediately after the division was made, perceiving that the members of the former sect unanimously cast their eyes upon Zosimus, whereby they seemed to designate him as their spokesman, and that their adversaries agreed on calling Erastus forth to carry their cause to victory, Heliodorus summoned the latter of these Virtuosi (because a greater number stood on his side)[h] to the debate, which, so far as I can recall, commenced in the following way.

"Most noble Gentlemen, had I not quite deeply considered how much we are naturally inclined readily to believe those things for which we avidly hope (like those two things besought with the greatest prayers of all—namely wealth and health—which the boasters of this famous Quintessence promise) I would indeed marvel that so many men, not otherwise gullible, could have been so easily cajoled into having both faith and hope in so utterly improbable a thing as that which is pleased to be called the Philosophical Stone. And we cannot grant such a Stone unless we likewise suppose that a grain of I know not what sort of a powder is sufficient to convert (and that in a moment) a mass of fluid, white, crude, volatile quicksilver, which exceeds it a hundred, a thousand, indeed perhaps ten thousand times in weight, into solid, malleable, highly refined, and yellow gold. Against the existence of this so very improbable and (if I mistake not) impossible thing, I would be able to adduce the authority of many learned men of nearly every age and nation, who condemn in writing (generally not without the applause of others) the empty pride of this alchemical pretension. In spite of all that (proceeds Erastus), I will not use this method; for I scarcely have the need to overwhelm my adversaries with the authority of so many philosophers, Scholastics, or lawyers when reason more than suffices to supply me with arguments. Therefore, since I recall the brevity suggested to us by our laudable custom, let me pass over those many things (which, however confidently enough they may be urged, and even perhaps by learned men, nonetheless I do not think it is sufficient that they be supported by such a fact), and be content to propose a few things that will seem to approach as near as can be to the acuteness of this Assembly. Indeed, for the sake of method I have resolved to reduce them to two main points.

The first is this: That the means of proof used by the Lapidists are not competent; and the other: That the operations attributed to the Philosophical Stone are rendered unbelievable by the strength of solid objections as well as by insurmountable difficulties."

"The first of these general points includes many particular considerations that I shall touch upon as they come to mind."

[h] Fragment 1, however, lists more "Friends of Chymistry" than adversaries.

Imprimis autem indicandum mihi est, quod non obstante ingenti Chymicorum strepitu, quem in suis historicis probationibus excitant de Metallorum per Projectionem Transmutatione, longe absim ab hac fide, tam a vero absimilia aut vera esse, aut sufficienti testimonio comprobata.

Commendantur autem favori nostro,[32] partim talium Chymicorum authoritate qui se *Adeptos* jactitant, partim a diversis relationibus quas de Transmutationibus, quasi ab hoc alterove, tali loco aut tali tempore factis accepimus. Sed neutrum horum argumentorum aliquid apud me ponderis obtinuit.[33]

Atque quantum ad prius; Cur ego crederem aliquid, aut penitus impossibile, aut ad minus stupendum et vix credibile, esse verum, eoquod ab uno alterove obscuri nominis Authore asseritur, licet illi forte haec, ut in Theatro Chymico compingeretur obtigit felicitas, non video. Aeque parum mihi constat Gebberum Adeptum lapidem philosophicum,[34] ac Regem Arabiae fuisse. Quis me [203] certiorem reddet, Basilium Valentinum, Roggerum Baconum, aut multos alios ^e Caenobitis illis^ qui inter Hermeticos Philosophos recensentur,[35] invitô paupertatis votô, opum plusquam Regiarum possessores fuisse, aut Paracelsum habuisse lapidem Philosophicum, quo tamen non obstante nescio quot paginas impenderet, ut in typis,[36] contra Germanicarum Civitatum aliquam, ob non persolutum sibi salarium, gravius debacharetur. Equidem si hi Viri in suis scriptis, inventa nobis vel experimenta quaedam, majoris precii, aut usus non ultimi, communicassent, tum sane, sine tanta praesumptione, fidem sibi Lectoris polliceri possent, cum etiam talia se habere secreta quae non revelarunt profiterentur. Iam vero, ut, postquam quis universa, Authorum horum alicujus, opera evoluit, neque ullum forte offendit experimentum, quod vel curiositatis vel utilitatis prae se ferret speciem, ut inquam hi nihilominus praesumerent, talem, se Nobilissimi hujus Arcani putare Dominos, quem non convicerunt quod majus quid quam triviale saperent, magnae id ipsorum foret stultitiae; longè autem majoris esset in illo, qui ineptâ credulitate, eorum praesumptioni patrocinaretur. Caeterum si quis tam facile sibi imponi patiatur, ut credat aliquem habere Lapidem Philosophicum, quia is in scriptis coram orbe se habere venditat, nihil impediet quo minus fidem adhibeat Artephio, graviter praetendenti, se non minus esse Adeptum ac ullum, quando, quod millesimum annum ageret, ope huius Philosophicae medicinae, dum librum suum scriberet profitetur.

"Particularly, I must declare that notwithstanding the enormous din that the chymists stir up in their historical arguments for the transmutation of metals by projection, I am far from believing that these accounts are either true or verified by sufficient testimony."

"They are commended to our favor partially by the authority of such chymists as style themselves adepti, and partially by the various relations of transmutations that we receive from this person or that, which were performed in such-and-such a place at such-and-such a time. But neither of these types of argument holds any weight with me."

"And furthermore, I do not see why I should believe anything to be true that is either entirely impossible or at least stupendous and scarcely credible just because it is asserted by one or another author with some obscure name that might perhaps have had the felicity to conceal what he contrived in the *Theatrum chymicum.*[i] It is as little evident to me that Geber was an adept of the Philosophical Stone as it is that he was the king of Arabia.[j] The more certainty with which anyone repeats to me that Basil Valentine, Roger Bacon, or the many other monks[k] who are counted among Hermetic Philosophers were, in spite of their vows of poverty, possessors of riches greater than that of kings, or that Paracelsus had the Philosophical Stone, in spite of which he nonetheless threatens (in I know not how many pages of print) one of the German States that did not pay up his salary, the more seriously he is deluded.[l] Indeed, had these men communicated to us in their writings some certain invention or experiment of greater value or of nontrivial use, then they could reasonably expect some belief from the reader, without so much presumption as they do when they declare that they have some secrets which they do not reveal. Now indeed, after one has gone through all the works of any one of these authors, and does not meet with a single experiment that presents any appearance of either curiosity or utility, if, I say, the authors still presume that such a reader will think them masters of this most noble Arcanum, though they have not convinced him that they know anything more than trivial, it is to their great folly; and there is far more foolishness in

[i] *Theatrum chemicum,* ed. Lazarus Zetzner, 6 vols. (Strasbourg, 1659–1661): a voluminous collection of alchemical tracts, many of them anonymous or pseudonymous.

[j] The *Summa perfectionis* is of thirteenth-century Latin origin, although its author was long identified as Jābir ibn-Ḥayyān. In many editions and references Geber was accorded the title "King of the Arabs."

[k] The very influential Basil Valentine corpus presents its author as a Benedictine monk of the early fifteenth century; they were written, however, after ca. 1600 and probably by several different authors, most notably Johann Thoelde. Roger Bacon (1214?-1294) was a Franciscan.

[l] Probably a reference to the story that in 1528, Paracelsus, then in Basel, was promised a large fee by a noble canon of Liechtenfels who suffered with severe stomach cramps, if he could cure him. With three pills of his laudanum, Paracelsus cured the canon, who then refused to pay, saying that the cure was too easy for such a high fee, and jested that Paracelsus had given him nothing but "three mouse-turds." Paracelsus brought a legal suit, but the judge awarded him only a trifling recompense, which put Paracelsus into a vindictive rage, and he soon thereafter left Basel for good. See Boerhaave, *Elementa chemiae,* 22.

Altera etiam via est, quâ Spagyristarum nonnulli, Transmutabilitatem inferiorum metallorum in aurum probatam dare attentarunt. Notio cui hi innituntur, haec esse videtur. Supponunt auro Vulgari, inesse certam partem sive portionem materiae, longè prae caeteris [205] nobiliorem, quâ (licet admodum sit exigua) separatâ, reliqua massa non colore solum, verum ipsâ etiam naturâ exuitur, aurumque esse desinit, in aliam degenerando substantiam, argentum sit illa licet, vel aliud quoddam album metallum, aut minerale fortè, cui nondum est, proprium nomen assignatum. Porro hi, (de quibus ajo) Chymici, supponunt separatam hanc seu extractam substantiam, quam veram auri tincturam, vel animam potius existimant, a crasso corpore expeditam, et artificiosè, alteri alicui conjugatam metallo, aptam esse quae vi aureae in se contentae naturae, aeque ac ingentis subtilitatis atque agilitatis[37] acquisitae per praeparationem, per quam manicis[38] compedibusque soluta ˆestˆ, invadat alia metalla et penetret, omniaque quibus ut genuinum aurum evaderent destituebantur, illis impertiat.

Quamvis autem haec, inter omnes quas mihi videre contigit Lapidistarum hypotheses, maxime plausibilis videatur, vereor tamen ne sit magis speciosa quam vera. Nam quod quaeso intelligibile, aut ad praxin reducibile, hucusque ab illis edocti fuimus experimentum, cujus exploratione factâ nobis persuadere possemus, in auro, aliquam portionem et nobiliorem multò, et ipsâ metalli mole puriorem, realiter existere; aut florem hunc animamve auri, actû esse a reliqua massa separabilem, cum in quantum eousque innotuit, aurum sit unum e[39] maximê homogeneis et firmis in rerum natura corporibus, ut et post tot torturas et sophisticationes, quibus adulterari potest, ad pristinum statum reducibile. Et fateor sanè, omnes illos conatus, quos vel ipsemet adhibui, vel[40] vidi adhibitos ad aurum *permanenter,* a sua natura alienandum, ad desideratum finem consequendum, minus efficaces fuisse. Non me fugit Chymicorum diversos, praetendere auri in sua tria principia Analysin, verum aeque mihi perspicuum est, non ess[e] a [207] talibus hominibus alienum, ea plenis venditare buccis, quae nunquam praestare potuerunt; quemadmodum etiam quosdam Chymicorum magis Ingenuos, narisque prae caeteris emunitioris, qualis est Angelus Sala Billichius etc, non alio talium Arcanorum jactatores, quam ignominiosô Impostorum Agyrtarumve titulo decorare. Praeterea quodsi praetensa illa ad *tria prima* analy sis foret possibilis, non tamen statim sequeretur, eâdem methodô, ex auro talem tincturam seu animam de qua jam loquimur, educi posse: siquidem non praetenditur quod huius separatio, reliquam metalli massam relinquat

him who (with absurd credulity) would defend these authors' presumption. Besides, whoever would allow himself to be so easily imposed upon that he should believe someone has the Philosophical Stone just because he has advertised so in his writings before the whole world will not have any difficulty in believing Artephius as well when he gravely asserts that not only is he an adept but also that when he began to write his book he had already lived a thousand years by the help of this philosophical medicine."[m]

"There is yet another way whereby some spagyricists attempt to show that the transmutability of lesser metals into gold is proven. The idea upon which these spagyricists are based seems to be this: They suppose that there is in vulgar gold a certain part or portion that is far more noble than the rest, which being separated (although it is very little), the residual mass is deprived not only of its color but of its very nature, and ceases to be gold, by being degenerated into another substance, which is allowed to be silver, or some other white metal, or perhaps a mineral to which no appropriate name has yet been assigned. Furthermore, these chymists of whom I speak suppose that this extracted or separated substance (which they believe is the true tincture, or rather, the soul of the gold), when it is freed from the gross body and joined by art to some other metal, is able (as much by the power of the golden nature contained in it as by the power of the enormous subtility and agility acquired in its preparation whereby it is freed from its shackles and fetters) to invade other metals, penetrate them, and impart to them all those things which they lacked so that they become genuine gold."

"Though this, among all the hypotheses of the Lapidists that occur to me, may seem the most plausible, I nonetheless fear that it is more specious than true. For I wish that they would teach us an intelligible experiment, or one reducible to practice, which when examined could convince us that there really exists in gold some portion both far nobler and more pure than the mass of metal itself; or that this flower or soul of gold is actually separable from the remaining mass, since it has been hitherto noted that gold is one of the most homogeneous and constant bodies in nature, so that even after all the tortures and sophistications wherewith it can be adulterated, it is still reducible to its original state. And I confess readily that all those attempts which I have either tried myself or seen others try toward *permanently* depriving gold of its nature have been ineffective in reaching that desired end. It does not escape me that various chymists allege the analysis of gold into its three principles, but it is equally clear to me that is not unknown for such men to boast brazenly of things that they could never perform; and also that some more ingenious chymists of more secure judgment than the rest (like Angelo Sala, Billich, and others) adorn boasters of such arcana with nothing

[m] In *De vita prolonganda,* Artephius claims to be 1,025 years of age at the time of its writing. Roger Bacon cites Artephius's longevity in *De secretis* (in *TC,* 5:854). On Artephius, see chapter 5, n. 126.

in forma hydrargiri et Salis, aut Mercurii et Sulphuris, sed quod potius relinquat eam in forma, alicujus albi metalli, naturae anomalae, et proportione majore, quam remanere posset, si educta anima, ad aliquod ex tribus hisce principiis esset referenda. Adiice, gratis sumi, flavedinem auri, distinctis quibusdam exiguis particulis, reliquum metallinae[41] materiae tingentibus proficisci; quia quantum nobis obvium est, color in auro, aeque ac in multis aliis metallis non procedit a particulari aut separabili aliqua portione, verum a contextu totius corporis ex particulis consistentis, quarum ad hunc modum concursus componit corpus, aptum, modificando lumini quod ad nostros reflectit oculos, ut pro modo requisito, queat in illis flavedinem producere.

Sed dato etiam, talem de qua nobis sermo est, ex auro posse educi Tincturam, nondum tamen video qua ratione queant Lapidistae evincere, illam posse vilius cum quo conjungitur metallum, in aurum convertere. Nam ex. gr. plumbum, est metallum per se completum, omnibusque praeditum qualitatibus, quaecunque requiruntur ad talem corporum speciem, cujus, plumbum esse natura designavit, constituendam: quantum enim ad Chymicorum suppositionem, quâ, destinatum a [209] natura ajunt, ut omnia metalla ad aurum tenderent et in eo terminarentur, eam jam olim[42] nullo subnixam fundamento ostendimus. Si igitur perpenderimus qualitates ad plumbum pertinentes, quod: scil: sit corpus crudum coloris subfusci, ante excandescentiam fusile, sustinendae cupellationi impar, specificè auro levius, volatile, in Spiritu aceti et aqua forti dissolubile etc., penitus improbabile videtur,[43] ut natura, plumbum tam aptum ad recipiendam tincturam,[44] et ad[45] constituendum cum illa aurum reddidisset, quam si ipsammet, quae de industria praeparatur, fecisset materiam cui ex suo instituto, anima haec uniri, eamque animare debuerat: sive ut corpus hoc modô, quo ego exposui plumbum esse qualificatum, habeat internam dispositionem,[46] ad tantum alterationem, vi exiguae quantitatis liquoris, ab auro separati, subeundam, quemadmodum est commutatio in corpus, a plumbo quoad omnes supradictas hisque plures alias qualitates diversum.

Tametsi igitur huius, plumbi animandi possibilitas verisimilis non sit, tamen, etiam hâc concessâ, nihil Lapidistarum causae accederet praesidii,

less than the ignominious title of imposters or quacks.[n] Moreover, even if that supposed analysis into three principles were possible, still it would not follow therefrom that by the same method such a tincture or soul like that of which we have just been speaking could be extracted from gold: since it is not alleged that a separation of the soul leaves the residual mass of metal in the form of Quicksilver and Salt, or of Mercury and Sulphur, but rather that it leaves the gold in the form of some white metal of an anomalous nature, and in a greater proportion than could remain if the extracted soul is referred to any of the three principles. Note now how it is freely assumed that the yellowness of gold arises from certain distinct, tiny particles that tinge the rest of the metallic matter; for it is quite obvious to us that the color in gold, as well as that in many other metals, does not proceed from any particular or separable portion, but actually from the texture of the whole body, which consists of particles whose concourse composes the body in such a way that it is fit to alter the light which it reflects to our eyes so that in the requisite way it can produce yellowness in them."

"But even granting that of which we have been speaking—that a tincture could be extracted from gold—I nevertheless still cannot see how the Lapidists can assert that this tincture can, when a baser metal is joined with it, convert that metal into gold. For lead, for example, is in itself a complete metal possessed of all those qualities required to constitute that species of bodies of which nature designed lead to be. As for the chymists' supposition that it is destined by nature that all metals should strive toward and terminate in gold, we have already shown this supposition to be groundless. If therefore, we attend to the qualities belonging to lead, namely, that it is a crude body of a dusky color, fusible before becoming red-hot, unable to withstand cupellation, less dense than gold, volatile, dissolvable in spirit of vinegar and aqua fortis,[o] etc., it seems utterly improbable that nature should have put forth lead so fit for receiving the tincture and constituting gold with it, as if that tincture which is prepared by industry had made a material by its own design to which the soul is fit to be united and to animate it; or that a body qualified in such a way as I have shown lead to be would have an internal disposition tending toward so great an alteration (by the power of a trifling quantity of a liquor separated from gold) that there would be an alteration into a body so differing from lead in all the aforementioned qualities and in many others."

"Even though there is no real possibility of such an animation of lead, nevertheless, even if the point were conceded, it would not help the Lapid-

[n] Angelo Sala (1576–1637) was a spagyric iatrochemist; his son-in-law Anton Gunther Billich (1598–1640) wrote a dissertation on the "vanity of chymical, hermetic, and spagyric medicines," *Thessalus in chymicis redivivus* (Frankfurt, 1640). Note that Boyle assails Sala's position in *Producibleness,* 148.

[o] Although lead is soluble in aqua fortis (nitric acid), it is not soluble in spirit of vinegar (acetic acid); presumably, Boyle was referring here to calcined lead (lead oxide), which is soluble in both acids.

eoquod esset quaedam (si ita loqui liceat) transpla^nta^tio aurificae tincturae ex uno metallino corpore in aliud, cum Chrysopoei nostri Quintessentiam suam,[47] praetendant posse plumbum nescio quot centenis vicibus se ponderosius, in aurum perfectum convertere.

His ita dictis, postquam Erastus, decumbendo, finem se Orationi imposuisse innuisset, diversa inter Antilapidistas eius amicos emicuere signa, quibus transactionem eius admodum sibi probari testarentur: exinde pars altera quae Zosimo assidebat, stimulavit eum, in defensionem Phylosophiae Chymicae consurgeret, cui post aliquam reluctantiae speciem assensus, filum orationis his verbis pertexuit.

[211] Nisi Nobilissimae hujus Societatis instituta, apologeticis interdicerent, profecto nimis multum juris mihi, aequè ac necessitatis suppeteret, solemnem instituere, eoquod me in controversiam immiserim, cui debitè ventilandae, parum idoneitatis mihi superesse confiteor. Existimo utcunque, posse me absque ulla legum nostrarum Violatione, fateri palam, dolendum esse, tam excellentium Philosophorum, quales sunt Adepti, causam, talis Hyperaspitis committi patrocinio, qui ne quidem spem alit, tantum abest ut honorem nactus fuerit, eorum accenseri ordini. Quia vero abest Pyrophilus, cujus praesentiam ardentibus expeto votis, minus me et causae et Auditorio injuriae illaturum censeo, si fecero periculum, numquid causa tam bona, mediante tàm vilis ac ego sum Patroni ore, queat se in sui defensionem arrigere, adversus tam formidabilem prout est Erastus, Antagonistam.

Aegrè haec absolverat Zosimus, ^cum Heliodorus^ fibulam ulteriori impositurus processui, Eum interrumperet, ajens, eodem Orationis tenore, quo se legum in his Colloquiis observari solitarum, tenacem profiteretur, transgressorem extitisse; quamobrem ex ipso peteret, supersedere supervacaneis hisce excusationibus dignaretur, atque quantociius sese ad multum desiderata argumenta accingere. Hac admonitione incitatus Zosimus, incunctanter eam consectatus, discursusque seriem hoc modo aggressus est.

Nisi saepius observassem, comitatam ab ignorantia superbiam, ad quas natura sumus proclives, plurimos nostrum eo flectere, ut Normarum nos instar reputemus, eamque ob rationem, entia rationis solummodo et mera impossibilia ^esse^ concludamus,[48] quicquid vel explicare nequimus vel perficere; mirarer sane Viros non ineruditos, nobilissimum [213] humanae indaginis productum, non obstantibus irrefragabilibus per quae fidei nostrae majorem in modum commendatur argumentis, quà hermeticam chimaeram reiicere.

ists' cause, because it would be only a certain (if one might so call it) transplantation of an aurific tincture out of one metallic body into another, while our chrysopoeians allege that their Quintessence can convert I know not how many hundred times its weight of lead into perfect gold."[p]

After Erastus had said these things, and then by sitting down signified that he had brought his speech to an end, various signs were obvious among his friends the anti-Lapidists whereby they made it known that they were completely satisfied with his treatment of the matter. Then the others who stood on Zosimus's side urged him to rise in defense of the chymical philosophy, and after some appearance of reluctance, he consented and wove the thread of his speech with these words:

"Had the regulations of this Most Noble Society not forbidden apologetics, certainly too many would come to mind by right as well as by necessity, because I have dispatched myself into this controversy though I confess that the ability of properly airing it exceeds me. I believe, however, that I can, without any violation of our rules, confess publicly that it must be suffered that the cause of such excellent philosophers who are adepti be committed to the patronage of such a Hyperaspes, who, lest he nourish some unwarranted expectation, he must declare that much is lacking from his having attained the honor to be counted among their rank. Indeed, because Pyrophilus is absent (whose presence I seek with ardent prayers), I the less account that I might render some injury to both the cause and the Assembly if I should hazard whether so worthy a cause, through the mouth of so mean a protector as I, can rise up in its own defense against even so formidable an opponent as Erastus."

Zosimus had scarcely finished this speech, and was about to set down a link to a further continuation of it, when Heliodorus interrupted him, saying that in the same course of speaking wherein Zosimus had professed himself steadfast in observing the accustomed regulations for these discussions, he had also revealed himself a transgressor of them. Wherefore Heliodorus asked that Zosimus might deign to desist from these overempty excuses and as quickly as possible restrict himself to the desired arguments. Stirred up by this admonition, Zosimus unhesitatingly adopted it and began the order of his discourse as follows.

"Had I not quite frequently observed that many of us are naturally inclined to bend toward the haughtiness attendant upon ignorance, so that we think ourselves the standards by which things are to be judged, and by the same reason conclude that whatever things we cannot explain or accomplish are merely hypothetical or mere impossibilities, I would marvel that men, who are by no means unlearned, reject the most noble product of human investigation as a hermetic chimera quite regardless of the irrefutable arguments by which it is strongly commended to our belief."[q]

[p] Erastus's argument rests upon the distinction between a "transplantation" of the golden soul into lesser metals and the projective transmutation afforded by the Stone; see chapter 3.

[q] Note the rhetorical parallelism between this opening statement by Zosimus and the corresponding part of Erastus's oration.

Antequam dimisero cui Erastus institit, Objectionem, quod scil: Chymici nullam adhucdum deposuerint hypothesin, per quam ratio, quâ Lapis Philosophicus transmutat metalla, clare aut ad minus intelligibiliter exponeretur, desiderabo a Nobilissima hac Societate, quatenus observare dignentur, quicquid hucusque a me in negotio huius Positionis, in quantum scil: illa relationem habet ad quaestionem facti, prolatum fuit, discursus solummodo gratiâ propositum fuisse, non vero ulla necessitate adacto, in causae defensionem adhibitum. Quamvis enim aliquis probatum daret nullum ex hermeticis authoribus talem proposuisse hypothesin, explicandae transmutationi idoneam, in qua Erastus aut fortè egomet Ipse acquiescerem, tamen non sequeretur, quod, quia id nondum per Chymicos, angustis jejunisque Peripateticarum Scholarum imbutos principiis, perfectum fuit, perficiendum non sit in posterum, quando profundior in Chymiam inspectio, rationalibus obtinget Philosophis, aeque principiorum fertilitate ac hermeticis experimentis instructis: interea vero, hoc sufficiat indicasse Erasto, argumentum eius, utcunque ^circa enodandam istam hypothesin^ casura res est, contra me nihil probare. Neque enim unquam dixi, propterea me lapidem Philosophicum admittere, quod clarè explicare possem, quomodo attributae et operationes queant perfici, sed quod sufficienti convictus essem testimonio, tales operationes, actù perpetratas fuisse: neque ulla incumbit necessitas, ut aliquis sit idoneus ad reddendam rationem Philosophicam, [215] modi quô tale vel aliud quid per alios vel per ipsummet perficitur, quae reapse perfici cernimus. Quemadmodum videmus quotidie, Braxeatorem, posse coquere cerevisiam, qui tamen nequeat explanare vel naturam fermentationis vel quomodo maceratum[49] hordeum ad pullulandum[50] perducatur, aut quomodo aqua pura, per additionem hujus hordei exsiccati in brasium, reducatur ad mustum, longe a priore sui natura diversum: ita imperitus aliquis Mechanicus, practicarum quarundam regularum ope, poterit conficere solarium, absque ulla cognitione aut explicandi facultate, illorum mathematicae scientiae principiorum, quibus Sciatericorum[51] ars innititur.

Brevibus, quodsi Erastus velit inficiari, Existentiam omnium illorum Chymicorum productorum, quorum naturae et effectuum, ipsimet Artifices, nequeunt distinctam dare descriptionem, debebit magnoperè illorum minuere catalogum, ut Chymicarum non entitatum augeat, et aeque eos repudiare, qui ex auro, fulminantem pulverem conficere suscipiunt, quam qui ex eodem metallo Transmutantem pulverem educere praetendunt.

"Before I discharge the objection that Erastus has urged (namely, that chymists have hitherto set down no hypothesis which explains clearly or at least intelligibly the means whereby the Philosophical Stone transmutes metals), I will desire from this most noble Society, so far as it may deign to observe, that whatever might be brought forward by me in the treatment of this position, namely, as far as it has a relation to a question of fact, has been proposed only for the sake of the discourse, and not at all employed towards the defense of this cause by the urging of any necessity. For howsoever much anyone might maintain that no trial from the hermetic authors has advanced a hypothesis capable of explaining transmutation (in which Erastus and perhaps even I myself might acquiesce), nonetheless, it still would not follow that just because no chymist steeped in the narrow and lean principles of the Peripatetic Schoolmen has yet given an explanation, that one could not be given in the future when a deeper investigation into chymistry is carried out by rational philosophers instructed in both fertile principles and hermetic experiments. Meanwhile, let it suffice to indicate to Erastus that his argument (whatever may befall concerning the resolution of this hypothesis) proves nothing against me. For I have never said that I accept the Philosophical Stone because I can clearly explain how its properties and operations can be performed, but rather because I have been convinced by sufficient testimony that such operations have actually been carried out. And no necessity requires that a person must be able to provide a philosophical explanation of the way in which this or that thing (which we perceive actually to be performed) is carried out either by himself or by others. In this way we see every day that brewers can make beer although they cannot explain either the nature of fermentation or how the soaked barley is brought to sprout; nor can they explain how pure water, by the addition of this barley after it had been dried out into malt, is reduced to a brew far different from its former nature. In the same way any unskilled mechanic can construct a sundial by the power of certain practical rules without any recognition of, or any ability to expound upon, those principles of mathematical knowledge wherein the art of sundial-makers rests."

"Briefly, if indeed Erastus wishes to deny the existence of all those chymical products of whose nature and effects the artificers themselves cannot give a clear description, he will have to diminish their catalog enormously that he might increase the number of chymical nonentities, and he shall have to repudiate as fully those who maintain that they can make a fulminating powder from gold as those who claim to draw a transmuting powder from that same metal."[r]

[r] Fulminating gold (*aurum fulminans*), an easily prepared and highly explosive gold compound, was well-known in the seventeenth century. Boyle mentions it elsewhere in several contexts; see, for example, *Some Physico-Theological Considerations about the Possibility of the Resurrection,* 18.

First Part
(Fragments 3–10)

FRAGMENT 3 [BP, 38:160][a]

The Sage Heliodorus ^who as you know is^[54] alwayes desirous to husband the Assembly's time, perceiving that the Discourses both of the Anti-Lapidists & their Opposers might be made more short & argumentative than by the stile they had been begun & hitherto carry'd on in, he guess'd they would prove ^the president,^ I say, after having very briefly but yet civilly taken notice to Erastus and Zosimus that they had both of them ^ingeniously^[55] perform'd the office of Advocates for their respective causes, He told him [*sic*] that the Company did not doubt but two such persons as they, knew how to Declaime, but that the scarcity of time, & the genius of that Assembly oblig'd him to desire[56] both the Antagonists to forbear insisting upon extrinsical arguments such as[57] those ^drawn from moral or^ political[58] considerations, and[59] in a close debate betake themselves to those Proofs that are more intrinsick, & properly Philosophical as being grounded on the nature & Physical Circumstances of the things in question.

This advice of the President was applauded by all the rest of the assembly that were not ^either^ heated by their zeal to their party or[60] ^ambitious^ to make a show of their various Learning & Eloquence. And therefore Erastus, thô not without som reluctancy, dispos'd himself to comply with Heliodorus & told him[61] ^&^ the Company that since they appear'd unwilling that he shold make the best use of some Arguments[62] that he thoght might deserv'd [*sic*] to be fully displai'd & ^if they were so^ wold scarce ever be fully answer'd, he wold imploy the weapons whose use was permitted[63] him, and[64] in pursuance of the method prescribd him, he ^& his Assistants^ wold content thimselves to make good[65] four Arguments against the Chrysopoeian assertion, all of them[66] drawn from the[67] Nature of the things in controversy. For[68] he declar'd[69] that he was altogether dissatisfy'd with Zosimus's opinion; 1st because of the great & ^in^credible chang[70] that must be made[71] in the pretended Transmutation of inferior metals in to Gold. *2ly* because of the permanent & as it were stubborn ^nature^ of the bodies in which this change is to be made *3ly* because of the small proportion of the aurific ^powder or^ Agent that is imploy'd to make a change in so great a quantity of indispos'd matter. [160*v*] And *lastly,* because of the very little time in which such[72] wonderful alterations are said to be performed.

Helliodorus being wel satisfied with this declaration turn'd himself to Zosimus to whom he said that he hopt the example of Erastus would invite[73] him also to plead for the Lapidists, in the directest & shortest way he could, & ^therefore chuse to^ imploy onely the[74] ^most proper^ arguments which are usual ^and^ those that the nature of the the thing in question dos suggest & require. To this notion Zosimus answerd with great respect that

[a] Margin: "T[ranscri]b'd."

he needed not be inducd by the example of an Adversary to do that for which he had such a motive as the Desire and Authority of Heliodorus. And therefore he promisd ˆin his owne Name & that of his Chymical Friendsˆ that they would reduce ˆalmost all they would insist on, in theˆ[75] behalf of the Lapidists to two ˆmaineˆ heads. And ˆas in the 1st partˆ[76] of his discourse he would answer the four arguments newly mentiond by Erastus, so[77] in the 2d he would manifest that what[78] his Antagonist had not been able to prove impossible to ˆNature &ˆ art, Experience has evinct to ˆhave been actually by Artˆ performd.[79]

FRAGMENT 4 [BP, 25:287–88][b]

Thô the 3rd and 4th Objections urg'd by [Erastus][c], *One,* as you may remember, drawn from the smallnes of the quantity that in Projection works upon a great deal of mettal, and the *other* from the celerity wherewith it performs its effects, seem by much the strongest that can be urg'd against Transmutation by the Elixir: yet I shall endeavour ˆto answerˆ them both together, because the things that I am to alleage ˆdo, asˆ to several of them, reach to both the objected difficulties, thô some do more directly regard the former of them, and others the latter.

I consider then in the first place, that there are 2 or 3 things that may justly keep us from Estimating the power of *all Agents,* whether Natural or Artificial, by the same measures by which men are wont to judge of the Efficacy of *Agents that act but at an ordinary* or moderate *rate.* For 1st, some Bodies may consist of Parts so minute so solid, so advantagiously shap'd, & endow'd with so much vehemence of motion, that[80] a small portion of a matter thus qualify'd, may operate more powerfully than a far greater quantity of ordinary matter can do; And this Energy will be much the greater, if the Body to be wrought upon be peculiarly qualify'd to receive its action; and especially if it be so dispos'd, as not only to admit readily, but powerfully promote by its own Concourse, the action of the Agent.

Of these things, continues Zosimus, 'tis not difficult to find severall Instances among the Productions of Nature, and also of Art, especially when they happily cooperate to the same effect. We see how little a Proportion of Runnet will serve to coagulate a great quantity of Milk. And I remember an Excellent Botanist learn'd of some Country people, a way of curdling milk only by rubbing an Herb (which I think it not fit to name,) on the inside of Pales that receiv'd it, which Trick they too often us'd, to be unsuspectedly reveng'd of their neighbours, who could not guess how so much milk should be spoil'd by undiscernable means. How small a Portion of fire will extent its Operation to a vast quantity of wood or other combustible matter, every

[b] Sheets endorsed on reverse (290): "C. 45 Relating to Dialogue of Transmutation of Metals." Obviously, a lacuna separates fragments 3 and 4, for we have missed Zosimus's response to Erastus's first two objections.

[c] A blank space has been left for the later addition of the name of the speaker. In this case, the speaker is obviously Erastus/Simplicius.

body knows, thô every Body dos not reflect upon it. There are Poysons that in a lesser quantity than that of a Pins head, may Operate on a mass of Blood of a large Animal, as is related by [288] Authors very credible, and has been confirm'd to me by what I have known of the sad effects of the Teeth of angry vipers both in Brut^e^s & men; thô the wounds their teeth made were sometimes so small, that I could very hardly perceive them. But, continues Zosimus, to give an Instance or two in Mineral Bodies; those that have not try'd it, (and there are very few that have,) will be surpriz'd to see how small a proportion even of crude Tin, will alter some mettals, and make Lead itself ^unfit^ to work as it should do, and is wont to do, upon the Cupel. And if you will beleive the unanimous tradition of Chymists, that the vapour of Lead will coagulate Quicksilver, and quite deprive it of its fluidity;[d] you must acknowledg that a very small proportion of matter, may have a ^notable^ operation even upon a mettaline Body. Nay divers of them pretend by this saturnine vapour to make Quicksilver malleable; ^of^ which, thô I deny not the Experiment to have somthing of pretty in it, yet I confess my Tryals have not convinc'd me.

FRAGMENT 5 [BP, 25:291–93][e]

To perswade you Gentlemen that great changes may be made even in such stubborn Bodys as Mettals by a very small quantity of matter, I shall give you a short account of an Operation that may seem to many of you, as it did to me rather more than less difficult and strange, than that of the Elixir it self. The story in short is this[:] having in my hands a very small Portion of a Powder (about which I am not at present to answer questions, and therefore hope I shall not be askt any)[f] from whence some Circumstances made me guess I might expect uncommon things: I invited an Experienc'd Physician who was also an Eminent Chymist to be present & assistant at an Experiment which I told him he would not think vulgar.[g] This curious Person gladly embrac'd the Motion and a private Place being appointed for our meeting I say Pyrocles[81] did at the appointed hour come thither alone as I found him when I came & brought with me a Guiney (a Piece of good Gold weighing about two drams) and my Powder together with a new Crucible, and a nice pair of Goldsmiths Scales, whilst the Crucible was kneeling[h] in the Fire, we weigh'd out the Powder which seem'd but light in Specie, and

[d] On these illustrations, see chapter 4. This curdling of milk with herbs is mentioned also in "Fluidity and Firmnesse," in *Certain Physiological Essays* (London, 1661), 218.

[e] Margin (pencil): "This belongs to the Dialogues about the transmutation of Mettals. [Zosimus] speaks."

[f] This refusal to entertain questions is similar to that in Boyle's "Incalescence" paper; see chapter 5.

[g] Compare this choice with that of "an experienced Doctor of Physick, very well vers'd in the separating and copelling of Metals," in *Anti-Elixir,* 282. This chymically minded physician might be Edmund Dickinson (1624–1707); below he is referred to as an "old man": Dickinson would have been in his fifties during the writing of this fragment.

[h] Amanuensis's error for the homonym *nealing* (= annealing).

was so little that its weight was inconsiderable even in that tender Ballance. The Doctor would have added some Niter and other salts to facilitate the Fusion of the Mettal but I would not consent to it for fear of making the Experiment less simple and certain and chose rather to blow longer, then not melt the Mettal per se; when it was brought to Fusion, I did with my own hand put the Powder[,] lap'd up in a small piece of Paper to keep it from scattering[,] into the Crucible where it quickly seem'd to have made some little change in the Mettal, but we kept it in good Fusion awhile longer, and then taking the Crucible from the Fire and suffering it to cool awhile, the Physician was greatly surpriz'd to see how much such a Mettal was alter'd in so short a time. This done, we weigh'd it again in the same Scales, and found that instead of having lost any thing it had gain'd by the Operation of the Fire, and my Opinion being that this Augment was made by a kind of Magnetism whereby many Parts of the Fire or Air, or both were detain'd by & imbody'd with some Parts of the Powder or some Mettaline Ones by its action made as it were magnetical. I had a great mind to try whether this accruing substance would be fixt, and according[ly] having weig'd our Mettal & cupel'd it carefully we found that we had remaining upon the Test a dark colour'd Heterogeneous Substance, to which I cannot give a name, because neither I, nor my Experienc'd Assistant could reduce it to any thing near [293] any Body we knew.[i] But it appear'd that whatever [it was,] it weigh'd between an hundred & fifty, and two hundred, not to say between two and three hundred times as much as did the Powder that we imploy'd at first, so that this wonderful Magnet did in a very short time obtain from the Air & Fire (for we could not suspect the Augment to proceed from any thing else,[)] at least between an hundred & fifty or two hundred times its own weight of a Substance fixt enough to indure the Cupel. This Experiment the Experienc'd Chymist has told as a great wonder to some of his nearest Friends, from whom I heard of it a long time after, and thô he were a Person that being an old man and an old Traveller was extreamly reserv'd, yet he was soe transported at the sight, that he confess'd to me, then what he repeated afterwards, that thô he had seen ----------[j] more than once, yet he never saw any thing that he thought so unaccountable.

FRAGMENT 6 [BP, 25:299][k]

The Learned Olaus Borrichius[l] related to me that he being at Amsterdam met with an Anabaptist who with much Intreaty shewed him a certain Powder communicated to him as he said by one ---------- who was a while after

[i] In cupellation gold or silver is melted in a bone-ash dish with an excess quantity of lead, and a current of air directed across the molten metal. Base metals, including the added lead, are converted to their oxides and blown away, leaving the noble metals pure. Boyle's "Heterogeneous Substance" was left on the dish ("the Test"), showing its "fixity."

[j] Perhaps "projection."

[k] Margin (pencil): "This belongs to the Tract of the Transmutation of Metals"; *verso* endorsed: "belonging to the Tract of Transmutation of Metals."

[l] On Boyle's contact with Borrichius, see chapter 5.

murther'd in the streets by some that had a mind to obtain his secret from him but could not[,] and that this Anabaptist bid him bring a piece of Silver which he did[,] the Piece being flat and about the bignes of the naile of a mans hand. That then this Chymist laid on it a little Powder that look'd like Powder of Brick and amounted as Olaus quoted to about a 5th part of the Silver, then having put the little Plate upon a live Coale with a small Pipe that Olaus lent him he blew a little upon the Powder which was very fusible and melted yet without melting the Silver, whereupon Olaus desired leave to look upon the little Plate, and found that the Powder seem'd to have pierc'd into it almost quite to the other side, and the Chymist blowing it upon the Coale again, and blowing it safely as before Olaus desir'd leave once more to take off the Plate before it began to melt and found it all transmuted into very lovely Gold except the very corner, wherefore the Chymist laying it the third time upon the Coale increas'd his blast soe that it melted into a small round Button which not only seem'd to be perfect Gold, but to have on one part of it a little red vitrum like a very small Ruby which the Anabaptist affirm'd would tinge much more Silver if it were added to the Button but Olaus sent it as 'twas to the now King of Denmark for a Present.

FRAGMENT 7 [BP, 10:10][m]

Though I confes that to be very plausible which[82] hath bin objected by Erastus[83] against the peice of gold that [Philetas][n] lately presented to the company yet ^his reasoning^[84] is not so convictive[85] as at first sight it seems to be ^for^ though it be true, that tis not likely[86] a man who pretends that the gold he shows was made by the elixir should have so little wit as to offer ^any other than^ fine gold, yet tis possible that a peice of mettal may have all the marks of true gold, & consequently may pas for such[87] in any mint in Europe, & yet be endowed[88] with some other properties, and those perhaps noble ones, whereby the skilfull may know it & distinguish it from common gold from which it may differ either in the richness of the colour, or a greater degree of softnes and malleability, or, as it sometimes happens in a power of fixing & tinging some little portion of the lead wherewith tis cupeld. Upon these or the like grounds I suppose it was, that Raymund Lully ^and some philosophers of his school^ speaking of gold made by projection do more than once tel the reader, that tis[89] *purius omni auro minerali,*[o] & which I

[m] Margin (pencil): "Zozimus speaks." On the redundancy of part of this fragment, see the introduction to this appendix.

[n] Name left blank in manuscript.

[o] I.e., "Purer than any mineral gold." A large alchemical corpus exists under the name of Ramon Lull, the Catalan mystic of the thirteenth century. The earliest members of the heterogeneous pseudo-Lullian corpus date from the late fourteenth century; see Michela Pereira, *The Alchemical Corpus Attributed to Raymond Lull* (London: Warburg Institute, 1989). The comment about the superiority of artificial precious metals is made in several places (see, for example, the *Practica* of the *Testamentum,* in *TC* 4:135–70, on 166, and the *Clavicula,* in *TC* 3:295–303, on 296–97). Such claims have been seen as a key juncture in arguments over the art-nature dichotomy; see William R. Newman, "Technology and Alchemical Debate in the Late Middle Ages," *Isis* 80 (1989): 423–45.

the less wonder at because the presant Emperours embasador at the English court a person of much candour & no way partial to the Chymists,[p] confessd to me, that at Vienna where the art of refineing is thought to flourish more than any where[,] if any gold[90] made by projection (as[91] was often done by ^famous^ fryar Wenceslawas)[q] was brought to the mint the officers would presantly by the signs of purity ^manifestly discernable^[92] in it say that it was not common but philosophick gold, which they were wont gladly to receive because it would admit of an allay which turnd to their profit[;] and I remember that a traveling virtuoso who showed a friend of mine (well known) [10*v*] to the famous royal society of London, a noble and surpriseing mettaline experiment presented me as a rarity a small lump that amounted to some drams of silver of an extraordinary whiteness or luster with an earnest desire that I would have it severly examind, where upon ^haveing^[93] shew it to a refiner noted for his skill ^without^[94] telling him anything but that I desired to have it strictly assayd he apeard surprised at the beuty of it, & tould me that it was no ordinary silver, & needed not be put upon the test to satisfy him that it was perfectly fine: notwithstanding ^which^ I caused it to be tryed by the publique saymester, among other peices of the same mettal that were brought him, & received from him a full approbation of it, this I gave notis of to the Gentleman (who had then a publique character) I received the presant from, & he then assured me that it was made by projection, & to satisfy me that it was no natural fine silver though it had all the marks required in mints, he imparted to me an easy way of distinguishing it, which when I soon after went about to try, I was by a very odd & unluckey accident deprived of the mettal it self, to all which contineous Zozimus[95] I shall add that I have had [a metal] (&[96] suppose that I ^yet^[97] have a little[98] fragment of it left) which is malleable[,] looks like gold,[99] both without & within disolves like it in aqua regia, & cupels like it, & yet as a[100] stricter though less obvious tryal hath assured me is not true gold nor belongs to any other sort of mettal of ^a^ known denomination.

FRAGMENT 8 [BP, 25:303][r]

On this occasion continues ---------- I remember a singularity which I presume the company will not be displeas'd to hear, and it is that I was once Possessor of a piece of Gold which was affirm'd by the Person it came from to me to have bin made by projection and given as a Present to that Person by the man that made it who yet was no Adeptus thô he was affirm'd to have some of the Powder. I am[101] too well acquainted with the warynes[102] of this assembly, to expect[103] they shold lay any weight upon those Circumstances of this story, that I had but by Hearsay, thô ^for^ my own part the

[p] Count Waldstein, ambassador from Leopold I to Charles II; compare this account with fragment 9 and the interview with Waldstein, text A in appendix 2.

[q] I.e., Wenzel Seyler, see chapter 4 and fragment 9.

[r] Margin: "T[ranscri]b'd"; *verso:* "Relation of a Lump of Gold said to be made by Projection."

knowledg I had of the person to whom the piece was presented makes me give credit to them. But that ˆforˆ which ˆIˆ scruple not to mention ˆthis Medall even toˆ this Company is, that thô the Person I had the Gold from, had try'd nothing with it, being loath to deface the fine, but ˆvery unusualˆ[104] figures that was stampt upon it, yet having without injuring it ˆcarefullyˆ made ˆaˆ Hydrostatical Examen of it, which I take to be at lest one of the surest ways of tryall I found to ˆmy no smallˆ[105] contentment that which perhaps you will not hear without som surprize namely that this Gold was indow'd with a considerably greater specific gravity or as some call it, Ponderosity, then the gold that I have met with either from Mints or among Refiners, for whereas ˆgood coin'dˆ[106] Gold dos often[107] weigh ˆbutˆ between 18 & 19 times as much as common water that is equal to it in Bulk ˆand I remember not to have found thatˆ even well refin'd Gold was Equivalent in weight to 20 times[108] ˆits Quantity of the same Liquorˆ, I found that this Gold amounted ˆto the weight ofˆ 25 times its bulk of water, so that whether it were made by projection or no, 'tis plain that Gold may be artificially brought to be in one of the chief ˆ& most radicalˆ Propertys of ˆthat Metalˆ[109] to be more perfect than the finest common Gold that[110] nature, assisted by vulgar Chymistry affords us.[s]

FRAGMENT 9 [BP, 25:419–21][t]

Thô says ---------- the Instances you have lately heard, contain an[111] Argument very favourable to the Chymical doctrine, yet I think I can strengthen it by ˆanˆ Instance yet more considerable ˆthan any of themˆ to the same purpose; ˆwithˆ[112] which I was lately supply'd by a very eminent Person Embassador of his Imperial Majesty, in the court of the King of great Brittain.[u] For having one day visted this Lord, whom I had found to be a Person of great vertue & honour, a serious man[113] and unfavourable enough to Chymists, I put him upon the discourse of what was said; to have been lately done at Vienna, upon which occasion his Excellency told me in the presence of another ˆInquisitiveˆ and very considerable Person[v] of the Imperial Court[114] that Wenchˆlˆeslaus the Augustinian Fryar,[w] that makes now so great a nois in Austria and German [*sic*], told him once, that he had found a particular by which mettals were transmuted ˆin a triceˆ but not by the way of projection. And I appearing desirous to know, not what his Excellency had heard, thô from Credible Witnesses, but what he had seen, he pursu'd his Relation by telling me. ˆButˆ when he seem'd Diffident of the truth of

[s] Compare Boyle's denomination of this gold as "more perfect" than natural gold with the classical chrysopoetic claim attributed to pseudo-Lull in fragment 7. The density of gold is 19.3; no metal has a density as high as 25. Note also the distinction between the capabilities of chrysopoeia and of "vulgar Chymistry."

[t] Margin: "T[ranscr]b'd."

[u] Count Waldstein; see above, fragment 7, and appendix 2, text A.

[v] Count Lamberg, see appendix 2, text A.

[w] Wenchleslaus = Wenzel Seyler; see chapter 4 and fragment 7. On "particulars," see chapter 3.

what the Fryar had said[115], this Person put into his hands a white Powder, which he told him, was a transmuting one ^and gave him directions how to use it^· But my illustrious Relator appearing backward to make tryal his Brother[116] ^since nominated to^ be a cardinal[x] (who was present) & had much more favourable thoughts ^of Chymists^ than he had, perswaded him to make tryal of the Powder presented him. In compliance with this Motion they two, permitting no Body, much les the Fryar to be present did, in the Embassadors or his Brothers hous weigh out as much of this powder, as serv'd to counterpoize a Ducate of Gold, and then put in them both together into a Crucible, encompass'd with a strong fire, they blew til they had melted both the Powder & the coyne into one lump, which being taken out, the Elder Brother ^confes'd^[117] [420] to me, that he was[118] surpriz'd to find this whole mass look like the purest gold, as indeed it was found to be, when curiously examin'd at the Emperors Mint, and when I ask'd[119] whether they had weigh'd the Augment of Gold, which the Ducat recev'd in this Operation, he answer'd that they had, & found to their wonder that they ^wanted^ not in the Lump one ful grain of 2ice the weight of the Ducat.

This, Gentlemen, ^continues ----------^ did I confess, put[120] divers thoughts into my head[121] ^especially^ as it seemd to argue that there may be some principle of metalline transmutations into Gold that differ from those, which mens speculations & Theorys, have been accustom'd to.

But adds ---------- still prosecuting his Discours, the Embassador's Relation ended not here, for he subjoin'd that soon after the Fryar, meeting him at Court, ask'd him if his Excellency had found his Experiment to succeed. To which the Embassador making a Reply, that intimated his suspicion, that Wench^l^eslaus had made that powder by disguising gold, or by convaying [with] the white substance[122] som aurific corpuscules, the Fryar seem'd troubl'd & somwhat nettl'd at this Jealousy, & finding by Enquiry that this Lord had not employ'd above one half of the Powder which was sufficient for at le[a]st two Experiments. He desir'd his Excellency, to take the remaining part of it,[123] which was yet in his own keeping, and melt it down with an equal weight of fine Silver. This, ^the^ Embassador going home, did by the help of his brother doe, & then they found that this Silver, which weighed a Ducat before, was so increas'd, as not to want a full grain of double its former weight of a metall that try'd at the mint, justify'd it self to be, what to the Eye it appear'd, ^exceeding fine Silver^· At this chang both the Brothers were much surpriz'd, and I as well by the relation of it, since it seems [421] plainly to show that there may be a matter so adiaphor^o^us or so strangely dispos'd as to be ^indifferently^ assimilatable, if I may so speak, into Gold or Silver,[124] by a bare comixtion with either of these 2 Mettals as if gold & silver even in their crude simplicity or natural state had in them a power of transmuting Body's into such metals as themselves are if they be

[x] Waldstein's brother was the archbishop of Prague.

^but^ of a nature dispos'd, to admit their actions and be perfectly subdued by them.

FRAGMENTS 10 [a: BP, 7:160–163*v*; b: 164][y]

Haveing received a couple of visitts from a forraigne Dr. of Phisick lately come into England, I thought my selfe obleidg'd to make him my acknowledgments for them at his owne Lodgeing where he receiv'd me very civily & after some complements told me that he was very glad to see me there at that time not only for his owne sake but for mine since he expected every Minute the presence of a Gentleman that haveing lately Travel'd into the Indies & even as far as China it selfe would probably prove no unwellcome company either to me or to him, to whome this Travailer tho but newly knowne to him[125] had been free & instructed in conversation, especially when he found by compareing notes that He & the Phisitian were of kinn & were both born in one Towne.[126]

We had not long been speaking before the stranger came in & gave me opertunity to take notice that I had seen him once before in the company of one of my acquaintance who it seems had told him who I was[,] for he very civily address'd himselfe to me with a Complement that 'tis not needfull & would be improper for me to repeat, wherefore I shall proceed to tell you that after the Civilities usuall on such occasions were exchang'd between me & those two Gentlemen whereof him at whose Lodgeing we mett by calling him the forreigner or Phisitian I shall distinguish from his Cozen the Stranger whom I shall also call the Traveller or the Virtuoso; after I say[,] some Mutuall Civillities had pass'd between us the Traveller told the Phisitian that if all things necessary were in readiness he came provided to make good his promise. And to obviate the scruple ariseing from the fear that what He the Traveller was about to do might hinder the forreigner & me from prosecuting our conversation, He (the Traveller) said that since I was come to that place so opertunely He would consent that I should if I thought fitt be a spectator of the Experiment he had promis'd to show His freind. I readily accepted & gratefully acknowledg'd so civill a profer & thereupon the strainger desired his freind to kindle a fire in a portable Iron furnace that stood in the Chimney & whilst [10b; 164*v*] that was doing [10a; 160*v*] He so spoke of the Experiment He was about to Make that I rightly concluded the designe of it was to turne Lead into runing Mercury in a very short time & by the help of a very small proportion of a Certaine powder that He had brought with him. [10b ends, 10a continues] Looking upon such an operation as a ^very^ rare Curiosity I was impatient to see a tryall made of it, & to satifie me the better it happened luckely that there had been an omistion to provide Lead enough & therefore the Travailer who was come to London in his way to another Country & spoke not a word of English desired me that in regard I had a footman as he knew by his Livery at the house dore I would

[y] Margins of 10a and 10b read: "Philoponus speaks"; 10b is a partial fair copy of 10a.

send him for some good unsofisticated Lead which he thought a Domestick of mine could not but know how to chuse. I sent therefore my ^servant^[127] for some of that mettall & two or three new Crusiples & the ^Virtuoso^[128] haveing melted some Lead & cast some graines of His powder upon it desired the Forraigner to blow lustily which he did but so unluckily, that some of the Coales upon which the Crusiple had been unskillfully made to lean suddainly broke or gave way whereby the Crusiple was overth[r]owne & the contained matter fell out amongst the Coals & Ashes before the operation was near halfe ended. At this mi[s]chance we were all three troubled & especially the[129] Travailer who said he had not brought over enough of the powder to repair losses but promised the forraigner to send him over some of it soon a[f]ter his arivall on the other side of the sea to convince both his friend & me that he did not pretend to a thing that he was not possesser of[,] affirming to us that his master could easily make a Considerable quantity of that powder & that to his knowledge it had divers times in very few hours turn'd Lead into Mercury. But, [161*r*] subjoyns he as willing to make me some amends for the late miscarage, since *Philopinus* has shar'd in our misfortune tis fitt we should make another tryall for his sake. Wherein if he please he shall be more then a bare spectator. Haveing[130] spoken to this purpose he desired me because his freind said I was vers'd in weighing things nicely that I would take some new Lead & weigh out some Drams of it which haveing done we putt it into a clean Crusiple which was plac'd in the furnace with more wariness then formerly. And then the Lead being strongly melted the Traveller opened a small peice of folded paper wherein there appear'd to be some grains but not very many of ^a^ powder that seemd ^somewhat^ transparent almost like exceeding small Rubies & was of a very fine & beautifull red. Of this he tooke carelessly enough, & without weighing it upon the point of a knife as much as I guest to be about a graine or at most betwixt one graine & two & then presenting me the haft of the knife he told me that I might if I pleas'd cast in the powder with my owne hand. But the fire by that time burning feircly & I haveing that morning had an accidentall indisposition in my eyes found them so Dazelld & offended by the glowing fire when I came very near it, that imagineing that the Experiment to be made was but the same that lately succeeded so ill I was unwilling to hurt my eyes & apprehensive lest the experiment ^might^[131] miscarry ^in my^ hands & therefore restoreing the knife to the Traveller I desired him to cast in the powder himselfe which he did whilst I stood by & lookd on. This being done he praid the ^Phisitian^[132] to blow the fire lustelly which the other ^if I misremember not^ did[133] ^with a pair of^ ordinary hand Bellows for about a quarter of an hour [161*v*] by my estimate dureing which time the Traveller & I sait Discoursing by the furnace minding the progress of the operation. The time newly mentioned being elaps'd the Virtuoso[134] told his freind that he supposd the pott had been long enough in that fire & therefore ordered him to take it out as soon as conveniently he could which being done & the Crusiple haveing been kept till it was cool enough to be managed without ^doeing harme^[135] we remov'd ^it^ to the window where in stead of runing Mercury I was surprizd to find a solid Body & my surprize was increasd ^when^ the Crusiple being inverted ^tho yett^ a little hott the Mass that came out & still retaind the figure of the lower part of the vessell apear'd very yellow & when I took it into my hand felt to my thinking manifestly heavier then so much Lead would have done. Upon[136] this, turning my eyes with ^a^ somewhat amazed look upon the Travellers face he smild & told me he thought I had suffitiently understood what kind of experiment that newly made was design'd to be which made me reflect upon some Circumstances that the preocupation I had that the bare repetion[*sic*] of the 1st experiment was the thing intended in the second kept me from considering & particularly I remembered that the powder first employd was of a Dusky Colour[137] not at all like the latter of a shineing red. Other Circumstances ^also^ occur'd to me that made me somewhat wonder at my owne mistake & the success of the ^second^ tryall made me regrett that I refus'd the obleidgeing offer that had been made me to cast in the powder of projection with my owne hand. And[138] I had the more cause to be troubl'd[139] at what I had omitted because the Traveller would probably[,] if I had not before appear'd willing to spare my eyes[,] have permitted me to act as well as look on[140] when he made another operation[141] wherein being willing to show me a Phaenomenon that[142] he justly thought would very much please me tho ^for some reasons^ I dare not now mention it[,] he desired me to weigh the yellow mettal on which when 'twas putt in a clean Crusiple and melted he cast that which I gues'd to be about two grains or less[143] of a powder which I could not certainly tell whether it were the very same for kind that he formerly usd or another like it but the effect of this powder which lessened not the weight of the mettall was strangly surprising to both the[144] Phisitian & ^to^ me whereat the Traveller seem'd not ill pleas'd. But tho I appear'd satisfied by the colour & ponderousness of the mettall (by which qualitys[,] exercise had made me a ^confident^[145] judge of what was or was not Gold when I pois'd it in my hand) that it was[146] the product of a real transmutation yett the Traveller press'd me to give [162*r*] my selfe a farther satisfaction by makeing some more rigorous examine of it. This to comply with him as well as my owne curiosity after a while I did. For goeing to a place where I ^had^[147] a say furnace & other accommodations I caus'd a servant of mine that knew nothing of the past story to Couple[z] ^[the] factitious mettall^ with severall times its weight of Lead, & part of it being mix'd with three or four times as much silver we dissolv'd in Aqua Fortis where it fell as Gold uses to doe ^so that we were^[148] satisfied both by Cupellation & quartation[aa] that it was true Gold but when I would[,] as I was in justice bound[,] have restor'd it to the owner he would not receive it unless at his importunity which 'twill easily be thought was not unwellcome

[z] I.e., cupel.

[aa] Methods of assaying gold; for an explanation of cupellation and quartation, see n. i above and n. n in the conclusion of this appendix.

to my Curiosity I would accept of above halfe of it ^the greatest^ part of which, for some of it I afterwards lost in making experiments with it[,] I yett have by me & carry always about me ^with that concerne for it that^[149] such a rarity deserves. I would have acknowledg'd the favour with a little present to the Person[150] I ow'd it to, but ^he^ absolutely refus'd it only ^he^ was content to accept of a small new fashioned Microscope with one Glass of which sort he had never seen[151] any before, & which was not worth a quarter so much Gold as he had presented me. He[152] seem'd also to be somewhat troubled that he could not leave with me a graine or two of his powder but excus'd it by frankly confessing that he was not the owner but a servant & ^young^ Disciple of an Adeptus who would exact a strict account of it. And he added that he durst not have done so much as to lett me be a spectator of what I had seen but that[153] his Master had not only in his hearing made a very favourable mention of me but had given him a liberty if not a command in case he should ^see^[154] me at London & meet with a fitt[155] opertunity he might give me the satisfaction to see a projection. And indeed before ^this virtuoso^[156] went away from London whence he made but to much hast he was so free with me as to ^take^ out of his pockett a Letter ^of instructions^ from his Master wherein he lett me read a [162*v*] permission he had to show me the transmutation of Mettalls. And by the sight of this Letter I had the means to Discover who[157] his Master was[158] whose name I had[159] met with in printed Gazetts where he was mention'd as a ^very rich man^[160] famous for extraordinary ^knowledge^[161] & who is now rais'd as his Disciple own'd to me to very eminent dignities which I wish he may long enjoy since I hear by severall ways that his Piety & Charity make very good use of them. In regard [that] the powder that was employ'd in the operations was not weigh'd I cannot tell precisely how many parts of Lead were transmuted by it but I remember the Gold weigh'd much above halfe an ounce & tho ^to my great trouble^ I so soon lost the Company of ^the Virtuoso^[162] that made it that I could[163] learne very little concerning Chymicall affairs whereof he declar'd he had not leave to speake freely yett I afterwards saw cause to thinke that one ^part of the powder^ would have gone as they speake or had its effect ^on^ some hundred parts of the Ignobler Mettall.

I have ^the less^[164] scrupl'd to[165] ^mention at the beginning of this Narative the experiment that miscaried because 'tis not only introductory to what follows but declares that a thing so little beleiv'd by many good Chymists[,] as the posible Mercurification of Mettalls was affirm'd to me upon his owne knowledge by such a Person that could transmute Mettals & I have not much scrupl'd to be very^ particular, (perhaps even to tediousness) in setting downe the Circumstances of the foregoing account because of the importance of the subject it relates to & the generall difidence that wary men are wont[166] upon no slight grounds to have of the Narratives that are made of projections by profest Chymists or those that have their relations from them who were generally by the cautious suspected more than I hope I shall be by them that know me[,] of Credulity[,] imposture or unskillfullness. I might if discretion did not forbid it say divers things[167] that are not common about this ^illustrious^ Adeptus & his ^young^ Disciple ^&^ thus much I shall [163] venture to add[:] that after our Traveller went away the esteem I had for the rarities committed to his trust was not only confirm'd but hightened partly by a strange experiment which I must not now mention that by his assistance I made with my owne hands when he was not within a 100 mile of the place [and] partly by what was told me by ^two^ strangers who[168] ^knew nothing of^ what I have been hitherto relating, that seperately told me of two severall transmutations that our Traveller had made in one towne of both which operations they were assur'd by the Persons themselves in whose sight & favour they ^were^ made whereof one ^whose name I know^ is at present a learned professor of Phisick in a university not very far of[f][bb] & partly because the Phisitian[169] already often mentioned who is himselfe ^a man of note &^ a scholar[170] has divers times assurd me that besides some other surpriseing things he saw our Traveller do he saw him make at the request of my relator[171] two other transmutations besides that which he & I were present at together one whereof was made upon Lead & the other (to show the Doctor that the vertue of the powder was not confind to that Mettall) upon Copper[172] and in this last the Phisitian for fuller satisfaction[173] would needs[174] have the operation try'd[175] on some of our English Copper farthings that he took out of his owne Pockett, which tho much more Difficulty melted then the Lead had been were no less really transmuted into Gold.

And 'tis not long since that being visited by an outlandish[cc] virtuoso ^that^[176] came to see [163*v*] the Country & brought with him marks of his being a ^Man^[177] of Quality, of fortune, & of parts, he told me when we began to converse freely[178] which he would not at first do, being a[179] Man of a reserv'd humour[,] that some months before he went to visitt & pass some time with a freind of his whom he nam'd to me that was then & is still receiver generall for the French King in one of the fairest Cities of france, which was not unknown to me by reason of the Curiosity that once lead me to see it.[dd] In the house of this Treasurer,[180] who has been long adicted to Chymistry, there ^lodg'd^[181] at the time by the Masters invitation a Person that ^to his host^ pretended to be an Adeptus & had gain'd Credit to his pretence by some projections that he had made in his presence, this Treasurer finding my relator difident as haveing never seen such an operation ingagd the Chymist to lett him be a Criticall spectator[182] to convince the virtuoso I had the story from and partly also to have a skilfull eye wittness, who would be able & forward to discover the fraud if he could find any in

[bb] Almost certainly Edmund Dickinson; see chapter 4.

[cc] I.e., foreign.

[dd] French king = Louis XIV; "one of the fairest Cities" presumably seen during Boyle's grand tour in 1641–1642; see Boyle, "An Account of Philaretus," printed in Hunter, *Robert Boyle by Himself*.

the operation. Upon this account the Virtuoso was present when the Alchymist made another projection by which he ^fairly^ turn'd a Mass of Lead into good Gold but tho the quantity of the transmuted Mettall was extraordinary for the Crusiple being large & haveing been almost full of Lead the Gold amounted to above 15 lbs, yett that which I cheifly minded in this Narrative & for which I here mention it is that my Relator tho convincd that the powder of projection was true & excellent[183] did by severall discourses find the possessor of it so very ignorant of the more secret parts of Chymistry that he firmly concluded that he never made or knew how to make that powder but gott it by some indirect means & accordingly warn'd the Treasurer to be upon his gaurd[;] who [the Treasurer] lately sent him word that the pretended Artist was fled from his house & that by sedulous inquiry that he ^found the imposter^ had stolen the powder with which he made so many projections from a Person[184] whose name, when I came to learn [it,] I found to be the very same & to be mentioned with some of the same titles & other Circumstances with the Philosopher that the Gentleman who show'd me projection own'd to me to be his Master, & the Person of Honour who told me this without knowing that I had heard of such a man, added that the same Philosopher had in his passage by such a town in france bestowed upon a curious and eminent person that still lives there & is my relators particular freind as much powder as being cast upon an imperfect mettall transmuted a mass of it worth 400 Ducats which amounts to above 300 & 50 lbs Sterling & for my further satisfaction gave me an address & recommendation to this Gentleman to send me if I desired it a full account of the whole transaction.

Second Part
(Fragments 11–16)

FRAGMENT 11 [RSMS, 199:121][a]

Whilst the President & some of the chief of the Assembly were entertaining the illustrious Stranger, & acquainting him with the Laws & Customs practis'd by their Society for the regulation of their Meetings, ^most of the^[185] Members sorted themselves into severall small Companyes, as Chance or Choice determin'd them. And I having observ'd that in one of these *Philetas, Cleanthes, Eleutherius* & three or four others[186] had associated themselves, with Intention not to loose so much time, as *Heliodorus* would be oblig'd to spend in[187] giving Answers to the curious Questions of the inquisitive Stranger; I desir'd leave, & easily obtain'd it[,] to[188] be an Auditor of their ^Conference^, which was[189] soon begun by *Eugenius;* who ^turning^[190] to *Philetas,* express'd himself in this manner. You will, I hope, pardon me if Charity as well as Curiosity puts me upon desireing, that whilst the prosecution of the main Controversy betwixt the Chrysopoeans

[a] Margin: "T[ranscri]b'd."

& their Antagonists is suspended, I may be inform'd upon what Grounds the Hermetical Philosophers, write [121*r*] so invidiously ^even^ of their Medicinal *Arcana,* and[191] appear so solicitous to conceal those noble Remedyes, which would be ^highly^ beneficial to Mankind.

Philetas was preparing to return *Eugenius* an Answer, when *Simplicius,* not having the patience to stay for it, address'd himself to[192] ^the same person^ & told him; I doubt ^*Eugenius*^ you have ask'd *Philetas* a puzzling Question; for I much fear it will not be easy, if at all possible, for him to render you good Reasons of a Practise against which[193] ^unanswerable^ ones may be alleag'd, ^on the account^ not only of Reason, but of Religion too. If you please, ^sayes^ *Philetas,* (a little surpris'd at this briskness) to let us know what those Reasons are, we shall be the better able to judge, whether[194] you have right to call them unanswerable. The rest of our little Company having joyn'd in seconding *Philetas's* Motion, *Simplicius,* who was fully perswaded of his own being in the right, stay'd not to be further press'd to begin his Discourse, but ^presently^ did it in the following Terms.[b]

FRAGMENT 12 [RSMS, 199:25–27][c]

I know saith Erastus that one of the ^grand^[195] proofs that is brought by the Lapidists for the real existance of the Philosophers stone is grounded on the Authority of those thy [*sic*] call the *Adepti*[196] ^as^ these are by them belived ^to be^ most profound Philosophers & consequently fit to have their doctrine and assertions intirely acquiest in. But I confess this argument is of no great weight with me who am so far from being in this point of the Chymists minde that instead of ^acknowledging^[197] the generality of the writers of the Theatrum Chymicum to be incomparable Phylosophers I question wether they deserve to be repeated Phylosophers at all. And to this opinion I have been lead by considering partly the matter of their writing and partly the stile.

And that I may not wander to variety of Authors I will at present pitch[198] upon Raimond Lully and that ^antient & numerous^ school or sect of Chymists whereof he is the master to which[199] school may be refer'd Parissinus[,] [25*v*] Ripley and many others that I shal not stay to name.[d]

[b] Note that Eugenius asks specifically about *medical* receipts; the following fragments do not directly address this topic. Thus either a significant lacuna exists here, or fragments 12–16 do not belong to this *Dialogue;* see the introduction to this appendix.

[c] Margin: "T[ranscri]b'd."

[d] On Lull, see appendix 1, First Part, n. o. "Parissinus" (Christopher of Paris), is an obscure alchemical author, proportedly thirteenth-century, whose best-known work is the *Elucidarius* (*TC*, 6:195–270). His attribution to the Lullian school is based upon his use of an alphabetical code like that in some of the pseudo-Lullian works. One of Boyle's alchemical correspondents promises to send Boyle a six-folio key to "Parisinus's code" (BL, 1:64). Parisinus is also the name chosen for the alchemical student in the surviving fragment of Boyle's dialogue on alchemical supernatural arcana; see appendix 3. George Ripley was a highly regarded English alchemist of the fifteenth century. The most popular of his many works is the *Compound of Alchymie,* better known as the *Twelve Gates;* it was first published at London in 1591, and

There[200] are 2 things (co^n^tinues *Erastus*) that keep me from looking upon our mystical Spagyrists as great Philosophers; the first is taken from the consideration of the materials that they do or ^do^ not employ in their writings, and the other from the ratiocinations & the stile that they make use of in them.

And first I do not perceive in their discourses any thing that argues that they were great masters, not ^to^ say so much as forward Apprentices in divers parts of learning, that are either necessary or very conducive to make ^one^ a good Naturalist, and much more to make him a profound Philosopher. I me^e^t with[201] in their writtings no documents of any egr^e^gious skill either in Geometry, Cosmography, Anatomy, Michanicks, Botanicks, Staticks, or Opticks. ^And,^ tho some of them pretend much to Astronomy and Astrology, yet I fear wee shall finde none ^of^ their discoveries or observations in the copious collections of the learned and diligent Ricciolus[e] or other Mode^r^ns[202] that have given us catalogues of the promoters of the Astral science.

[26*r*] And when haveing observed how many ^usefull^ parts of[203] Philosophy our Adepti do not appear to be acqu^ai^nted with, I come to consider what kind of philosophy 'tis that they do most commonly[204] (not to say solely) employ, I cannot but think them to have either less learning or less judgement then their disciples ascribe to them, for the Philosophic notions they are wont almost at every turn to make use of, are taken from ^the^ narrow[,] precarious & barren doctrine ^of the Peripateticks^ about the 4 Elements & their imaginary first Qualities. Which doctrine thô the more judicious of the later Aristotelians themselves have[205] been asham'd of, and therefore thought fit to mince and disguise it, yet the more antient of our *Adepti* (and after them most of their disciples) ^were^[206] involved in ^the^ profound ignorance of the times they lived in, and thought this doctrine of their adversaries the schoolmen so fine a piece of learning that they[207] adopted it, and affected so much to seem ready in it that in some books you shal meet with it, either employ'd or suppos'd in allmost every page or C^h^apter thô often times they wrest it so manifestly [26*v*] or apply it so unfitly[208] that they seem resolv'd rather to bring it in upon ^any^ terms, then[209] forbear any occasion they can meet with or create to make use of it, forgeting[210] that known sentance of the Citharaedii Ridetus c^h^ordâ qui semper oberrat eâdem[.][f] one would expect that such illuminated philoso-

---

numerous times thereafter in various languages, but the best edition is Elias Ashmole's, in *TCB*, 107–93. On Boyle's reading of Ripley, see chapter 5.

The following paragraph repeats the thought of the preceding paragraph, and Boyle does not provide the critique of the Lullian school prefaced in the present paragraph.

[e] Giovanni Battista Riccioli (1598–1671), who compiled several collections of astonomical data, including *Almagestum novum* (Bologna, 1651), which included a *Chronicon astronomorum vel astrologorum* (1:xxvi–xlvii).

[f] Horace, *Ars poetica* 356; "ut citharoedus Ridetus, chorda qui semper oberrat eadem [Like that lyre-player Ridetus, who always blundered on the same string]."

phers as our *Adepti* are said to be sh^o^uld by the light of their chymical fires make rare discoveries of noble and usefull Th^e^or^i^es as well as Experiments, and should adorn their writtings with such instructive notions and hints as should inrich the understanding of their readers, thô not their purses.[g] And since their Publishing books to the world, and the preference they therein give themselves before the rest of mankinde, do not argue them to be despisers of fame or ^haters^[211] of applause, one would think that out of kindness to their own reputation ^if not to their readers^ they would take ^now & then,^ what they may often easily finde, occasions to give the world some choise *specimens* or tasts[212] of their egregious skill in several parts of Philosophy, if really [27*r*] they were such masters of it as they would have us belive ^them^· But I am so far from finding in their books either clear notions, or firm deductions, or those Coruscations of Philosophical light that the Titles would make one expect that I finde ^there^ for the most part a confused *chaos* of dark & indigested fancies, and som scraps of Philosophy, which as far as they are intelligible are but mean[,] narrow[,] superficial and in a word[,] Plebe^i^an.

FRAGMENT 13 [RSMS, 199:27*v*–29*v*][h]

Nor are the things they[213] employ for Illustrations, wont to be either learned or very apposite: for they are usually taken from the most ^obvious^[214] things that occurr in nature, & the most trivial Operations of what I can scarce vouchsafe to call *Art.* A man need not be a vulgar Philosopher, much less an *Adeptus* to be stor'd with such slight Comparisons: to be a Plowman or a Gardner may abundantly serve the turn. Not that I despise all *Similes* that are borrow'd from familiar things, merely because they are so; but I cannot think they argue any deep inspection into the dark Mysteries of Nature, nor[215] any great Judgement ^neither^, unless they be very apposite to the thing to be declar'd by them; And this the Illustrations of our Lapidists are most commonly very far from being. For they usually borrow their similitudes to explicate the Transactions that occurr between Mineral Bodies, ^from^[216] what passes between men & women, or ^in^ the Nutrition & Propagation of Plants, which are Bodies of quite differing kinds from Metals, these being Inanimate Creatures & those living ones, so that in the Chymists [28*r*] Phrase & Estimate, these belong to the Mineral, & those to the vegetable or the Animal Kingdom. How ^for instance^ can it make me understand the manner how Gold or Silver, that are dead & inorganical Bodies, are produc'd, and much less how either of them is propagated, to tell me a story of the Conjunction of *Gabricius* & *Beia*, of a red man & a white woman,[i] that are suppos'd to be Animals[217] of differing Sexes, & furnish'd with all the Organs (which are both many & various & exquisitely contriv'd) that are necessary or usefull to the Propagation of the *Species?* I

[g] Compare this line to *Anti-Elixir,* 278, and *Origine of Formes and Qualities,* 367.

[h] Margin: "T[ranscri]b'd."

[i] These terms are common *Decknamen* of ingredients for the Philosophers' Stone, not, as Erastus claims, in the constitution of metals.

shall add that in many of their Illustrations some things are suppos'd & slily impos'd upon the unwary Reader, that a Philosopher ought not without clear proof to admit: as when the Lapidists pretend to explain some things either by the Sperm of Metals or by the Life or Soul of gold, & ^by^ other like precarious suppositions. In short many of the Comparisons they make use of to explain things, are so unfit for that purpose, that themselves need Explications, & they rather darken than illustrate the Subjects they are so incongruously apply'd to.

But perhaps our Hermetic Sages are at [28*v*] lest good Logicians, ^&^ thô their Explications be but mean, are wont to make cogent Inferences. But *thô* I have sometimes employ'd my self in making Algebraical Calculations, & examining Geometrical Demonstrations, which are Exercises that require, & after some practise are wont to confer, a more than ordinary patience & application of mind; yet I confess that[218] I have often met with Argumentations in the Lapidists Books, that all my attention would not help me to discern the connexion or consecution of. And I sometimes thought, that 'twas no less Difficult to understand the force of their Ratiocinations, than the Process of their Elixir. Nor I think it ought to serve their turn on this occasion, as they will have it do upon some others; ^To say that^ they are not oblig'd to write clearly. For what ever excuse that may be when they deliver their *Arcana,* he that uses an Argument is suppos'd to employ it to convince another, & therefore must propose it so as that he may discern the Consecution of it, & if that be not discernable, he may safely conclude the Argument is [29*v*] to be judg'd either sophistical or vain and illusory.

These are concludes *Erastus,* some of the things that keep me from looking upon the Chrysopoean Writers, with ^near^ that veneration which their Disciples give them. And I confess that when I meet ^even in their^[219] applauded Books, such trivial Notions of Philosophy, such incongruous Illustrations, & such weak or undiscernable Inferences, I cannot look upon them as very profound, & much less as the only, Philosophers. And cannot but wonder, either that Authors no better qualify'd should so arrogantly pretend to Medicines capable of purifying & emproving mens Bodies as well as ^the^ imperfect Metalls, or that the Possessors of such improving *Arcana* should in their Writings leave so little footsteps of extraordinary, not to say of so much as ordinary, knowledge: & upon this account I hope I shall be excus'd if I do not look upon their authority as a weighty Argument to perswade me of what is in it self so unlikely as that they had the Philosophers Stone.

FRAGMENT 14 [RSMS, 199:46–51][j]

There is one thing perticularly (continues[220] ----------) that I can by no means tollerate in the way of wrighting affected by those that ^pas for^[221] Adepti in the Theatrum Chymicum and that is their gross hypocrisy. For most of these wrighters after they have frequently cald the^ir^ reader their son[222] & ^made^ solemn professions[223] & protestations (and that perhaps

[j] Margin: "T[ranscri]b'd."

not without interposeing the sacred name of God) that they will disclose to him their secrets clearly & candidly, & ^after they have^[224] by those artifices[225] made him gape wide to receive what they promise him they are not ashamed to[226] [47*r*] to put him off with riddles insteed of instructions, & delude him by sayings that are either so dark as not to be ^at all^ intelligible, or which is wors so deceitfully expressd as to make him think he understands[227] a truth that they never meant to teach him, so that insted of the radient light they promisd him they put him off either with shaddows that have much obscurity & no solidity, or deceitfull light that like Ignes fatuii[k] mislead insted of guideing him. Nay adds ---------- the wrighters I speak of are so habitualy unsincere that they employ [47*v*] this delusory way of wrighting not onely when they treat of the most secret & ^essential^[228] poynts of their great Arcanum such as are their matter & the first preparations of it but when they discource of things ^of lesser moment^ where they ^might^[229] exercise their candour & benevolence with out at all discovering the grand secrets of their work.

I can not[,] pursues ----------[,] but think it unworthy of Christians or of Philosophers & even of men to abuse ingenious or at least studious persons that never Injurd or [48*r*] provokd them, under a pretence of no less than paternal kindness. And they must little know or little value the pleasures & the troubles of the mind to make it their sport to wrack the brains of sedulous enquirers after truth with professions of haveing ^candidly^ deliverd it in expressions so dark & so ambiguous as to produce ^even^ in thier[230] attentive readers nothing but anxiety or mistakes. And this way of wrighting is so much affected by these hermetick hypocrits that it occurs almost in every ^leaf of their books^[231] & it is nausiously employd [48*v*] on so many occasions, that their protestations of sincerity are little less numerous then their riddles, & those indulgent fathers do as much oppress their childrens patience as deceive their expectations. Were it not much better that[232] these pretended instructers would plainly[233] & with^out^ tedious circumlocutions tell their reader that they dont mean to discover their secrets to him[234], or at least that they do not intend to do it plainly but leave him to guess at the sense of [49*r*] what they deliver as well as he can, for why should they do so fraudilent & inhuman a thing as under a pretense of great kindness to raise a longing expectation that they never meant to answer.

I know subjoyns ---------- that this is said in their defense that sometimes they declare they are not to be litteraly understood. But beside that this excuse is not applicable to all of them, besides this I say since these declarations are very unfrequent & since their professions of kindness are ^very^ often subsequent (as well as precedent) to such [49*v*] intimations,[235] what Reader that ^being^ himself[236] an honest man is on that score disposd to[237] believe the Author ^to be^ so, can think other then that by the subsequent protestations of kindness he meant to ^releas[e]^[238] the right ^(such as it

[k] I.e., will-o'-the-wisps.

was) that^ he could pretend to[239] by the intimation he had given long before that he meant to wright sometimes obscurely. Indeed ---------- the first or second person that made use of this deceitfull way of wrighting might with some colour [50*r*] though not justify yet excuse his style by[240] ^aledging^ that he thought[241] mens sagacity or good fortune might bring them to understand his[242] dark sayings & reap some benefit by[243] the incouragements he gave them to suppose they understood his meaning[;] but after the universal experience of divers ages has made it but too manifestly apear that these pretended paternal instructions of the adepti have bin so little understood ^even^ by those that studyed them with great application of mind and proceeded to action upon the incouragements [50*v*] given them have[244] generaly lost their expectations & many of them their time, their money, & their health[;] after this long experience I say[,] to continue to alure innocent & studious persons with professions of great kindness to[245] think themselves well instructed by what the author knows will a hundred to one delude his reader & make him fall into a precipis is a peractis that can scarce be too much condemd, and ^though something could^[246] be pretended to[247] ^to make it consistent with^ justis yet I see nothing that can [51*r*] reconcile it to sincerity ^much less^ kindness.

FRAGMENT 15 [RSMS, 199:52–55][l]

Methinks sais ---------- not without some fierceness in his looks & the tone of his voyce you make very bold with the hermetick Philosophers,[248] & too much forget the veneration their ^profound^ knowledg chalenges & the gratitude due to these great instructors of the curious & benefactors of mankind that have[249] many times built hospitals for the poor & oftentimes made their disciples more rich then princes.[m]

There[250] are certain rules answers ---------- of justis[,] equity & decency that are[251] to be observd ^thô it were but^ for our ^own^ sakes when we[252] speak of the authors of books; and those bounds [52*v*] I ^am still^[253] carefull not to transgress when I speak of the most censurable of the Authors of the Theatrum Chymicum.[254] But beyond what those rules exact I confess I know no obligation I have to treat them with more respect & gratitude then other wrighters. For as for their learning [Erastus][n] has already shown that it is not transcendent,[255] but rather inferior to that of ^a great^ many other wrighters[256] & though they are pleasd (such is their modesty) to stile themselves philosophers exclusively to others yet why We should therfore think them so I confess I do not see. If they were[257] realy (which few of this judicious company believe them to have bin) possessors of the Elixir yet what is that to[258] us that never are the better for their [53*r*] wealth and have[259] ^not on that^ score any reason to prefer them to many rich bankers

[l] Margin: "T[ranscri]b'd to the break" (in Robin Bacon's hand).

[m] Nicholas Flamel and his wife Perenelle are the most renowned for having built hospitals, churches, and almshouses with their transmuted gold.

[n] Space for name left blank; obviously Erastus from fragment 12.

& merchants[260] by whose ^wealth^ none but a few of their ^own^ nere relations or intimats were enrich'd, & there is no more cause[261] why I should prefer Bernard Trevisan for instance; or Tepheus[o], to Galileo or sir Francis Bacon, as that we should respect the memory of Cresus more than that of Ptolomeus Philadelphus or that of Crassus more than that of Cicero or Mecenas. And to speak freely continues --------- we are so far from being bound to treat these invidious & deceitfull Adepti with a peculiar respect & kindness that where tis well if we forbear treating them with indignation for ^supposing themselves to have bin^[262] posessors of such wealth & such [53*v*] medicines as they ^boast^[263] of, we are very little beholden to such ill naturd persons as might without impoverishing themselves do men a great deal of good[264] & yet would not do any. And as for mankind in general they have bin solicitous not to deserve well, so from those in particular that cultivate the study of nature they scarce deserve common civility, since they have exercisd none in wrighting of them[;] for in a hundred places of their books where they speak of Physitians, or of the Chymical students of natures secrets that have not hit upon their matter & their way of managing it, they aford them no better Epithets then Ignoramus's [54*r*] fools, Imposters, ^or^[265] Sophisters though many of these were ^not onely more sincere & usefull persons[,]^ & considering learning in general[,] better scholars & perhaps better naturalists then themselves, & 'tis certainly a great want not onely of civility but of common equity to insult over studious & industrious men for not haveing bin succesfull in very difficult atempts, especialy since many of those who were misled by the dark & envious wrightings of these that dispose them, and these insolent adepti are perhaps the onely sort of wrighters in the world that think a man cannot be mistaken ^or make a paralogisme^ about an abstruse subject without being a sophister or a Duns. how much more like a good man & a philosopher did the great Hypocrites[p] who [54*v*] alone has done mankind more servis then all the adepti put together, speak of ingenious men who misd of the[266] mark where he sais

[8 cm left blank]

and as for the great good that --------- intimates that they have done the world though we should grant that here & there some of them (for we hear not of many) have, ^as tis said of the Fairies, dropt some money in places where they but pasd & quickly vanishd^ now & then privately relievd ^the wants of^ an indigent person or rectified the mistakes [55*r*] of an erroneous ^Chymist^,[267] or ^even^ built some almshouse yet generaly speaking the good they have done in the world by their private charitys is much inferiour, to the harm they have done by their mischievous writings[268] by which (& they might ^fore^se[e] it would prove so) ^to a very few that have bin bene-

[o] Bernard Trevisan, or Bernard of Trier, was a fifteenth-century alchemist and author of the *Epistola ad Thomam de Bononia,* a chrysopoetic work popular in the seventeenth century. The phrase "or Tepheus" is the amanuensis's error for the similarly sounded name Artephius (on whom see chapter 5, n. 126).

[p] I.e., Hippocrates.

fitd^ hundreds have ^bin^ ruind, & thousands bin damnified, either in poynt of estate or health or reputation or ^other mens concerns^[269] so that we may safely say that the serch of their great medicins have made[270] incomparably more sick men then the medicins themselves have curd, & they have fild far more hospitals than ever they have built.

FRAGMENT 16 [RSMS, 187:6v--8][q]

As to the Argument propos'd in favour of the Books printed by the Adepti you speak of, I am content so far to allow of the allegation as to grant that these Books have been the occasion of ^som^ mens lighting upon some things, not unusefull to ^physicians &^ Apothecarys[271] and perhaps to some minerallogists too, by exciting their Endeavors by the Proposal of such rich & desirable recompences as would make ^the greatest^[272] costs and Labours in pursueing them a cheap price for them. But when I have granted thus much I must own, that I think the writings of the Adepti collected in the Theatrum Chymicum to be Books for which the Publick is more justly warranted to complain of the Authors then oblig'd to thank them; and that upon the whole matter they have done more harm than good, if it were but for this reason, [7*r*] that they ^have^ hinder'd ^or retarded^ ^a great deal more^[273] of solid & useful knowledg then they have either imparted or procur'd. For thô I[274] grant that men in seeking with great industry cost & obstinacy for what they call the Philosophers Stone, have some of them now & then chance to light on, an useful (or if you please) a noble Experiment or two: ^yet^[275] besides that many more seekers have wholly lost their ^time^[276] and costs &[277] hopes ^& oftentimes^ which is worse their healths & reputations too, in following those alluring but unfaithfull guides, what can these few & Casual hits amount to in comparison of that rich harvest ^of Experimental knowledg^ that might have been justly expected, if those great Charges, labors, & In^de^fatigable Industry had been imploy'd in a regular & methodical way by analysis & all ^other^ kind of Chymical and mechanical [7*v*] Operation[278] to discover the true nature^s^ of Mettals & Minerals & the wayes of preparing[279] both them ^&^ vegetables & Animals for the use of Physicians[,] Apothecarys, Goldsmiths, ^Refiners^[280], Say-Masters, Mint Men[,] Painters, & especially Myne workers and Mineralogists. To all which I doubt not but for less charges & endeavors, than the envious & deluding obscurity of the Adepty's writings have[281] defeated, would have ^make^ [*sic*] Chymistry a great Benefactor ^to mankind^ & convinc'd the world to its ^great^ advantage, how much ^true^ knowledg is likely to be more advanc'd ^by a rational &^ orderly way of investigating truth by the Instruments ^that^ Chymistry[282] affords than by attempts that have no solid ground or Inductions[283] To proceed upon[,] but instead of pursuing the Interpretation of Nature, prosecute that of Ridles.

[q] Margin: "NN speaks" (penciled), "T[ranscri]b'd."

Conclusion

FRAGMENT 17—[*Anti-Elixir;* in the interest of continuity I have placed the printed "Publisher's Note" after the conclusion of the *Dialogue.*]

[1] After the whole Company had, as it were by Common Consent, continued silent for some time, which others spent in Reflections upon the Preceding *Conference,* and *Pyrophilus,* in the Consideration of what he was about to Deliver; this *Virtuoso* at length stood up, and Addressing himself to the rest, "I hope, *Gentlemen,* sayes he, that what has been already Discoursed, has Inclin'd, if not Perswaded you to Think, [2] That the Exaltation, or Change of other *Metals* into *Gold,* is not a thing Absolutely Impossible; and, though I confess, I cannot remove all your Doubts, and Objections, or my own, by being able to Affirm to you, That I have with my own hands made Projection (as *Chymists* are wont to call the Sudden Transmutation made by a small quantity of their Admirable *Elixir*) yet I can Confirm much of what hath been Argued for the Possibility of such a sudden Change of a Metalline Body, by a Way, which, I presume, will surprize you. For, to make it credible, that other Metals are capable of being Graduated, or Exalted into Gold *by way of Projection;* I will Relate to you, that *by the like way,* Gold has been Degraded, or Imbased."

The Novelty of this *Preamble* having much surprised the Auditory, at length, *Simplicius,* with a disdainful Smile, told *Pyrophilus,* "That the Company would have much thanked him, if he could have assured them, That he has seen another Mettal Exalted into Gold; but, that to find a way of spoiling Gold, was not onely an Useless Discovery, but a Prejudicial Practice."

Pyrophilus was going to make some Return to this *Animadversion,* when he was prevented by *Aristander;* who, turning himself to *Simplicius,* told him, with a Countenance and Tone that argued some displeasure; "If *Pyrophilus* had been Discoursing to a Company of Goldsmiths, or of Merchants, your severe Reflection upon what he said would have been proper: but, you might well have forborn it, if you had considered, as I suppose he did, that he was speaking to an Assembly of *Philosophers* and *Virtuosi,* who are wont to estimate Experiments, not as they inrich Mens Purses, but their Brains, and think Knowledge especially of uncommon things very desirable, even when 'tis not accompanyed with any other thing, than the Light that still attends it, and indears it. It hath been thought an Useful Secret, by [3] a kind of Retrogradation to turn Tin and Lead into brittle Bodies, like the Ores of those Metals.[a] And if I thought it proper, I could shew, that such a change might be of use in the Investigation of the Nature of those Metals, besides the practical use that I know may be made of it. To find the Nature

[a] Presumably Boyle refers to the calcination of these metals to their oxides, and the fusion of those oxides (with sand) to provide high-quality glass.

of Wine, we are assisted, not only by the methods of obtaining from it a Spirit; but by the ways of readily turning it into Vinegar: the knowledge of which ways hath not been despised by Chymists or Physitians, and hath at *Paris,* and divers other places, set up a profitable Trade. 'Tis well known that divers eminent *Spagyrists* have reckon'd amongst their highest *Arcana* the ways by which they pretended, (and I fear did but pretend) to Extract the Mercury of Gold, and consequently *destroy* that Metal; and 'twere not hard to shew by particular instances, that all the Experiments wherein Bodies are in some respects deteriorated, are not without distinction to be rejected or despis'd; since in some of them, the Light they may afford may more than countervail the Degradation of a small quantity of matter, though it be Gold it self. And indeed, (continues he) if we will consider things as Philosophers, and look upon them as Nature hath made them, not as Opinion hath disguised them; the Prerogatives and usefulness of Gold, in comparison of other Metals, is nothing near so great as Alchymists and Usurers imagine. For, as it is true, that Gold is more ponderous, and more fix'd, and perhaps more difficult to be spoiled, than Iron; yet these qualities (whereof the first makes it burthensom, and the two others serve chiefly but to distinguish the true from the counterfeit) are so balanced by the hardness, stiffness, springiness, and other useful qualities of Iron; that if those two Metals I speak of (Gold and Iron) were equally plentiful in the World, it is scarce to be doubted, but that Men would prefer the [4] more useful before the more splendid, considering how much worse it were for Mankind to want Hatchets, and Knives and Swords, than Coin and Plate? Wherefore, (concludes he) I think *Pyrophilus* ought to be both desired and incouraged to go on with his intended Discourse, since whether Gold be or not be the Best of Metals; an assurance that it may be degraded, may prove a Novelty very Instructive, and perhaps more so than the Transmutation of a baser Metal into a Nobler. For I remember it hath long pass'd for a Maxim among Chymical Philosophers, That *Facilius est aurum construere quam destruere:*[b] And whatever becomes of that, 'tis certain that Gold being the closest, the constantest, and the least destructible of Metals, to be able to work a notable and almost *Essential change* in such a Body, (though, by detereorating it) is more than to work a *like change,* (though in popular estimation for the better) in any Metal less indisposed to admit alterations, especially in such an one as *Pyrophilus* intimates, by telling us, that 'twas made by *Way of Projection* and consequently by a very small proportion of active matter; whereas the destructions that vulgar Chymists pretend to make of Gold, are wont to be attempted to be made by considerable proportions of Corrosive *Menstruums,* or other fretting Bodies; and even these, Experience shews to be usually too weak to *ruine,* though sometimes they

[b] "It is easier to make gold than to destroy it." Boyle also quotes this in the *Sceptical Chymist,* 177, and in *Origine of Formes and Qualities,* 363, where he attributes it to Roger Bacon.

may much *disguise* the most Stable Texture of Gold. *Cuncta adeo miris illic complexibus haerent.*"[c]

Pyrophilus perceiving by several signs that he needed not add any thing of Apologetical to what *Arristander* had already said for him, resumed his Discourse, by saying, "I was going, Gentlemen, when *Simplicius* diverted me, to tell you That looking upon the Vulgar Objections that have been wont to be fram'd against the possibility of Metalline Transmutations, from the Authority and Prejudices of [5] *Aristotle,* and the School-Philosophers, as Arguments that in such an Assembly as this need not now be solemnly discuss'd; I consider that the *difficulties* that really deserve to be call'd so, and are of weight even with Mechanical Philosophers, and Judicious Naturalists, are principally these. *First,* That the great change that must be wrought by the *Elixir,* (if there be such an Agent) is effected upon Bodies of so stable and almost immutable a Nature as Metals. *Next,* That this great change is said to be brought to pass in a very short time. *And thirdly,* (which is yet more strange) That this great and sudden alteration is said to be effected by a very small, and perhaps inconsiderable, proportion of transmuting Powder. To which *three* grand difficulties, I shall add *another* that to me appears, and perhaps will seem to divers of the new Philosophers, worthy to be lookt upon as a *fourth,* namely, The notable change that must by a real transmutation be made in the Specifick Gravity of the matter wrought upon: which difficulty I therefore think not unworthy to be added to the rest, because upon several tryals of my own and other men, I have found no known quality of Gold, (as its colour, malleableness, fixity, or the like) so difficult, if not so impossible, to be introduc'd into any other Metalline Matter, as the great Specifick Gravity that is peculiar to Gold. So that, Gentlemen, (concludes *Pyrophilus*) if it can be made appear that Art has produc'd an *Anti-Elixir,* (if I may so call it) or Agent that is able in a very short time, to work a very notable, though deteriorating, change upon a Metal; in proportion to which, its quantity is very inconsiderable; I see not why it should be thought impossible that Art may also make a *true Elixir,* or Powder capable of speedily Transmuting a great proportion of a baser Metal into Silver or Gold: especially if it be considered, that those [6] that treat of these *Arcana,* confess that 'tis not every matter which may be justly called the Philosophers Stone, that it able to transmute other Metals in vast quantities; since several Writers, (and even *Lully* himself) make differing *orders* or *degrees* of the *Elixir,* and acknowledge, that a Medicine or Tincture of the first or *lowest* order will not transmute above ten times its weight of an Inferior Metal."[d]

[c] "Thus all together its properties cling there in tight embraces." Slightly misquoted from Giovanni Aurelio Augurello's poem *Chrysopoeia* (*TC,* 3:197–244, on 208), originally published in Venice, 1515. Boyle quotes the same line in the *Sceptical Chemist,* 56, as "Cuncta adeo miris illic compagibus haerent." In either case, Boyle retains the sense of the steadfastness of gold's properties as expressed by Augurello.

[d] On the orders of the Stone (here, "Medicine or Tincture"), see the discussion of multiplication in chapter 3.

Pyrophilus having at this part of his Discourse made a short pawse to take breath, *Crattippus* took occasion from his silence to say to him, "I presume, *Pyrophilus,* I shall be disavowed by very few of these Gentlemen, if I tell you that the company is impatient to hear the Narrative of your Experiment, and that if it do much as probably make out the particulars you have been mentioning, you will in likelyhood *perswade most* of them, and will certainly *oblige* them *all.* I shall therefore on their behalf as well as my own, sollicite you to hasten to the Historical part of a Discourse that is so like to gratifie our Curiosity."

The Company having by their unanimous silence, testified their approbation of what *Crattippus* had said; and appearing more than ordinarily attentive,[e]

As I was one day abroad, saith *Pyrophilus,* to return visits to my Friends, I was by a happy Providence (for it was beside my first Intention) directed to make one to an Ingenious Foreigner, with whom a few that I had received from him, had given me some little acquaintance.

Whilst this Gentleman[f] and I were discoursing together of several matters, there came in to visit him a stranger, whom I had but once seen before; and though that were in a promiscuous company, yet he addressed himself to me in a way that quickly satisfied me of the greatness of his Civility; which he soon after also did of that of his Curiosity.[g] [7] For the *Virtuoso,*[h] in whose Lodgings we met, having (to gratifie me) put him upon the discourse of his Voyages; the curious stranger entertained us an hour or two with pertinent and judicious Answers to the Questions I askt him about places so remote, or so much within Land, that I had not met with any of our English Navigators or Travellers that had penetrated so far as to visit them. And because I found by his discourse that I was like to enjoy such good company but a very little while, (since he told me that he came the other day into *England* but to dispatch a business which he had already done as far as he could do it, after which he was with speed to return, as (to my trouble) he did to his Patron that sent him) I made the more haste to propose such Questions to him, as I most desired to be satisfied about; and among other things, enquiring whether in the Eastern parts he had travers'd, he had met with any Chymists; he answered me that he had; and that though they were *fewer,* and more *reserved* than ours, yet he did not find them all less *skilful.* And on this occasion, before he left the Town to go aboard the Ship he was to overtake; he in a very obliging way put into my hands at parting a little piece

[e] Fragment 17c (draft) reads simply: "The Company being all silent & appearing more then ordinarily attentive . . ."

[f] Fragment 17e, the Latin translation, reads, "vir omnigenâ literaturâ clarus, in cujus diversorium casu conveneramus [the man, renowned for all kinds of learning, in whose lodging we met by chance]."

[g] The draft reads here: "he addresed himselfe to me with soe much ^unaffected^ sivility that I soon perceived he was as much a ^Gentleman [replaces *Courtier*] as after I found he was^ a traviller." On the possible identity of this "stranger" and Georges Pierre des Clozets, see chapter 4.

[h] Latin translation reads "generosus ille peregrinus [the foreign gentleman]."

of Paper, folded up; which he said contained all that he had left of a rarity he had received from an Eastern *Virtuoso,*[i] and which he intimated would give me occasion both to Remember him, and to exercise my thoughts in uncommon Speculations.

The great delight I took in conversing with a Person that had travelled so far, and could give me so good an account of what he had seen, made me so much resent the being so soon deprived of it, that though I judg'd such a *Vertuoso*[j] would not, as a great token of his kindness, have presented me a trifle, yet the Present did but very imperfectly consoal me for the loss of so pleasing and instructive a Conversation.

[8]Nevertheless, that I might comply with the curiosity he himself had excited in me, and know how much I was his Debtor, I resolved to see what it was he had given me, and try whether I could make it do what I thought he *Intimated,* by the help of those few *hints* rather than *directions* how to use it, which the parting haste he was in (or perhaps some other reason best known to himself) confin'd him to give me. But in regard that I could not but think the Experiment would one way or other prove Extraordinary I though fit to take a Witness or two and an Assistant[k] in the trying of it; and for that purpose made choice of an experienced Doctor of Physick, very well vers'd in the separating and copelling of Metals.[l]

Though the Company (says *Heliodorus*) be so confident of your sincerity and wariness, that they would give credit even to unlikely Experiments, upon your single testimony; yet we cannot but approve your discretion in taking an Assistant and a Witness, because in nice and uncommon Experiments we can scarce use too much circumspection, especially when we have not the means of reiterating the tryal: for in such new, as well as difficult cases, 'tis easie even for a clear-sighted Experimenter to over-look some important circumstance, that a far less skilful by-stander may take notice of.

As I have ever judged, (saith *Pyrophilus*) that cautiousness is a very requisite qualification for him that would satisfactorily make curious Experiments; so I thought fit to imploy a more than ordinary measure of it, in making a tryal, whose event I imagined might prove odd enough. And therefore having several times observed that some men are prepossessed, by having a particular Expectation rais'd in them, and are inclined to think that they *do see* that happen which they think they *should* see happen; I resolved to obviate this prejudication as much as innocently I could, [9] and (without telling him any thing but the *truth,* to which Philosophy as well as Religion

[i] The draft reads "from his ˆeasternˆ patron," and the Latin translation agrees more closely with the draft than the published text, reading: "quantum residuum erat de κιμήλοι suo eoo patrono acceperat [as much as was left of an earth (reading κιμήλοι as κιμωλία, a kind of earth) he had received from his Eastern patron]."

[j] Latin reads instead "virum, reconditae sapientiae laudem summam meritissimum [a man, most worthy of the highest praise for his recondite wisdom]."

[k] Draft and Latin both read simply "a witness and an assistant."

[l] Edmund Dickinson? See n. t below.

obliges us to be strictly loyal) I told him but thus much of the truth, that I expected that a small proportion of a Powder presented me by a Foreign *Virtuoso,* would give a Brittleness to the most flexible and malleable of Metals, Gold itself. Which change I perceiv'd he judged so considerable and unlikely to be effected, that he was greedy of seeing it severly try'd.

Having thus prepared him not to look for all that I my self expected, I cautiously opened the Paper I lately mentioned, but was both surprized and troubled, (as he also was) to find in it so very little Powder, that in stead of two differing tryals that I designed to make with it, there seem'd very small hope left that it would serve for one, (and that but an imperfect one neither). For there was so very little Powder, that we could scarce see the colour of it, (save that as far as I could judge it was of a darkish Red) and we thought it not only dangerous, but useless to attempt to weigh it, in regard we might easily lose it by putting it into, and out of the Balance; and the Weights we had were not small enough for so despicable a quantity of matter, which in words I estimated at an eighth part of a Grain: but my Assistant, (whose conjecture I confess my thoughts inclin'd to prefer) would allow it to be at most but a tenth part of a Grain, Wherefore seeing the utmost we could reasonably hope to do with so very little Powder, was to make *one* tryal with it, we weighed out in differing Balances two Drams of Gold that had been formerly English Coyn, and that I caused by one that I usually imploy[m] to be *cupell'd* with a sufficient quantity of Lead, and *quarted,*[n] as they speak, with refin'd Silver, and purg'd *Aqua fortis,* to be sure of the goodness of the Gold: these two Drams I put into a new Crucible, first carefully neal'd, and having brought them to fusion by the meer action of the fire, without [10] the help of Borax, or any other Additament, (which course, though somewhat more laborious, than the most usual we took to obviate scruples) I put into the well-melted Metal with my own hand the little parcel of Powder lately mentioned, and continuing the Vessel in the fire for about a quarter of an hour, that the Powder might have time to defuse it self every way into the Metal, we poured out the well-melted Gold into another Crucible that I had brought with me, and that had been gradually heated before, to prevent cracking. But though from the first fusion of the Metal, to the pouring out, it had turn'd[o] in the Crucible like ordinary Gold, save that once my Assistant

[m] Latin reads: "ab amanuensi quem occupatum tenere consuesco [by an amanuensis I usually employ]." An assistant who is both amanuensis and skilled in chymical operations may be Frederic Slare, who was working for Boyle in the 1670s and was admitted F.R.S. in 1680. See Boas Hall, "Frederic Slare, F.R.S. (1648–1727)"; Hunter, *Guide,* xxxv.

[n] The Latin translation inserts a definition of quartation: "quartatione (nam sic ea chymicis operatio audit, qua aurum cum triplo argenti confunditur, et in aqua forti id quod est auro alienum solvitur) [by quartation, for that is what chymists call that operation in which gold is melted with thrice its weight of silver, and then in aqua fortis whatever is not gold is dissolved away]."

[o] Draft reads "melted and turnd." In this case, "turning" refers to the appearance of the molten metal when the crucible containing it is rotated. Different metals and their alloys have different viscosities in the molten state, and a good operator can make judgments regarding the identity and purity of a metal on these grounds, i.e., how it "turns."

told me he saw that for two or three moments it lookt almost like an Opale; yet I was somewhat surpriz'd to find when the matter was grown cold, that though it appear'd upon the Balance that we had not lost any thing of the weight we put in, yet in stead of *fine Gold,* we had a lump of Metal of a dirty colour, and as it were overcast with a thin coat, almost like *half vitrified Litharge;* and somewhat to increase the Wonder, we perceived that there stuck to one side of the Crucible a little Globule of Metal that lookt not at all yellowish, but like course Silver, and the bottom of the Crucible was overlaid with a vitrified substance, whereof one part was of a transparent yellow, and the other of a deep brown, inclining to red; and in this vitrified substance I could plainly perceive sticking at least five or six little Globules that lookt more like impure Silver than pure Gold. In short, this *stuff* look'd so little like *refin'd,* or so much as *ordinary, Gold,* that though my Friend did much more than I marvel at this change, yet I confess I was surpriz'd at it my self. For though in some particulars it answered what I lookt for, yet in others, it was very differing from that which the Donor of the Powder had, as I thought, given me ground to expect. Whether the cause of my disappointment [11] were that (as I formerly intimated) this *Virtuoso's* haste or design made him leave me in the dark; or whether it were that finding my self in want of sufficient directions, I happily pitcht upon such a proportion of Materials, and way of operating, as were proper to make a new Discovery, which the excellent Giver of the Powder had not Design'd, or perhaps thought of.

I shall not at all wonder, saith *Cratippus,* either at your Friends amazement, or at your surprize, if your further tryals did in any measure confirm what the *superficial change* that appeared in your Metal could not but incline you to conjecture.

You will best judge of that (replies Pyrophilus) by the account I was going to give you of what we did with our odd Metal. *And First,* having rubb'd it upon a good Touchstone, whereon we had likewise rubb'd a piece of *Coyn'd Silver,* and a piece of *Coyn'd Gold,* we manifestly found that the mark left upon the Stone by our Mass between the marks of the two other Metals, was notoriously more like the Touch of the Silver than to that of the Gold. *Next,* having knockt our little lump with a Hammer, it was, (according to my prediction) found brittle, and flew into several pieces. Thirdly, (which is more) even the insides of those pieces lookt of a base dirty colour, like that of Brass or worse, for the *fragments* had a far greater resemblance to *Bell-Metal,* than either to Gold or to Silver.[p] To which we added this *fourth,* and

[p] A chemical explanation of the anti-Elixir may be hazarded here. The gold's loss of malleability and color after fusion with a very small amount of material accords well with the action of antimony on gold. Molten gold absorbs antimony extremely readily, even from the vapor state, thereby becoming brittle and white. One part of antimony in 1,920 of gold has been shown sufficient to render it brittle (Carl Hatchett, "Versuche und Beobachtungen über verschiedene Legirungen ges Goldes," *Neues Allgemeines Journal der Chemie* 4 [1804]: 50–92, on 55–56). Antimony was of course a popular material for chymists, so the anti-Elixir may

more considerable, Examen; that having carefully weigh'd out one dram of our stuff, (reserving the rest for trials to be suggested by *second thoughts*) and put it upon an excellent new and well-neal'd Cupel, with about half a dozen times its weight of Lead, we found, somewhat to our wonder, that though it turn'd very well like good Gold, yet it continued in the fire above an hour and an half, (which was twice as long as we expected) and yet [12] almost to the very last the fumes copiously ascended, which sufficiently argu'd the operation to have been well carried on; and when at last it was quite ended, we found the Cupel very smooth and intire, but ting'd with a fine Purplish Red, (which did somewhat surprize us) and besides, the *refined Gold,* there lay upon the cavity of the Cupel some dark-coloured recrements, which we concluded to have proceeded from the deteriorated Metal, not from the Lead. But when we came to put our Gold again into the Balance, we found it to weigh only about *fifty three Grains,* and consequently to have lost *seven;* which yet we found to be fully made up by that little quantity of recrements that I have lately mention'd, whose Weight and Fixity, compared with their unpromising Colour, did not a little puzzle us, especially because we had not enough either of Them, or of leisure, to examine their nature. To all which circumstances, I shall subjoin this, that to prevent any scruples that might arise touching the *Gold* we imploy'd, I caused a dram and a half that had been purposely reserv'd out of the same portion with that that had been debased; I caused this (I say) to be in my Assistants presence melted by itself, and found it (as I doubted not but I should do) fine and well-coloured Gold.

I hope you will pardon my curiousity, saith *Arristander* to the Gentleman that spoke last, if I ask why you take no notice of the effect of *Aqua fortis* upon your *imbased Metal?* Your Question, replies *Pyrophilus,* I confess to be very reasonable, and I am somewhat troubled that I can answer it but by telling you that we had not at hand any *Aqua fortis* we durst relie on; which yet I was the less troubled at, because heretofore some *tryals* purposely made had inform'd me, that in some *Metalline Mixtures* the *Gold* if it were much predominant in quantity, may protect another Metal; (for instance *Silver*) from being *dissolved* by that *Menstruum,* though not from being at all *invaded* by it.

[13]There yet remain'd, saith *Heliodorus,* one examen more of your odd Metal, which would have satisfied me, at least as much as any of the rest, of its having been notably imbas'd; for if it were altered in its *specifick gravity,* that quality I have always observ'd (as I lately perceiv'd you also have done) to stick so close to Gold, that it could not by an additament so inconsider-

---

easily have been antimony-containing. Boyle's proportion of anti-Elixir to gold (1 to 1,000) is within the limits of action of antimony upon gold (1 in 1,920). It must be admitted, however, that this impurity should have been removed by the subsequent cupellation and ought not to have produced so notable a change in density.

able in point of bulk, be considerably altered without a notable and almost Essential change in the texture of the Metal.

To this pertinent discourse, *Pyrophilus,* with the respect due to a person that so worthily sustain'd the dignity he had of presiding in that *choice company,* made this return: I owe you, Sir, my humble thanks for calling upon me to give you an account I might have forgotten, and which is yet of so important a thing, that none of the other *Phaenomena* of our Experiment seem'd to me to deserve so much notice. Wherefore I shall now inform you, that having provided my self of all the requisites to make Hydrostatical Tryals, (to which perhaps I am not altogether a stranger) I carefully weighed in water the *ill-lookt* Mass, (before it was divided for the coupelling of the above mentioned dram) and found, to the great confirmation of my former *wonder* and *conjectures,* that in stead of weighing about nineteen times as much as a bulk of water, equal to it, its proportion to that liquor was but that of fifteen, and about two-thirds to one: so that its *specifick gravity* was less by about 3⅓; than if it had been pure Gold it would have been.

At the recital of this *notable circumstance,* superadded to the rest, the generality of the Company, and the President too, by looking and smiling upon one another, express'd themselves to be as *well delighted* as *surpriz'd;* and after the murmuring occasion'd by the various whispers that pass'd amongst them, was a little over, *Heliodorus* address'd himself to *Pyrophilus,* and told him, I *need not,* and therefore *shall* [14] *not,* stay for an express order from the Company to give you their hearty thanks: for as the *Obliging Stranger* did very much gratifie *you* by the Present of his *Wonderful Powder,* so you have not a little gratified *us* by so *candid* and *particular* a Narrative of the effects of it; and I hope (continues he) that if you have not yet otherwise dispos'd of that part of your *deteriorated Gold* that you did not cupel, you will sometime or other favour us with a sight of it.

I join in this request, said *Crattippus,* as soon as he perceived the President had done speaking, and to facilitate the grant of it, I shall not scruple to tell *Pyrophilus* he may be confident that the *Degradation* of his *Gold* will not *depreciate* it amongst Us: since if it be allowable for Opinion to stamp such a value upon *Old Coyns* and *Medals,* that in the Judgment of good Antiquaries, a rusty piece of *Brass* or *Copper,* with a half defaced Image or Inscription on it, is to be highlier valued than as big a piece of well-stampt *Gold;* I see not why it should not be lawful for Philosophers to prize such a lump of *depraved Gold* as yours, before the *finest Gold* the Chymists or Mintmasters are wont to afford us. And though I freely grant that some old *Copper Medals* are of good use in History, to keep alive by their *Inscriptions* the memory of the taking of a Town, or the winning of a Battel; though these be but things that almost every day are some where or other done, yet I think *Pyrophilus's* imbas'd Metal is much to be preferr'd, as not only *preserving* the *memory,* but *being an effect* of *such* a Victory of Art over Nature, and the conquering of such generally believ'd insuperable difficulties, *as* no Story that I know of gives us an example of.

As soon as ever *Crattippus* had made a pawse, *Pyrophilus* to prevent complimental discourse, did in few words tell the President, That his part had been but that of a Relator of matter of Fact, and that therefore he could deserve but [15] *little thanks* and *no praise* at all; though a good measure of both of them were due to the *Obliging Virtuoso* that had given him the Powder; and in that, the opportunity of complying with his duty, and his inclination, to serve that learned Company.

These Gentlemen (saith *Arristander*) are not persons among whom modesty is either *restrained* from expressing it self, or *construed* according to the Letter; and therefore whatever you have been pleas'd to say, the Company cannot but think its self much obliged to you; and I know the obligation would be much increas'd, if you would favor us with your reflections upon the extraordinary Experiment you have been pleased to relate to us.

If, replies *Pyrophilus,* I had had wherewithal to repeat the Experiment, and vary it according to the hints afforded me by the first tryal, I should be less unfit to comply with *Arristander's* motion; but the *Phaenomena* are too new and too difficult for me to attempt to unriddle them by the help of so slender an *information* as a person so little sagacious as I could get by a *single tryal;* and though I will not deny that I have had some raving thoughts about this puzzling subject, yet I hope I shall easily be pardon'd, if I decline to present crude and *immature* thoughts to a Company that so well *deserves* the most *ripe* ones, and can so skilfully *discover* those that are not so.

I confess, saith *Heliodorus,* that I think *Pyrophilus's* wariness deserves not only to be *allow'd* but *imitated;* and therefore by my consent the further discourse of so abstruse a subject, shall be deferr'd till we shall have had time to consider seriously of *Phaenomena* that will be sure to *imploy* our most speculative thoughts, and I fear to *pose* them too: only we must not forget that *Pyrophilus himself* ought to be not barely *allow'd,* but *invited* to draw before we rise, what Corrollaries he thinks fit to propose from what he hath already delivered.

[16]The inference, saith *Pyrophilus,* I meant to make, will not detain you long; having for the main been already intimated in what you may remember I told you I design'd in the mention I was about to make of the now-recited Experiment. For without launching into difficult Speculations, or making use of disputable Hypotheses, it seems evident enough from the matter of Fact faithfully laid before you, that an Operation *very near,* if not *altogether* as strange as that which is call'd *Projection,* and in the difficultest points much of the same nature with it, may safely be admitted. For our Experiment plainly shews that Gold, though confessedly the most homogeneous, and the least mutable of Metals, may be in a very short time (perhaps not amounting to many minutes) exceedingly *chang'd,* both as to *malleableness, colour, homogeniety,* and (which is more) *specifick gravity;* and all this by so very inconsiderable a proportion of injected Powder, that since the Gold that was wrought on weighed two of our English drams, and consequently an hundred and twenty grains, an easie computation will assure us that the Medicine did thus powerfully act, according to my estimate, (which was the modestest) upon near a thousand times, (for 'twas above nine hun-

dred and fifty times) its weight of Gold, and according to my Assistants estimate, did (as they speak) *go on* upon twelve hundred; so that if it were fit to apply to this *Anti-Elixir,* (as I formerly ventur'd to call it) what is said of the true *Elixir* by divers of the Chymical Philosophers, who will have the virtue of their Stone increas'd in such a proportion, as that at first 'twill transmute but *ten* times its weight; after the next rotation *an hundred* times, and after the next to that *a thousand times,* our Powder may in their language be stil'd *a Medicine of the third order.*[q]

The Computation, saith *Arristander,* is very obvious, but the change of so great a proportion of Metal is so wonderful and unexampled, that I hope we shall among other things [17] learn from it this lesson, That we ought not to be so forward as many men otherwise of great parts are wont to be, in prescribing narrow limits to the power of Nature and Art, and in condemning and deriding all those that pretend to, or believe, uncommon things in Chymistry, as either Cheats or Credulous. And therefore I hope, that though (at least in my opinion) it be very allowable to call Fables, Fables, and to detect and expose the Impostures or Deceits of ignorant or vain-glorious Pretenders to Chymical Mysteries, yet we shall not by too hasty and general censures of the sober and diligent Indigators of the *Arcana* of Chymistry, blemish (as much as in us lies) that excellent Art it self, and thereby disoblige the genuine Sons of it, and divert those that are indeed Possessors of Noble Secrets, from vouchsafing to gratifie *our* Curiosity, as we see the one of them did *Pyrophilus*'s, with the *sight* at least, of some of their highly Instructive Rarities.

I wholly approve, saith *Heliodorus* rising from his seat, the discreet and seasonable motion made by *Arristander.*

And I presume, subjoins *Pyrophilus,* that it will not be the less lik'd, if I add, That I will allow the Company to believe that as *extraordinary,* as I perceive most of you think the *Phaenomena* of the lately recited Experiment; yet I have not (because I must not do it) as yet acquainted you with the Strangest effect of our Admirable Powder.[r]

The Publisher to the Reader.

Having been allowed the Liberty of Perusing the following *Paper* at my own Lodging; I found myself strongly tempted, by the Strangeness of the things mention'd in it, to venture to Release it: The knowledge I had of the *Author's* Inclination to Gratifie the *Virtuosi,* forbidding me to despair of his pardon, if the same disposition prevail'd with me, to make the Curious Partakers with me of so Surprising a Piece of *Philosophical* [ii] *News.* And,

[q] On "orders," see chapter 3.

[r] The text ends with a typically Boylean teaser referring to "unmentionable" knowledge. What this "Strangest effect" was apparently caused some discussion: "Some students in Hermetic Philosophy have conjectured, from a variety of passages in this narration . . . that this was, in reality, the true elixir, and that the powder employed in the experiment might possibly be recovered with some addition, as well as part of the gold. This, they imagine, is hinted at in the very last words relating to the concealed wonders of this extraordinary powder." *Biographia britannica,* 2d ed. (London, 1778–1793), 5:507, n. Q.

though it sufficiently appear'd, that the insuing Conference was but a Continuation of a larger *Discourse;* yet, considering, that this Part consists chiefly, not to say only, of a *Narrative;* which (if I may so speak) stands upon its own legs without any need of depending upon any thing that was deliver'd before; I thought it was no great Venture, nor Incongruity, to let it come abroad by it self. And, I the less scrupled to make this Publication, because I found, that the Honorable Mr. *Boyle*[s] confesses himself to be Fully Satisfied of the Truth, of as much of the Matter of Fact, as delivers the *Phoenomena* of the Tryal; the Truth whereof was further Confirm'd to me, by the Testimony, and Particular Account, which that most Learned and Experienc'd *Physitian,* who was Assistant to *Pyrophilus* in making the Experiment, and with whom I have the honor to be Acquainted (being now in *London*) gave me with his own Mouth, of all the Circumstances of the Tryal. And, where the [iii] Truth of that shall be once Granted, there is little cause to doubt, that the Novelty of the thing will sufficiently Indear the Relation: especially to those that are studious of the *Higher Arcana* of the *Hermetick Philosophy.*[r], most of the *Phoenomena* here mention'd, will probably seem wholly new, not only to vulgar *Chymists,* but also to the greatest part of the more knowing *Spagyrists,* and *Natural Philosophers* themselves: none of the *Orthodox* Authors, as far as I can remember, having taken notice of such an *Anti-Elixir.* And, though *Pyrophilus*'s Scrupulousness (which makes him very unwilling to speak the utmost of a thing) allowes it to be a Deterioration into an Imperfect Mettal onely; yet, to tell the truth, I think it was more Imbas'd than so; for the part left of it (and kept for some farther Discoveries) which I once got a sight of, looks more like a *Mineral,* or *Marchasite,* then like any *Imperfect Mettal:* and therefore this Degradation is not the same, but much greater, [iv] than that which *Lullius* doth intimate in some places.[v] These Considerations make me presume it will easily be granted, That the Effects of this *Anti-Philosophers Stone,* as I think it may not unfitly be call'd, will not only seem very *strange* to *Hermetick,* as well as other *Philosophers,* but may prove very *Instructive* to Speculative Wits; especially if *Pyrophilus* shall please to acquaint them with that more odd *Phoenomenon,* which he Mentions darkly in the Close of this *Discourse.*

[s] The manuscript reads quite differently here, omitting much of the rest of the sentence and running into the next: "I found that Mr. B. who is very well acquainted with the person that tryed the Experiment, confeses himselfe to be fully satisfied of the truth of the matter of fact & where that shalbe once granted . . . "

[t] This assistant may be Edmund Dickinson; "now in London" implies that he was not always resident there, and Dickinson moved to London at about this time (see n. l, above, and chapter 4, n. 34).

[u] This sentence continues in the manuscript: "for I have heard more then ordinary Chymists wonderingly acknowledge that they never saw nor read nor heard of any Experiment paraleld to this: Tho one of those skilfull persons confest at length to me that he had more then once seen projection."

[v] This sentence and the preceding one exist only in the published text; the following sentence in the manuscript begins: "And I presume it will easily be granted . . ."

Textual Notes

1. The following name, *Leanthes*, has been deleted.
2. nonulli
3. Replaces *primum.*
4. *illa*
5. [adnectend]um . . . dividium
6. Postremo
7. per naturam et Actum/Actionem, productarum analogiâ. Clearly, the translator has misread *art* as *act.*
8. contuitu illius, inspectando illo
9. pluris
10. *Ipsius*
11. Replaces *circa.*
12. continet
13. experiri
14. habitudines, proprietates
15. particulares
16. *quam ali*
17. "intra se" altered from "in specie."
18. *ventilata*
19. expendamus, discutiamus
20. obicem p[rop]onere
21. congruum
22. adhaerere
23. adjuvare
24. [metalli]cis
25. restricturos
26. Virtuosos
27. factam divisionem, factam trichotomiam
28. tuendam (to replace the idiom *in arce collocandam*)
29. major stetit numerus
30. Elixir
31. sufficientes
32. fidei [nostr]ae
33. meruit
34. [lapid]is [philosophic]i
35. numerantur
36. scriptis
37. activitatis
38. vinculis
39. Final *x* deleted.
40. *adhibitos*
41. [metalli]c[ae]
42. in antecessum
43. Unclear in MS, probably *videbit* written over as *videtur.*
44. *reddidisset*
45. *recipiend[am]*

46. habitudinem
47. Elixir [su]u[m]
48. inferamus
49. humectatum
50. germinationem
51. ars Sciaterica
52. This numeral appears in the margin; I have inserted it in its logical place in the text.
53. In 2b, this marker (inconsistently) bears parentheses, while the succeeding Roman numeral does not. I have indicated here the consistent outline form of 2a.
54. Replaces *being*.
55. Replaces *handsomly*.
56. *that*
57. *were more proper for*
58. *& moral*
59. *betake themselves to*
60. *persuaded*, then *willing*
61. *in*
62. Replaces *Arms*.
63. Replaces *left*.
64. *without*
65. Altered from "*himself to declare*."
66. *grounded on the nat*[ure]
67. *very*
68. *first*
69. *himself ag*[*ainst*]
70. *'d*
71. *by*
72. Replaces *so*.
73. Replaces *right*; dictation error?
74. Altered from *these*.
75. Insertion replaces *what he had to say on the*.
76. Insertion replaces *in one part*.
77. *haveing*
78. *the*
79. Altered from *be performable by Art*.
80. Replaces *not*.
81. Name entered later in pencil.
82. Replaces *that*.
83. Name added in pencil.
84. Replaces *it is*.
85. "his . . . convictive" replaces *it is not so unquestionable*.
86. *that*
87. *in judgment of goldsmiths & even of refiners*
88. *perhaps*
89. *more pure*
90. *were*
91. Replaces *that*.
92. Replaces *conspicuous*.

93. Replaces *I.*
94. Replaces *not.*
95. Name entered later in pencil to fill a space initially left blank.
96. *keep*
97. Replaces *still.*
98. *of it*
99. *disolves like it in*
100. *more*
101. Replaces *know.*
102. Replaces *cautiousnes.*
103. *that*
104. Replaces *extravagant.*
105. Replaces *be my great.*
106. Replaces *the purest.*
107. *not*
108. Replaces *as much water.*
109. Replaces *Gold.*
110. *vulgar*
111. Altered from *haveing contain a very considerable.*
112. Replaces *by.*
113. Replaces *Person.*
114. "of another . . . Court" replaces *great Lord of that Country.*
115. *to him*
116. *who was who was present*
117. *from whom I have these part*[iculars]
118. *not a little*
119. *how much*
120. Replaces *suggest.*
121. Replaces *mind.*
122. Replaces *by hiding into it some.*
123. Replaces *the Powder.*
124. *either*
125. The copy 10b accidently omits this clause beginning "to him, to whome . . . knowne."
126. 10a reads "in the same towne."
127. Replaces *Boy.*
128. Replaces *strang Travailer.*
129. Replaces *particularly the.*
130. Altered from *And Upon my* haveing.
131. Replaces *should.*
132. Replaces *forraigner.*
133. *yet useing no other*
134. Replaces *Traveller.*
135. Replaces *offence.*
136. *the*
137. *not very*
138. *And tho my by omissions I had spard my eyes*
139. *that my wilingness to spare my eyes had made*
140. *at another*

141. *which he*
142. *some reasons forbid me not now mention*
143. *which I*
144. *forraigner*
145. Replaces *competent.*
146. *really transmuted*
147. Replaces *knew I could command.*
148. Replaces *it being.*
149. Replaces *as.*
150. Replaces *him.*
151. Replaces *saw.*
152. *told me to*
153. *he had heard*
154. Replaces *mett* [meet].
155. Replaces *a conveniency and.*
156. *he*
157. *was*
158. *& I perceiv'd*
159. *heard of*
160. Replaces *Person.*
161. Replaces *abilities & bounty.*
162. Replaces *him.*
163. *not handsomely procure answers to*
164. Replaces *not.*
165. *be very*
166. *to have*
167. *about*
168. Altered from *a stranger who was one to*
169. Replaces *forraigner.*
170. *& by profession a Doctor of Phisick*
171. *(who is his K tho unacquainted with him was found to be his Kinsman)*
172. *wherein*
173. *gott him to*
174. *employ*
175. Replaces *transmutation made.*
176. Replaces *now.*
177. Replaces *Person.*
178. *for being a reser*[*ved*]
179. *reserved Person*
180. Altered from *this Treasurers house.*
181. Replaces *lived.*
182. *partly*
183. *found the possessor*
184. a Person replaces *such a one.*
185. Replaces *divers.*
186. "Three or four others" replaces *two or three Virtuosi.*
187. Replaces *informing the curious Str*[*anger*].
188. *make one of their number*
189. *Eugenius*

190. *addressing himself*
191. *seem*
192. "Conference which was" replaces *Discourse which Eugenius.*
193. *very strong*
194. *they deserve*
195. Replaces *chife.*
196. *since*
197. Replaces *admiring.*
198. MS reads *picth.*
199. ˆ*antient & numerous*ˆ
200. After the false start *I observe*, the same hand continues, but using different ink—"T[ranscri]b'd" in margin.
201. *no documen*[*ts*]
202. *of*
203. *usefull*
204. *are wont*
205. *though*[*t*]
206. Replaces *being.*
207. *embraced it*
208. Replaces *impertinently.*
209. *deny themselves*
210. *their*
211. Replaces *enemies.*
212. Replaces *samples.*
213. *usually*
214. Replaces *trivial.*
215. *so much as*
216. Replaces *by.*
217. *endow'd with* [illeg.]
218. *all my att*[*tention*]
219. Replaces *in those.*
220. Replaces *says.*
221. Altered from *they call.*
222. Final *s* deleted from Reader and son.
223. Altered from *solemnly profest.*
224. Replaces *when.*
225. *they have*
226. ˆ*disap*[*point*]ˆ *delude him*
227. *more then*
228. Replaces *poy*[*nts*] ˆ*& import*[*ant*]ˆ *& fundamental.*
229. Replaces *may.*
230. *most*
231. Replaces *page.*
232. *for*
233. *tell their readers*
234. Modified from the plural *readers . . . them.*
235. *these egregious dissemblers have no right*
236. *so much*
237. *judge*

238. Replaces *divest himself of.*
239. *(such a[s it was)*]
240. *saying*
241. *that*
242. *riddles*
243. *his*
244. *hin a*
245. *engage*
246. Replaces *what ever may.*
247. *shew that tis reconcild it to*
248. Replaces *sages.*
249. *made*
250. "T[ranscri]b'd" written in the margin in pencil.
251. *fit*
252. *dis[course?*]
253. Replaces *shall be.*
254. Altered from *the most blameable wrighters of the Th[eatrum chemicum*].
255. Replaces *far from wonderfull.*
256. *of nature*
257. Replaces *be.*
258. Replaces *those.*
259. *no*
260. *of*
261. Replaces *reason.*
262. Replaces *whereas if they were.*
263. Inserted "boast" is in the hand of the transcriber Robin Bacon, replacing *brag.*
264. *which they*
265. Replaces *&.*
266. Modified from *their.*
267. Replaces *one.*
268. Modified from *wrightings.*
269. The insertion, in Robin Bacon's hand, replaces *otherwise.*
270. Replaces *occasiond.*
271. *& physicians*
272. Replaces *all.*
273. Text originally read *have much more* hinder'd *the encrease of* ^solid &^ useful knowled[*ge*].
274. *do*
275. Replaces *did.*
276. Replaces *Toils.*
277. *Expectations*
278. *made in appropriated furnaces had bee[n*]
279. *for*
280. Replaces *Mint Men.*
281. *defeat[ed] render'd*
282. *even vulgar*
283. *nor steadily pursue & clear prospect*

APPENDIX 2

INTERVIEW ACCOUNTS OF TRANSMUTATION AND PREFACES TO BOYLE'S OTHER CHRYSOPOETIC WRITINGS

Part I: Interview Accounts

The following documents record interviews Boyle conducted with visitors from Vienna concerning transmutations performed at the imperial court of Leopold I. The first manuscript contains two accounts of Wenzel Seyler, one from Franz Josef Lamberg (1637–1711), and the other from Karl Ferdinand Waldstein (Leopold's envoy to Charles II, 5 June 1677 to 9 March 1679).[a] The latter interview is dated "26 June," and the year must be 1677 or 1678. Further information from Waldstein and Lamberg appears in fragments 7 and 9 of the *Dialogue on Transmutation.*

The identities of the interviewee and projector in the second account are not so clear. It may be another account of Seyler, but if so, it is strange that Boyle omits his notorious name. The relator, an "Ingenious Gentleman . . . chosen to be one of the examiners of the Transmutation," may be Waldstein again (but it is odd Boyle would not be more specific as he was in the first account) for Waldstein was an inspector of Seyler's transmutations. Boyle's informant may instead be Johann Joachim Becher, another examiner of Seyler's projections. Alternatively, this transmutation of specifically *tin* (rather than lead) into gold may be the transmutation performed upon tin at the imperial court described by Wilhelm Schröder, F.R.S., in a letter written from Prague on 7 October 1674 (N.S.).[b] J. J. Becher was again examiner at that projection.

These two accounts were originally part of a much larger gathering of papers, presumably a collection of transmutation accounts. The interviews with Lamberg and Waldstein bear original pagination of 21–24, while the fair copy of the account from the "Ingenious Gentleman" is 27–28.[c] None of the remaining pages seems to have survived elsewhere in the archive.

Text A: Accounts of Wenzel Seyler from Count Lamberg and Count Waldstein [BP, 25:273–76]

Receiveing this Afternoone the Honor of a visit from the Count of Lamberg[,] son of the Lord High Steward to his Imperiall Majesty,[d] after some

[a] Maddison, *Life,* 168n.

[b] Schröder to [Royal Society?], BL, 5:69.

[c] The juxtaposition of these documents was first noted by Hunter, "Alchemy, Magic and Moralism," 402 n. 45.

[d] Franz Josef von Lamberg, son of Johann Maximillian von Lamberg (d. 1682); Franz Josef was apparently in England in the company of Waldstein, envoy from Leopold I to Charles II.

discourse of other Curiosities I desired he would be pleased to give me some Information that I might trust to about the Famous Fryer at Vienna Wenceslaus de ^Reinberg^ who has been soe much famed for having made projections at Vienna, from whence this Nobleman now comes.

To this request he answer'd, that he thought he could in some sort satisfy me, because Wenceslaus ^de Reinberg^ was his very perticuler Acquaintance, & profess'd soe much kindness to him, that he openly said ^that^ if he would communicate his skill to any, it should be to Him. But after other Passages of less moment, when I enquir'd more punctually what it was he knew him to have performed, whether he did really make projections, he told me that he had made severall at Vienna, not only in the presence of diverse other Persons of Note, but of this Relator Himself. And when I desired to know upon what Mettalls, & upon what quantity the Transmutation was made; he replyed *that* he could make them upon any Mettall, but those he saw, were made upon Lead & upon Tinn; that in the Experiments he was present at, the projection was made upon about --------- of Mettall at a time which came out very fine Gold, & endured the due *examens* of the Saymen or refiners. When I asked how much Powder he imployed at a time, his Lordship told me, that he tooke it out of a Box that was about the bredth of the Palme of [274] his hand and that he guess'd it to be about 2. or 3. graines: and to the question what Colour it was of he design'd it by the Colour of one part of a wrought Cushion that lay by us, which was of a red more inclineing to browne then common Cinnabar is wont to be. And when I enquired how long the operation lasted, I was answer'd that 'twas about a quarter of an houre or less after the powder was cast in: the Mettall not being poured out of the Crucible, but suffer'd to settle there, & then was knock'd out in a Lump of such a shape as the lower part of the Crucible had given it. He added that this *Adeptus* had to his knowledg of late paid threescore thousand Florens of 3th ----------[e] (of Debts that he had contracted by borrowing and otherwise) dureing his stay at Vienna, [margin: I forgat to adde that he told me the Adepty had resign'd backe to the Emperor the pension his Imperiall Majesty had settled on him] & that he now keeps Coach & horses,[1] Lackays, and other Equipage suitable to the degree of a Baron to which he has been advanc'd by the Emperour. He told me also that this Artist had left in the hands of the Relators Father, to keep safe for him about halfe a pound of Powder that passd for such as was formerly described, but the Relator knows not whither [*sic*] it was all a meere powder of Projection or that mingled with somewhat else, the latter of which I am induced to thinke the most probable, by what I learn'd from a publicke Minister, that comeing from the Court of Vienna, & passing for Vienna, was pleased to give me a visit.[f] Lastly the Count of Lamberg told me that the

[e] "Florens" altered from *Florence;* the text here is unclear.

[f] Possibly J. J. Becher, who in his account of Seyler published at Boyle's request (*Magnalia naturae*) remarked that when Seyler's store of stolen powder of projection began to run low, he adulterated it with red lead and cinnabar.

Adeptus affirmd, that [275] with the same powder he can by the way of ordering it, transmute some of the Ignobler Mettalls into Silver as well as into Gold: and thô himselfe had not been an Eye witness of that yet it had been done in the presence of diverse others & perticulerly of the Relators brother. The same day most of these perticulers were confirmed to me, by my having met with a Gentleman well skill'd in Chymistry, who having been also an acquaintance of this Adeptus, has formerly told me upon his own knowledge some such Passages as are here above related, & some others that I have not now time to set downe.

June 26. This Evening discourseing with the Count of Wallest^e^in his Imperiall Majesties extraordinary Envoye to his Majesty of great Brittaine about Fryer Wencel, he told me that he had seen him make Projections, and that this Adeptus was free enough to gratify his friends, & the Curious when they exprest stronge desires to be satisyed by ocular Demonstration. He confirm'd to mee what the Count of Lamberg had told mee of his having about halfe a yeare since paid all his debts, which amounted to diverse thousand of pounds Sterling and his having return'd his Imperial Majesty thankes for the Pension he had formerly received at his Court, with a[2] Declaration that now He intended to live & worke at his own charge, in his own House, where he is now married; and lives with his wife in a splendid, if not also in a profuse[3] way both as to his Table & other exam ----------. The Count also told me that being [276] once in need of a considerable summ, the Adeptus had borrowed of the Relator's brother, (I suppose the Archbishop of Prague) a convenient place in his house to make a considerable projection; which he did upon 12. or 16. pound of matter at a time ^which^ he turn'd all into pure Gold. He added, that oftentimes without staying to send for Lead or Tinn, he cut off a good piece of a plate of Metall, as Pewter, and cast that into the Crucible to be transmuted. Lastly he told me that the Adeptus once in the presence of the Emperour, the Count Wallestine being by; tooke a thin plate of silver, & having first cut off a Corner of it to shew that it was pure silver, he lightly rub'd upon some part of the rest a little of this Powder somewhat wetted to make it sticke, and which was in colour much like Cramosi[g] then having with a paire of Tonges or Flyers [*sic*] held of the Plate over kindled Coales, till it grew hot, but did not melt; the tincture penetrated the silver so far, as to turne great part of it into Gold. Of which transmuted plate, the Relator brought a peice with him into England, together with the Corner of Silver that had been before by the Operation cut off[,] both which peices the Relator some dayes since presented to his Majesty of Great Brittaine.

[g] I.e., a kind of crimson cloth.

Text B: A Projection Account from Vienna
[BP, 25:277–79, draft; BP, 25:307–8, fair][h]

Being yesterday in Company with an Ingenious Gentleman soe expert in Metalline affairs that he was chosen a while since by the Emperor to be one of the examiners of the Transmutation that was undertaken to be made before His Imperial Majesty, I desired this Gentleman (when no other Company was in the roome) to tell me freely whether it were true that I had heard of him (who did not love to talke of it) viz. that he had seen a projection not long agoe. To which question he frankly told me that he had and that it was shewne him in such a place by such a Person whom he nam'd to me, & whom I had some Cause to thinke a possessor of extraordinary Arcana in re metallica.[i] Upon this Confession I employed the time that was allowed me to converse with this stranger, in proposeing him some questions to which he civilly return'd Answers to this purpose.

1. That the projection he saw was made with a red powder taken up on the point of a knife & cast[4] upon the melted metal.

2. That, before the Medicine was cast in, the Metall was carefully scumm'd.

3. That the Mettall which was made use of in this Experiment was Tinne, which Mettall the Adeptus most commonly chose to employ when he would make any Transmutation.

4. That the fire was not extraordinary strong but only sufficient to bring the Tinn to a very good fusion.

5. That there was some Commotion made in the melted Metall upon the ingress of the Powder into it, but noe fulmination, or flame, that he tooke notice of. [308]

6. That the transmutation was made in as little time as he had related the story to me, which I guess'd to be about halfe a quarter of an hour or less, before I ask'd him the particular questions.

7. That he told me ^he^ saw an other time a projection made on the same Metall with an imperfect Medicine, which was longer in working before it effected the desired change, & transmuted but (if I mistake not) about 80 times its own weight. But,

8. That by the projection mention'd all the while, there was so great a proportion of Metall transmuted with a small quantity of powder, that the Tinn that was turn'd into a perfect Gold amounted to eight pound weight.

9. That he did not remember the proportion of the Tinn to the Medicine

[h] Margin: "Febr."; endorsed: "Papers relating to the transmutation at Vienna, told him by Count Lamberg and Count Wallensteyn." This endorsement seems to be the hand of William Wotton and need not pertain strictly to this account, as this sheet appears to have been the outermost sheet of a bundle.

[i] "In metallic affairs": in transcribing the draft, Robin Bacon missed the line "& whom I had some Cause to thinke."

employ'd to transmute it but had taken a note of it, which he intimated he would looke out for me; & as far as I can guess by his Discourse it was about one upon 2000.[5]

Part II: Prefaces to Chrysopoetic Processes

Boyle's real output of writings and processes relating to chrysopoeia must have been considerably greater than the present archive indicates. The following four documents are all preambles or explanatory notes to collections of chrysopoetic processes; while the texts published here do survive, there is no trace of the "annexed Paper," "following Process," or "foregoing parts," to which they make reference. Although one might conjecture that these items fell prey to the apologetic purges alluded to previously, it is more likely that they were taken early on by persons anxious to carry out the processes for personal financial gain. This theory could explain the survival of these truncated documents: while a purge of alchemical material would have disposed of both introductions and processes, a hopeful transmuter would have been interested solely in the laboratory directions and left behind Boyle's philosophical musings upon them.

All these manuscripts probably date from toward the end of Boyle's life and show Boyle's desire to leave behind a record of his transmutational pursuits; the "Hermetic Legacy" he alludes to is perhaps the clearest example. In spite of this communicativeness, Boyle is solicitous that the processes he reveals will not "threaten the welfare of States." Besides reinforcing the importance of chrysopoeia to Boyle, several texts contain valuable retrospective expositions of Boyle's motivations behind his pursuit of chymistry generally. While I have discussed these documents in chapter 6 (to which I refer the reader), I have not exhausted their usefulness to historians, and so, I believe, they deserve to be somewhat better known.

Text C: Preamble to Boyle's "Hermetic Legacy" [published in Works, *1:cxxx and More,* Life, *228–30; manuscripts: BL, 1:107–8 and 130–31]*[i]

I confesse you are not the onely Person among my Friends to whom it hath seem'd somewhat strange, that I, who have spent many of my Thoughts, some of my money, and, what I value far more, of my time too, upon Chymistry, as well as diverse other parts of Learning; have not been taken notice of to have found any *Particulars,* as Chymists speak, or other Luciferous experiments upon Mettals and Minerals: nor have pretended to be

[i] The copy at 130–31 is endorsed in a contemporary hand (possibly Boyle's): "To Mr. N., The [illeg.]." The last page bears a note in a later hand, "2. halfe sheets L. P. / 1. halfe sheet, lett. B," and a memo in the hand of Henry Miles, "Laid aside being transcribed into Materials for the Life."

Possessor of those difficult and Compounded Experiments, that are magnify'd by Chymists as Excellent ---------- Hermetick Arcana.[k] But sir, since I find *You* in the list of those that have made the newly mention'd Reflection, I am content to give You ^what I should not give any that I do not highly Regard; viz.^ such a summary account of my Comportment, as may at least lessen your wonder at it. I must inform you then, that when among other studies, I apply'd my self to the cultivating of Natural Philosophy, I soon perceiv'd that some insight into Chymical operations, was, thô not absolutely necessary, yet highly conducive to the true knowledg of nature, and especially the Indagation of several of her most abstruse Mysteries. On this score I was induc'd to make a nearer Inspection into Chymistry, then *Virtuosi* are wont to think it worthwhile to doe; and I did not repent me of my labour. But as I cultivated Chymistry not so much for it self, as for the sake of Natural Philosophy ^&^ in order to it so most of the Experiments I devis'd and pursu'd, were generally such as tended, not to multiply Chymical Processes [130*v*] or gain the reputation of having store of difficult and elaborate Ones, but to serve for Foundations, and other usefull Materials, for an Experimental History of nature, on which a Solid theory may in process of time be superstructed. For this purpose I judg'd that plain and easy Experiments, and as simple, or as little compounded as may be[,] would *caeteris paribus* be the fittest; as being the most easy to be try'd (and if need be, repeated,) and to be Judg'd of, both in relation to their causes & to their Effects. And for these Reasons, thô I had by me ^a^ not inconsiderable number of more Compounded & Elaborate Processes, some of which I had made, and others I receiv'd as great Secrets from noted Artists; I purposely forebore to mention any number of them in my Writings about Physicks; being desirous rather to Increase knowledg, than make an Ostentation of any that I thought would puzle most readers more than it would instruct them. This Sir, I hope, will appear to you a fair account of your not finding my Physical Discourses larded with long & intricate Processes; some of which may, I willingly grant, produce notable Effects, and for that reason are valuable; but are lesse Fit than far more simple ones, to discover the Causes of things, which yet is the chief scope of a Naturalist as such. And to those that think it strange that among my other Experiments about mettals and Minerals, I have not produc'd those gainful Ones that Chymists call *Particulars;* it may I hope, suffice to represent, that being a Batchelor, and through Gods Bounty [131] furnish'd with a Competent Estate for a young^er^ Brother, and freed from any ambition to leave my Heirs rich, I had no need to pursue Lucriferous Experiments, To which I so much prefer'd Luciferous Ones, that I had a kind of Ambition (which I now perceive to have been a vanity) of being able to say that I cultivated Chymistry with a Disinterest[ed] mind, neither seeking, nor scarce careing for any ^other^ ad-

[k] On particulars, see chapter 3, 77–80.

vantages by it, than those of *the* Improvement of my own knowledg of Nature, *the* gratifying the Curious and the Industrious, and *the* Acquist of some useful helps to make Good & Uncommon Medicines.

If I may be allow'd to judge of Courses by the success, the Entertainment that the Publick has been pleas'd to give my Endeavours to serve it, will not make me repent of the way I have made choice ^of^ to do it in. But however, since I find myself now grown old, I think it time to comply with my former Intentions to leave a kind of Hermetick Legacy to the studious Disciples of that Art, and to deliver candidly in the annexed Paper, some Processes Chymical & Medicinal, that are less simple and plain than those barely Luciferous ones I have been wont to affect, ^&^ of a more difficult & elaborate kind, than those I have hitherto publish'd, and more of kin to the noblest Hermetick Secrets, or as Helmont styles them, *Arcana majora*. Some of these I have made & try'd, Others I have (thô not without much difficulty) obtain'd by Exchange, or otherwise, from those that ^affirm they^[l] knew them to be Real, and were themselves Competent Judges, as being some of them Disciples of true Adepts, or otherwise[m] admitted to their Acquaintance & Conversation. Most[6] of these Processes are clearly enough deliver'd, and of the rest there is plainly set down without deceitful terms, as much as may serve to make what is Literally taught, to be of great utility; thô the full & compleat uses are not mention'd, *partly* because in spite of my Philanthropy, I was ingag'd [131*v*] to secrecy as to some of these uses, and partly because (I must ingeniously[n] confess it) I am not yet, or perhaps ever shalbe, acquainted with them my self. The knowledg I have of your great Affection for the Publick good, and your particular kindnes for me, invites me among the many *Virtuosi* in whose friendship I am happy, to intrust the following Papers in Your hands; earnestly desiring you to impart them to the Publick, Faithfully and without Envy, *verbatim* in my own Expressions, as a Monument of my good affections to Mankind, as well in my Chymical Capacity as in the others, wherein I have been solicitous to do it service. I am with sincere respect.

Sir,
Your most Faithful & most Humble Servant
Robert Boyle[o]

Text D: Preamble to a Collection of Particularia *[BP, 17:43]*[p]

Having in another[7] writing had occasion to discourse somwhat largely[8] of several things that concern Lucriferous Experiments in[9] general ^it^ may

[l] Insertion in 1:107–8 only.

[m] BL, 1:130–31 reads "and others" in place of "or otherwise."

[n] Probably a mistake for "ingenuously."

[o] This full closing occurs only in 1:107–8, where it is in pencil and autograph; the other copy closes simply, "I Ro: Boyle," again in autograph.

[p] Margin: "T[ranscri]b'd."

now suffice ^to add som things^ that relate to those[10] Particulars as Chymists are wont to call them, to which this Paper is to serve for a short Preamble.[q]

I ^have^ elsewhere[r] declar'd for what Reasons I, who had[11] neither wife[,] children, or ambition to grow rich, was not only not solicitous to procure Lucrative Experiments but was not forward so much as[12] to accept of[13] them, unles[14] I judg'd them also useful for the increase of knowledg or the cure of Diseases, but afterwards thô late considering that ^since^ most particulars worke notable changes on the mettals or other Bodys that afford them, & therefore may probably be apply'd to the discovery of unobvious Truths, or the Preparation of good medicines, I set aside from time to time some ^of the^ Experiments communicated to me that were affirm'd by the Imparters to be capable of exalting[,] ripening[,] or at least of separating from inferiour metalls & minerals some portions of one or both of the two noblest, nor did I take them all upon trust having had[15] tryals successfully enough made of more than one of them. But by misfortune they were for a very long time[16] mislaid & which was worse I missd them whilst I had the opportunity of Time & Furnaces to ^examine them.^[17] But haveing late recover'd a small collection ^of^ these Processes,[18] which I receiv'd, almost all of them from persons of [44*r*] uncommon skil & credible men ^who affirm'd[19] they know them to be true & who^[20] but present [*sic*] them freely to me, as testimonys of their kindnes or gratitude:[21] I think it not amiss to impart them ^first^ to you, & some of your Spagyrick Friends, and[22] then perhaps to the Publick to[o]. For thô[23] ^I acknowledge that^ several of these Experiments are more curious than Lucrative, yet they are not all so, and[24] ^I have several^ Inducements to communicate them. The first to gratify & convince diverse virtuosi[25] whom the frequent mistakes & frauds of Chymical Imposters have brought to think that there is really no[26] gold or silver to be obtain'd by any Chymical Art from[27] the baser mettals, or from Minerals. And these ingenious men being once satisfy'd that[28] ^those richer mettals^ may be obtain'd from Bodys that the vulger Processes can get no such matter from, wil be incourag'd to make ^many new[29]^ Tryals ^to the same purpose^ since they know it to be possible that some may succeed. 2[ly] Particulars[30] ^depending^ usually upon some uncommon wayes of operation upon minerals & metals, they are like much to help a sagacious person[31] ^in^ the discovery of the nature of mettals & minerals. 3[ly] Thô it be true that[32] the greatest number of Particulars are not considerably Lucrative unles made in great quantitys, yet there are some[33] ^that^ being skilfully wrought, even in small quantitys may enable a poor and industrious Artist ^especially if he be a single man^ to get a Livelihood, thô not to grow rich. And 4[ly] those particulars that are thought but very mean in case they be not wrought in great quantitys may be for that very Reason more beneficial to

[q] On particulars, see chapter 3; the "other writing" may be Text C.

[r] Possibly Text C.

the Publique, than divers that are more rich & easy to be conceal'd; for these meaner particulars requiring many hands, Materials & Instruments to carry them on with profit ^wil^[34] set many poor people at work[35] ^&^ thereby releive ^great^ numbrs[36] enabling them ^or at least assisting them^ to get maintenence for themselves & their distrest familys if any Industrious Artist shalbe so lucky as I wish many a one may[,] as to make a[37] considerable profit by any of these Processes.

I do not condemn the custom of Chymists when they pretend to teach Lucriferous Experiments thô in me I fear ^to pass^ for[38] canting (a style I am not fond of) if I should exhort those Industrious Artists that shalbe so happy[39] ^which^ I wish many of them may be, as to make a considerable profit, by any of these Processes to remember that the Poor has ^in som measure^ share[40] with them.

Text E: A Preamble to an Augmentation of Gold [BP, 25:283–85, fair; BP, 38:157, draft][s]

Having been acquainted with more than one or two persons that have exercis'd their Industry upon the following work I had the Curiosity to inform my self somewhat solicitously about it, and partly by Answers obtain'd to my Questions, and partly by reading a Book written by the first or reputed Author of the Process, and partly by some things that I had now and then opportunity to see about the Progress & Operations of the work. I so instructed my self, that I thought I was able to frame a Mental Process of it that would succeed. The chief Lights I had were from the Authors Brother, who is look'd upon as Possessor of several of his Brothers Manuscripts, and from an ingenious and dexterous Chymist, who wrought part of the work, with the last nam'd Gentleman, and afterwards wrought it by himself and perhaps with more skill. That the thing is true, for the main, I am very prone to beleive. For some of the Powder made according to the Process, thô it were very unripe and fugitive, yet being presented me by a *Virtuoso* of quality, to whom it had been given, we found a real augmentation of gold.[t] And not long after the dexterous Chymist I lately mention'd, made an Experiment with some that he had prepar'd himself and show'd me from time to time in its Progress towards maturity; and in this Tryal my scruple was prevented, for this Powder was not projected on gold, as the former had been, but upon silver carefully separated from gold as well as copper, and yet the quantity of gold prov'd to be the same of the Augment before mention'd. But these Powders, whereof I have yet a Dose by me that was part of the Powder that made the first Tryal, is very unripe, and we found that of (by my guess) near 4. parts or 5. of it would presently fly away in a strong fire.

[s] The draft reads in the margin: "T[ranscri]b'd"; and in pencil: "Philaretus speaks."

[t] Here and below, the fair copy uses standard alchemical symbols for the metals. These symbols fill spaces left in the manuscript at the time of writing; the draft preserves the spaces.

[285] When I was perswaded by these Inducements as to look upon the following work as Real for the maine, I committed the following Process to Paper, and then had the approbation of One, who had wrought it himself, that I understood it rightly. Which that it made me the more solicitous to obtain, was not that any profit was to be made by it in the State, I saw it try'd in, But 1st because it seems a very curious work, being different from all those I have met with among Chymists, and so may excite their Curiosity to work by wayes and practise [unknown] for ought yet appears, even by the Adepti. 2ly Because I take this Process to be a thing very improvable by a Sagacious Artist, who may carry it on much further than 'tis yet brought to, especially by diversifying & applying it. 3ly Because at a certain stage of the Operation it may if one please be broken off, and the already produc'd stuff, may be apply'd to great emprovement of ----------. 4ly Because this work is not of that benefit as may threaten the welfare of States, if it should fall into unworthy hands. Lastly, That in the Progress of this work, there is a certain stage, or nick of time, at which part of the Powder being taken away is by a very slight change turn'd into an Excellent Medicine which I have not without some wonder known to be a strong Sudorifick in the Dose of a single Grain, and yet the Patients being ask'd, complain'd of not being weaken'd by the Copious Evacuation.

Text F: Explanatory Note on Boyle's Attempts Toward the Elixir [BP, 38:158–59][u]

You will perhaps think it strange that in the foregoing part of this paper I[41] mention[42] my having consented to Tryals made upon different matters: since 'tis said to be the unanimous doctrine of the Adepti, that there[43] can be but One true matter[44] of the grand Elixir, & that whoever works upon any other is either an ignorant or deluded or a Sophister ^if not also^ and Imposter. But to this I answer 1. That my chief designe being to attain good medicins ^[45]there^ may be menstruums & other Instruments drawn from more than one matter, that may be capable either of becoming, or which is more easy, of preparing noble Remedys: as may appear by[46] Solvents mention'd by approv'd[47] Authors and particularly by Helmont, who makes a[48] vast difference between his Alcahest & his volatiliz'd Salt of Tartar, thô he inculcates that This however it be far inferior to the other is a noble menstruum that corrects all vegetable poisons, & dissolves ^even^ Quicksilver & silver and prepares them into powerful Medicines, and, in short, performs all that a Physician or a Chirurgeon needs to wish[49] in point of remedies. 2. That the Hermetic Philosophers are not so unanimous ^as is generally beleeved by Spagyrists^ in confining them that would make[50] the great work[v] itself, to one[51] matter, nor[52] to[53] ^one^ determinate way of working on it. This[54] I could countenance by the authority of some that are allow'd[55] to have bin

[u] Margin: "T[ranscri]b'd."

[v] I.e., the Philosophers' Stone.

true Adepti. But I scruple not to say, that it weighs more with me, that I have had the opportunity to be confirm'd in my opinion, by the autority or confessions of some[56] Hermeticks now alive whose candor & experimented skil I can more confidently relye on, as men that are well acquainted with the grounds of the receiv'd doctrine, and likely to have successively try'd new wayes of attaining their ends; especially since, Thirdly, if one of the Ancient Adepti that rejects all but one matter, shold revive,[57] and deliver to me his opinion in never so express & confident terms; I shold[58] thô with due respect to his[59] happy skil, presume to tell him, that I must ^take the Liberty to^ dissent from him. For thô I think it very possible to pronounce safely that[60] this or that ^determinate matter^ handled according to this or that method, will never produce the[61] ^Elixir, yet^ when we speak of inanimate & inorganical bodys, I think it very[62] difficult (if at all possible) to prove an Universal Negative. Since to doe it, one must[63] intimately know the nature of all the[64] subjects that are capable to be wrought upon ^for such a purpose^ & the extent or the power of[65] ^the^ combinations & of all the ways of[66] preparation that can be imploy'd upon such a subject, & that I think is more than any Chymist in the world ever [158*v*] did know or can reasonably be suppos'd likely to know. Solomons knowledg was the admiration of his own & aftertimes and besides the[67] ^natural & political^ advantages he had to excell other men, he was supernaturally assisted to surpass them. But thô these[68] unmatch'd abilitys of his were soe much[69] about the discovery of natural things that he gave the world a History (which none is mentioned to have done before him) of Animals & Plants yet this great Naturalist thô aided by the Skil & long Experience of the Tyrians then the greatest Navigators in the world[70]seems by the great slowness of his Fleets voyages to Ophir to have known no better wayes of sailing than by attending to Heel & Promontorys & ^perhaps^ the flight of Birds[71] & other such indications and which was anciently the seamans chief guide[,] the Pole-Star. And yet he that shold have dogmatically pronounced that[72] ^any of^ of these was the ^pilates^ true guide & that there could be no better ^wold^ have bin exceedingly mistaken, as may appear by the[73] Magnetick Needle or[74] ^marriners^ compass which is[75] reputed to have bin found less than 4 ages ago by one Flavius[76] of Almalphi a seaman that[77] before this lucky hit[78] liv'd an obscur'd person & was never afterwards faign'd for Learning or Sagacity. What[79] great surprizing changes may be[80] produc'd in natural Bodys[81] skilfully wrought on[82] by sagacious Naturalists furnisht not only with Chymical, but also with[83] Mechanical, & other Philosophical parts of knowledg; I may elsewhere purposely give Instances of. And indeed when I consider the vast extent of Natures workes & ^the wonderfuly^[84] various[85] ^wayes^ & methods that may be employ'd to alter them & the strange faecundity[86] that the marriage of nature & Art may be ^cap^able[87] ^of arriving at;^ I cannot but think[88] ^the man that is^ forward to assert Universal Negatives as to the changes of forms & qualities among inanimate bodys is much fitter to prove[89] that he ^himselfe^ is confident than that his doctrine may be confidently relyd on. In short I think an Adeptus may be[90] safely credited in what upon Experience he delivers affirmatively as that such a Body is the true matter that being handled in a due method will produce the Philosophers Stone, for he may have learnt[91] by his own[92] tryals that what he sayes is true. But I allow myself to distrust his authority, thô not his veracity[,] in what he teaches negatively when he tells me, that there is no other matter ^nor^ noe other way of managing it ^successfully^ than that he mentions or hints, and ^all[93] the respect I can on this occasion^[94] ^pay him is^ to beleive that he beleives himself & that indeed there is no other matter or method ^of^ elaborating it that is known to Him.[95] ^But^ I think [that] ought not to hinder an inquisitive man vers'd in chimical affairs from working upon more matters & in more manners than one, in case he be invited to it, either by[96] strong Philosophical[97] probabilitys grounded on[98] the nature of things[99] ^or^ analogous Experiments or by the faithfull communication of some Friend whose honesty & skill he has just reason to be[100] confident[.] This Advertisment Sir I take to be of that Importance to a virtuoso qualify'd like you, that I[101] ^thoght^[102] myself bound to endeavor to remove a prejudication that[103] by discouraging probable attempts may hinder the increase of knowledg by several discoverys that may be hop'd for in the industrious pursuit of them[,] & this Advertisment I chiefly intend for You & those other generous Inquierers that[104] [159] are ^not very^ solicitous that the Processes they attempt shold[105] afford them Gold, provided ^they make^ notable discoverys of truth & furnish them with extraordinary & noble Medicins.

Textual Notes

1. Altered from *houses.*
2. *the*
3. Altered from *profit.*
4. *into the*
5. In the draft, "2000" is added later in pencil; in the fair copy, the numeral is followed by "two thousand" in a different hand.
6. *Some*
7. *Paper*
8. *warily of the* ^several^ things that in a general way
9. *a*
10. *^particulars as^ the Chymists usually call'd*
11. Altered from *have.*
12. Altered from *backward enough* ^but even^.
13. *those them I might easily have got*
14. *somewhat they*
15. *successful*
16. *I miss'd*
17. Sentence altered from: "I laid them up soe carefully that for a very long time I miss'd them whilst I had the Opportunity of time & Furnishes [amanuensis' dicta

tion error for *Furnaces*] to make tryal of them. Not having ˆofˆ lately thô somwhat unexpectedly recov[ered]"

18. *I think it not amiss to*
19. *them to be true & who*
20. *who did not sell them*
21. *who affirm'd they know them to be true and who did not*
22. *if you*
23. *p[er]sons sev[er]al*
24. *my*
25. *from*
26. *change*
27. *any of*
28. *Gold and silver*
29. *several new*
30. *being ˆconsistˆ*
31. *from*
32. *most Par[ticulars]*
33. *p[ro]fitable enough if*
34. Replaces *must.*
35. *when*
36. *of people*
37. Altered from *any.*
38. Altered from *twould look like,* then *twould pass.*
39. *as*
40. Originally read *has som share ˆof itˆ with them in the least.*
41. *onse make*
42. *of*
43. *is not ˆneither Matterˆ*
44. *that is capable to be*
45. *that*
46. *divers*
47. *sp*[agyrists?]
48. *great*
49. *for*
50. *over*
51. *determinate*
52. *weigh*
53. *a*
54. *ma*[tter]
55. *that*
56. *now*
57. *& affirm to me in express terms*
58. *presume*
59. *great skill or at least*
60. *this or that determ*
61. *mixture*
62. *bold*
63. *know*
64. *bodys*

65. *skilful*
66. *pr[e]paring*
67. *great & peculiar*
68. *^his^*
69. *employd*
70. *did by the long time*
71. *or*
72. *one or more*
73. *divers directive vertue of*
74. *seamans*
75. *thô not found out until abt four Cent Ages ago, & consequently many after*
76. *a sea*[man]
77. *only*
78. *was*
79. Preceded by *^And^*.
80. *effe*[cted]
81. *handled*
82. *with*
83. *other Phys*[ical]
84. *almost uncomprehensively*
85. *forms that Art can superinduce in them I cannot but*
86. *^of forms & qualities^*
87. *to produce*
88. *those that are*
89. *that the Assertor*
90. *^much more^*
91. *^so much^*
92. *experience*
93. *that*
94. *I* [illeg.] *doe in this case is*
95. *which*
96. *good*
97. *gro*[unds]
98. Altered from *upon.*
99. *&*
100. *persuaded*
101. *think*
102. *I shold*
103. *by the*
104. *aim at*
105. *make*

APPENDIX 3

ROBERT BOYLE'S DIALOGUE ON THE CONVERSE WITH ANGELS AIDED BY THE PHILOSOPHERS' STONE [POSSIBLY A FRAGMENT OF "OF THE SUPERNATURALL ARCANA PRETENDED BY SOME CHYMISTS"]

A DIALOGUE AMONG TIMOTHEUS, ARNOBIUS, ELEUTHERIUS, AND CORNELIUS
[BP, 7:134v–150]

Arguments for the Negative

That there is great danger of and in mistaking an evill spirit for a good one.

That the worshipping of good Angels[a] is forbidden as a peice of Idolatry

That most of the stories own of witches and apparitions are fictitious or unfitt to be credited.

That there is no rationall account to be given of the pretended intercourse between good spirits and the Adepti and especially that tis unaccountable how a red powder should draw and procure such a familiarity.

[135]

Arguments for the affirmative

That the matter of fact may be probably made out (wherein the stories of witchcraft may be helpfull)

That tis not difficult for the Adepti to discriminate good spirits from evill ones.

That tis granted Angels ought not to be worship'd nor do the Adepti worship them.

That tis not likely that aire and fluid parts of the world should be destitute of spirits.

That tis not likely that there should be so few spirits or so few orders of them as is commonly presumd.

[a] Here and throughout the MS reads *Angle* for *Angel*, which I have silently corrected for readability's sake.

That there may be some spirits of a sociable nature with men and that we know very little of the nature, communities, laws, Politicks and government of spirits.

That ^therefore^ there may be congruities or magnatisms capable of inviting them which we know nothing of.

That therefore the unaccountableness objected is no sufficient Argument.

That such persons as the Adepti may well be supposd to be under a peculiar conduct and to have particular priviledges.

[136]

TIMOTHEUS: You are well met gentlemen and since I saw you coming all three together out of Parisinuss' house give me leave to aske you whether any of you has taken notice of the new sett of books that he has lately added to his library.

ARNOBIUS: I can easily answer you this question Timotheus for when I came in I found him in his study and was somewhat surpris'd to finde there a collection of I know not how many old books about the Phylosophers-stone in one of which he was so attentively reading that I had leasure to se[e] what the book treated of, before he was aware of my looking over his shoulder.

TIMOTHEUS: I presume then that you cannot but wonder as well as I do, that a man versd in rationall and experimentall Phylosophy and who neither loves money nor wants as much [137] of it as will supply him not onely with necessaries but conveniences should with so much application of minde peruse those dull Authors that deserve to be left in as great an obscurity as they[1] affect in their aenigmaticall writings: and that he should so industriously seek the Philosophers-stone of which he has no need, onely that he may be able to make more gold then he would dare own to have and could perhaps tell what to do with.

ELEUTHERIUS: If I mistake not gentlemen you mistake the scope and drift of Parisinus's studyes, for by some occasionall but free discourse I have had with him I am induc'd to guess that in his Hermetick attempts he has higher aims then the acquiring of the skill to Transmute inferior mettalls into gold.

ARNOBIUS: I confess to you gentlemen that what Eleutherius does but suspect concerning Parisinus has bin long my opinion, for I think his ambition reaches higher then gold or anything that gold can purchase.

TIMOTHEUS: But what higher aime can he have then that, which seems the top of humane prosperity and the utmost thing that mankind in this life is capable of arriving at. [138]

ARNOBIUS: I might tell you and truely too, that I think a deep incite[b] into the adept Phylosophy (as I finde the masters of it sometimes call it) may be a

[b] Amanuensis's error for the homonym *insight*.

more desirable thing as it enables men to make admirable medicines for the sick then as it enables them to make silver and gold[;] but I shall answer you more freely and directly by telling you that I am apt to think the maine reason why Parisinus so much vallues and cultivates the study of Hermetick Phylosophy, is that he has a higher aime in it then you suppose and hopes that the acquisition of the Philosophers-stone may be an inlett into another sort of knowledge and a step to the attainment of some intercourse with good spirits.

TIMOTHEUS: This surprising answer Arnobius does indeed remove my first wonder but it onely changes the object of it and very much increases the degree. For having allways lookt uppon Parisinus both as a [139] rationall person and as a good christian I cannot but exceedingly admire that he should think it either possible or lawfull for him, by the help of a red powder which is but a corporeall and even an inanimate thing to acquire communion with incorporeall spirits[,] and since I perceive he has some what opend himself to you and Eleutherius give me leave to desire you to lay before him when you shall have opportunity, such dissuasives as are proper to withdraw him from prosecuting a study that may exceedingly both indanger his conscience and blemish his understanding.

ARNOBIUS: I shall readily comply with your desire Timotheus if you please to acquaint me with the arguments you would have me imploy and convince me of the validity of them.

TIMOTHEUS: I will not burden you with many Arnobius especially because I can offer you three or four of such weight that they may well excuse my forbearing to name any more.

And first since an Apostle tells us that even the divell can transforme [140] himself into an Angel of light tis very dangerous to have, much less to procure the conversation of Angels for fear wee should by the deluding arts of spirits farr more knowing then men, be trapand into the fatall mistake of a divill for an Angel of light.[c]

But if it were granted to Parisinus that he could be sure that the spirits whose converse he aspires to, will prove such as belong to ^the^ Coelestiall court and not such as (though in disguise formes) assend out of the bottomeless pitt, yet how will he acquit himself of an idolatricall commerce with spirits which though nobler creatures then men are yet but creatures; and though good themselves may be the occasions of great mischiefe to us which saint Paul was well aware of when he carefully forewarnes the collosians to beware of the worshipping of Angels which though it seemd an act of great humility in them that exercisd it as it shewd ^that^ they ownd themselves [141] farr inferior to the spirits they worshipd and as it argued them to be deeply sensible of the vast distance between God and them which made them think it sauciness to presume to address them-

[c] Probably a reference to 1 John 4:1; "Beloved, do not trust every spirit but put the spirits to a test to see if they belong to God."

selves immediately to him (and therefore made them desire to be introduced by the mediation of his heavenly courtiers) yet this seeming humility is by the Apostle censurd as irreligious:[d] God having appointed us not Angels but our saviour for a Mediator who pertaking of our own humane nature and having in it dy'd for us, will be more ready to help us, as being God as well as Man he is more able to recommend and answer our requests then the Angels them selves can rationally be supposd to be.

[Blank space (ca. 8 lines) left in MS.]

The two foregoing Arguments have represented the aspiring to have commerce with spirits as a thing dangerous [142] and unlawfull. But now I shall add another dissuasive which though it were single would deserve to be prevalent for I dare appeal to all judicious men whether they do not think it an ambition ^fitter^ for some whimsicall Enthusiast then for so learned a man as Parisinus to hope that if he could turne silver into gold, he might make Angels themselves quit their heaven to court his company as if those glorious and happy spirits were like covetious mortalls greedy of gold and sollicitous to attend men, that are their inferiors, onely because they are well stord with that thick clay.

ELEUTHERIUS: I confess to you Arnobius that it seems somewhat unaccountable and therefore incredible that a little powder that is as truely corporeall as pouder of post[e] or of brick should be able to attract or invite incorporeall and intelligent beings that have neither need nor use of gold[,] to converse familiarly with those that [143] perhapps by chance or fraud have made themselves possessors of a few drams or ounces of Transmuting pouder[;] for what affinity or congruity can there be betwixt the stupid and inanimate Elixir and a rationall and immortal spirit that these happy beings should delight to hover about it, and for its sake should quite contrary to the custome if not also to the politie of those blessed spirits [to] discover themselves in a sensible way to the chymist that carries it about him and converse familiarly with a sinfull and perhaps too, an ignorant mortall from whose conversation what advantage can we suppose they can expect to reap? [144][2]

ARNOBIUS: I shall not Timotheus deny your arguments to be very considerable, nor shall I refuse to press them upon Parisinus if after this conference is over you shall persist to have them proposd unaltered and unaccompanied with auxiliary ones.

ELEUTHERIUS: I know not upon what account you interpose this condition.

ARNOBIUS: I do it,[3] Eleutherius, because that how ever weighty these arguments may be, I know they are not new to Parisinus who has sometimes taken notice of them to Cornelius and to me in the discourse we have had with him about conversing with spirits.

CORNELIUS: In which discourses he had sayd so much to weaken the argu-

[d] Col. 2:18.

[e] I.e., sandstone.

ments we are spekeing of that to make them prevalent they must[,] according to the hint given by Arnobius[,] be garded and seconded by considerations that Timotheus has not yet mentiond.

TIMOTHEUS: But may we not then know what answers Parisinus returns to the allegd arguments that we may discerne whether those arguments[4] be cogent enough of themselves or require to be other wise proposd and backt with auxilliary proofs.

ARNOBIUS: What Parisinus returns to the arguments we are speaking of is not so much by way of direct answer distinctly accomodated to such of them as by making such a discourse in favour of his own Hipothesis that intellegent men ^may^ discerne it to contain the grounds of answering those and divers other objections.

ELEUTHERIUS: I perceive it will be then requisite to beg of [145] You to acquaint us what his Hipothesis or discourse was.

ARNOBIUS: I will tell you the substance of as much as I remember of what he lately sayd to me upon this subject in the presence of Cornelius who I hope will prompt me if I have forgotten or mistaken any thing for on this occasion I pretend to make but a repetition of Parisinus opinions and arguments not a declaration of my owne but because it may be tedious or intricate to speak allways of a third person I will for brevitys sake when Circumstances invite me, discourse as if the opinions I seem to maintain were my own but I must once for all premise at the very begining that the discourse of Parisinus was not one continued thing with out any variety of contexture but consisted of severall distinct propositions or considerations which all of them together made up the discourse that contaynd, and as he thought justified, his Hipothesis.

I shall then with him proceed by steps in my discourse and begin with representing that the matter of fact (namely that men may have converse with good spirits) may be probably made out and though this seems to be the difficultest point of all and therefore fitt to be reserved for the latter end of our discourse yet I choose to mention it in the first place because if it can be made out twill much facilitate your assent to the subsequent part of my discourse and allow me to shorten it very much.

TIMOTHEUS: I readily grant Arnobius all that you say except that the matter of fact may be probably made out. [146]

ARNOBIUS: I am glad Timotheus that you take notice that I imployed the word *probably* because that assures me that you will not expect that so rare and industriously conceald an intercourse as that between men and spirits should not be capable of Demonstrative proofs.

Two mediums there are[5] by which this intercourse may be made probable.

And first that there are witches and magicians I presume Timotheus will not deny for besides the almost universall consent of mankind the scripture its selfe in divers places declares no less. Tis one of the mosaick laws that the Israelites should not suffer a witch to live and not to mention what feats the Eguptian sorcerers did before Pharoh the story is well known of the witch at Endor whom saul so Criminally and fatally consulted the night before his death which was expressly foretold him by the apparition that he took for Samuell[;] and in the new testament, not to mention Elymas the sorcerer we read of Simon the magician and his haveing by magicall feats bewitcht the Samaritans so as to make them style and beleive him the great power of God.[f]

CORNELIUS: Lett me add that though the scripture were silent our own times afford us undeniable instances of the converse of evill men with evill spirits and I my selfe know severall persons divers of them Physitians and men of parts and not superstitious but rather the contrary who have had converse with spirits even in visible shapes and more then [147] One of them against his will as themselves have complained to me.

ARNOBIUS: If then there be an intercourse betwixt men and Demons the grand difficulty that is urged against such conversations from the great disparity in nature between a man and a spirit falls to the ground for the[6] difficulty in point of Physick or Phylosophy rises from the distance between the humane nature and that of an unbodyed spirit as such and so consists in Physicall not in morall qualifications and it is not more inconceivable in Phylosophy that a good then that an evill spiritt should converse with men though for some other reasons (to be toucht hereafter) the visits of the former sort are much more unfrequent in there aparisions to men then those of the latter this may be further strengthend by what I shall have occasion to say under another head.

And therefore I shall now proceed to my second Medium which consists of some instances of Spirits whom wee have more reason to look upon as good than bad that have conversd with men.

There is in the Daemonology of the famous and learned Bodine among divers other remarkable passages a memorable story of a good spt. or Genius that for many years familiarly assisted a freind of his from whose own mouth he had the relation[,] doing him as occasion requird very many good offices as by diverting and instructing him by warning him of dangers and assisting him in them.[g] [148]

ELEUTHERIUS: You put me in mind of a gentleman that I once had some acquaintance with who haveing been bred a Physician & being an excellent Mathematician was much cajold and at last perverted by a famous writer who by many both friends and foes is thought the subtilest Atheist

[f] On the Mosaic prohibition, see Lev. 20:27; on Pharaoh's sorcerers, Exod. 7:10–12; Saul and the witch of Endor, 1 Sam. 28:7–20; Elymas, Acts 13:6–12; Simon Magus and the Samaritans, Acts 8:9–24.

[g] Jean Bodin (1530–1596), best known as a writer of political tracts, also wrote the extremely popular *De la Demonomanie des Sorciers,* first published in 1580 at Paris, and reissued many times in various languages. The text is a lengthy and rambling collection of supernatural events involving spirits.

of our times.[h] This Gentleman when I knew him was not only a profest Christian but a very knowing and devout one & when one day I askt him in free discourse betwixt us two alone how he came to be converted from Atheism in which I knew he had been instructed by so great a master he told me that he had met with an eminent person whose name & place of residence he told me who was able to performe by skill in naturall phylosophy much more than he himselfe could ever do and yet desspisd the world & the preferments and advantages he was possesd of in it because of the great impressions that were made on his minde by his converse with spirits both good and bad though the company of the latter was very unwellcome and uneasy to him and of his conversation with good spirits which yet he kept very secret he gave divers externall proofs to this Gentleman for whom he had a paternall love as ^lookeing on him as^ his spirituall sonne whom he had rescued[7] from Athesim [*sic*] and made a sincere convert.

[149] This same Mathematician being for his various learning & travels much in favour with a sovereign Prince whose name I ^must^[8] not tell you and who is still alive this Prince was lately soe free with him as to confess to him that when he first came to the government of his state he found it in soe odd a condition that (though he be known to be a Prince of great parts) he could not tell which way to turn himself to avoyd the dangers & inconveniences that on all sides seemd to threaten him[;] but whilst he was one night walkeing alone very pensive in his bed chamber and musing on those things there suddenly appeard to him a venerable old man who told him that he came to advise him in the streights that perplext him and thereupon haveing lay'd before him as it were a scheem of the state of his affairs he spent a great while in particularly counselling him how he should behave himself in such and such conjunctures of circumstances assuring ^Him^ that if he folloed his advice he should happily extricate himselfe out of the difficulties that besett him [150] Him which accordingly came to pass the Prince haveing attaind a very prosperous condition and appearing more confident then was expected for a man of his abilities[,] that some gathering clouds that afterwards threatned his state with a storme would happily blow over[.] This Prince has about him a parson of extraordinary learning both in Divine and Humane things especially Physicks[,] Mathematicks and Easterne languages who to confirme my relator[9] in the beleif of a deity and a providence has solemnly assurd him that he himselfe has often had conversation with spirits and that sometimes invisible-shaps severall of which he[,] who is a very pious as well as judicious man is satisfied to be at least[10] no Devils.

[h] Almost certainly Thomas Hobbes.

Textual Notes

1. MS reads *their.*
2. The balance of the document is in scribal hand different from the preceding; corrections and insertions are in the hand of Thomas Smith.
3. *not*
4. altered from *argues.*
5. *are* inserted over illegible crossing out.
6. "for the" repeated.
7. Replaces *redeem.*
8. Replaces *can.*
9. Altered from "relation."
10. *last*

WORKS CITED

ARCHIVAL SOURCES

Archives départmentales du Calvados, Caen
Bodleian Library, Ashmole Manuscripts
British Library, Additional Manuscripts
British Library, Sloane Collection
French Protestant Church Archives, London
Public Record Office, London, State Papers
Royal Society Library, Boyle Letters (BL), 6 vols.
Royal Society Library, Boyle Papers (BP), 46 vols.
Royal Society Library, Miscellaneous Manuscript Series (RSMS)
University of Glasgow, Ferguson Collection
University of Sheffield, Hartlib Papers
Wellcome Historical Medical Library

PRINTED SOURCES

Alexander, H. G., ed. *The Leibniz-Clarke Correspondence.* Manchester: Manchester University Press, 1956.

Arnaud, E. R. *Introduction a la chymie.* Lyons, 1650.

Ashworth, William B. "Natural History and the Emblematic World-View." In *Reappraisals of the Scientific Revolution,* edited by David C. Lindberg and Robert S. Westman, 303–31. Cambridge: Cambridge University Press, 1990.

Athenaeus. *Deipnosophistae.* In *Athenaeus,* translated by C. B. Gulick. Loeb Classical Library. London: William Heinemann, 1930.

Aubrey's Brief Lives. Edited by Oliver Lawson Dick. London: Secker and Warburg, 1950.

Austin, H. D. "Artephius-Orpheus." *Speculum* 12 (1937): 251–54.

Barlet, Annibal. *Vray et methodique cours de la physique resolutive, vulgairement dite Chymie . . . pour connoistre La Theotechnie Ergocosmique, c'est a dire, L'Art de Dieu, en l'ouvrage de l'univers.* Paris, 1653.

Basilius Valentinus. *Chymische Schrifften.* 2 vols. Hamburg, 1677. Reprint, Hildesheim: Gerstenburg Verlag, 1976.

———. *Triumph-Wagen Antimonii.* Leipzig, 1604.

Baumann, E. D. *Francois de la Boë Sylvius.* Leiden: Brill, 1949.

Becher, Johann Joachim. *Opuscula chymica rariora.* Nuremberg, 1719.

———. *Magnalia naturae: or, The Philosophers' Stone lately expos'd to Publick Sight and Sale.* London, 1680.

———. *Physica subterranea.* Edited by Georg Ernst Stahl. Leipzig, 1738.

———. *Tripus hermeticus fatidicus.* Frankfurt, 1689.

Beguin, Jean. *Tyrocinium chymicum.* London, 1669.

Beretta, Marco. *The Enlightenment of Matter.* Canton, MA: Science History Publications, 1993.

Berman, David. *A History of Atheism in Britain: From Hobbes to Russell.* London: Croom Helm, 1988.

Bernard Trevisan. *Responsio ad Thomam de Bononia.* In *BCC,* 2:399–408.
Berthelot, Marcellin. *Collection des anciens alchemistes grecs.* 3 vols. Paris, 1887–1888. Reprint, London: The Holland Press, 1963.
Bianchi, Massimo Luigi. "The Visible and Invisible: From Alchemy to Paracelsus." In *Alchemy and Chemistry,* 17–50.
Biblioteca chemica curiosa. Edited by by J. J. Manget. 2 vols. Geneva, 1702. Reprint, Sala Bolognese: Arnaldo Forni, 1976.
Biographica britannica. 2d ed. 5 vols. London, 1778–1793.
Birch, Thomas. *The Life of the Honourable Robert Boyle.* London, 1744.
Bligh, E. W. *Sir Kenelm Digby and His Venetia.* London: S. Low, Marston and Co., 1932.
Blomberg, William. *An Account of the Life and Writings of Edmund Dickinson.* London, 1739.
Boas, Marie. See under Hall, Marie Boas.
Boerhaave, Hermann. *Elementa chemiae.* 2 vols. Leiden, 1732.
Bonus, Petrus. *Margarita preciosa novella.* In *BCC,* 2:1–80.
Borrichius, Olaus. *Olai Borrichii Itinerarium 1660–1664.* Edited by H. D. Schepelern. 4 vols. Copenhagen: Danish Society of Language and Literature, 1983.
Boulton, Richard. *The Works of the Honourable Robert Boyle, Esq. Epitomized.* 4 vols. London, 1699–1700.
Boyle, Robert. *Animadversions upon Mr. Hobbes's Problemata de vacuo.* London, 1674.
———. *Certain Physiological Essays.* London, 1661.
———. *Christian Virtuoso.* London, 1690.
———. *Considerations touching the Style of the Holy Scriptures.* London, 1661.
———. *Excellency and Grounds of the Mechanical Hypothesis.* London, 1674.
———. *The Excellency of Theology, Compar'd with Natural Philosophy.* London, 1674.
———. *Experimenta et observationes physicae.* London, 1691.
———. *An Historical Account of A Degradation of Gold Made by an Anti-Elixir.* London, 1678.
———. *History of Cold.* London, 1665.
———. *A Hydrostatical Discourse occasion'd by some Objections of Dr. Henry More.* London, 1672.
———. *Invitation to Communicativeness.* In *Chymical, Medicinal and Chyrurgical Addresses,* edited by Samuel Hartlib. London, 1655.
———. *Mechanical Origine of Qualities.* London, 1675.
———. *Medicinal Experiments.* London, 1692.
———. "Of the Incalescence of Quicksilver with Gold." *Philosophical Transactions of the Royal Society of London* 10 (1676): 515–33.
———. *Of the Reconcileableness of Specifick Medicines to the Corpuscular Philosophy.* London, 1685.
———. *Origine of Formes and Qualities.* Oxford, 1666.
———. *Origine and Virtues of Gems.* London, 1672.
———. *The Producibleness of Chymical Principles.* Oxford, 1680.
———. *Sceptical Chymist.* London, 1661.
———. *A Sceptical Dialogue about Cold.* London, 1674.
———. *Seraphic Love.* London, 1659.

———. *Some Considerations about the Reconcileableness of Reason and Religion.* London, 1675.

———. *Some Considerations Touching the Usefulnesse of Experimental Natural Philosophy.* Oxford, 1663.

———. *Some Physico-Theological Considerations about the Possibility of the Resurrection.* London, 1675.

———. *Things Above Reason.* London, 1681.

———. *The Works of the Honourable Robert Boyle.* Edited by Thomas Birch. 5 vols. London, 1744.

Brennan, Katherine Stern. "Culture and Dependencies: The Society of Men of Letters of Caen from 1652 to 1705." Ph.D. diss., Johns Hopkins University, 1981.

Brock, William H. *The Norton History of Chemistry.* New York: W. W. Norton, 1992.

Brown, Harcourt. *Scientific Organizations in Seventeenth-Century France.* Baltimore: Johns Hopkins University Press, 1934.

Burnet, Gilbert. *Dr. Burnet's Travels.* Amsterdam, 1687.

Butterfield, Herbert. *The Origins of Modern Science.* New York: Macmillan, 1951.

Calvet, Antoine. "Les *alchimica* d'Arnaud de Villeneuve à travers la tradition imprimée." In *Alchimie: Art, histoire, et mythes,* Textes et Traveaux de Chrysopoeia I, edited by Didier Kahn and Sylvain Matton, 157–90. Paris: S.E.M.A., 1995.

———. "Alchimie et joachimisme dans les *alchimica* pseudo-Arnaldiens." In *Alchimie et Philosophie à la Renaissance,* edited by Jean-Claude Margolin and Sylvain Matton, 93–107. Paris: Vrin, 1993.

———. "Le *De vita philosophorum* du pseudo-Arnaud de Villeneuve." *Chrysopoeia* 4 (1990–1991): 34–79.

Casaubon, Meric. *A True and Faithful Relation.* London, 1659.

The Character of Town-Gallant. London, 1675.

Chymia Jenensis: Chymisten, Chemisten, und Chemiker in Jena. Jena: Friedrich-Schiller-Universität Jena Verlag, 1989.

Claveus. See DuClo, Gaston.

Clericuzio, Antonio. "Carneades and the Chemists: A Study of the *Sceptical Chymist* and Its Impact on Seventeenth Century Chemistry." In *RBR,* 79–90.

———. "From van Helmont to Boyle: A Study of the Transmission of Helmontian Chemical and Medical Theories in Seventeenth-Century England." *British Journal for the History of Science* 26 (1993): 303–54.

———. "A Redefinition of Boyle's Chemistry and Corpuscular Philosophy." *Annals of Science* 47 (1990): 561–89.

———. "Robert Boyle and the English Helmontians." In *Alchemy Revisited,* edited by Z.R.W.M. van Martels, 192–99. Leiden: Brill, 1990.

———. "Le Trasmutazioni in Bacon e Boyle." In *Francis Bacon: Terminologia e fattura nel XVII secolo,* edited by Marta Fattori, 29–42. Rome: Edizioni dell'Ateneo, 1984.

Clulee, Nicholas H. *John Dee's Natural Philosophy: Between Science and Religion.* London: Routledge, 1988.

Coelum philosophorum. Dresden and Leipzig, 1739.

Colie, R. L. "Spinoza in England 1665–1730." *Proceedings of the American Philosophical Society* 107 (1963): 183–219.

Collesson, Jean. *Idea perfecta philosophiae hermeticae.* In *TC,* 6:143–62.

Cope, Jackson I. *Joseph Glanville: Anglican Apologist.* St. Louis: Washington University Studies, 1956.
Crosland, Maurice. "Chemistry and the Chemical Revolution." In *The Ferment of Knowledge,* edited by G. S. Rousseau and Roy Porter, 380–416. Cambridge: Cambridge University Press, 1980.
———. *Historical Studies in the Language of Chemistry.* New York: Dover Publications, 1978.
Cudworth, Ralph. *True Intellectual System of the World.* London, 1678.
Davis, Edward B. "Newton's Rejection of the 'Newtonian World View.'" *Science and Christian Belief* 3 (1991): 103–17.
Davisson, William. *Philosophia Pyrotechnia.* 4 vols. Paris, 1633–1635.
Dear, Peter. "Totius in verba: Rhetoric and Authority in the Early Royal Society," *Isis* 76 (1985): 145–61.
———. "Trust Boyle." *British Journal for the History of Science* 28 (1995): 451–54.
De auro potabile. In *TC,* 6:382–93.
Debus, Allen G. "And Boyle Stood on the Shoulders of Whom?" *Isis* 57 (1966): 125–26.
———. "The Chemical Debates of the Seventeenth Century: The Reaction to Robert Fludd and Jean Baptiste van Helmont." In *Reason, Experiment, and Mysticism,* edited by M. L. Righini Bonelli and William R. Shea, 19–48. New York: Science History Publications, 1975.
———. *The Chemical Philosophy: Paracelsian Science and Medicine in the Sixteenth and Seventeenth Centuries.* 2 vols. New York: Science History Publications, 1977.
———. *Chemistry, Alchemy and the New Philosophy.* London: Variorum Reprints, 1987.
———. *The English Paracelsians.* London: Oldbourne Press, 1965.
———. "Fire Analysis and the Elements in the Sixteenth and the Seventeenth Centuries." *Annals of Science* 23 (1967): 128–47.
———. *The French Paracelsians.* Cambridge: Cambridge University Press, 1991.
———. "The Significance of the History of Early Chemistry." *Cahiers d'Histoire Mondiale* 9 (1965): 3–58.
———, ed. *Science, Medicine, and Society in the Renaissance.* New York: Science History Publications, 1972.
de Clave, Estienne. *Le cours de chimie.* Paris, 1646.
———. *Les vrais principes et éléments de la Nature.* Paris, 1641.
Dickinson, Edmund. *Epistola ad Theodorum Mundanum, Philosophum adeptum.* Oxford, 1686.
Dobbs, Betty Jo Teeter. *Alchemical Death and Resurrection: The Significance of Alchemy in the Age of Newton.* Washington, DC: Smithsonian Institution Libraries, 1990.
———. *Foundations of Newton's Alchemy, or "The Hunting of the Greene Lyon".* Cambridge: Cambridge University Press, 1975.
———. *Janus Faces of Genius.* Cambridge: Cambridge University Press, 1991.
———. "Studies in the Natural Philosophy of Sir Kenelm Digby." *Ambix* 18 (1971): 1–25; 20 (1973): 143–63; 21 (1974): 1–28.
DuClo, Gaston. *Apologia chrysopoeiae et argyropoeiae.* In *TC,* 2:4–80.
———. *De recta et vera ratione progignendi lapidis philosophorum.* In *TC,* 4:388–413.
———. *De triplici praeparatione auri et argenti.* In *TC,* 4:371–88.

DuClos, S. C. "Dissertation sur les Principes des Mixtes naturals." In *Memoires de l'Academie Royale des Sciences, 1666–1699,* 4:1–30. Paris, 1733.

Egidius de Vadis. *Dialogus.* In *TC,* 2:81–109.

Erastus, Thomas. *Disputationum de medicina nova Philippi Paracelsi.* 3 vols. Basel, 1572.

Evelyn, John. *Diary and Correspondence.* Edited by William Bray. 4 vols. London, 1854.

Feingold, Mordechai. "When Facts Matter." *Isis* 87 (1996): 131–39.

Ferguson, John. *Bibliotheca chemica.* 2 vols. Glasgow, 1906.

Figala, Karin. "Zwei Londoner Alchemisten um 1700: Sir Isaac Newton und Cleidophorus Mystagogus." *Physis* 18 (1976): 245–73.

Figala, Karin, and Ulrich Petzold. "Alchemy in the Newtonian Circle: Personal Acquaintances and the Problem of the Late Phase of Isaac Newton's Alchemy." In *Renaissance and Revolution: Humanists, Scholars, Craftsmen, and Natural Philosophers in Early Modern Europe,* edited by J. V. Field and F.A.J.L. James, 173–92. Cambridge: Cambridge University Press, 1993.

Fisch, Harold. "The Scientist as Priest: A Note on Robert Boyle's Natural Theology." *Isis* 44 (1953): 252–65.

Flamel, Nicholas. *Exposition of the Hieroglyphicall Figures.* Translated by Irenaeus Orandus. London, 1624. Reprint, Garland Publishing, 1994.

Foxcroft, H. C. *The Life and Letters of Sir George Saville.* 2 vols. London, 1898.

Frank, Robert G., Jr. *Harvey and the Oxford Physiologists: A Study of Scientific Ideas and Social Interactions.* Berkeley and Los Angeles: University of California Press, 1980.

French, Peter J. *John Dee: The World of an Elizabethan Magus.* London: Routledge & Kegan Paul, 1972.

Fulton, John F. *A Bibliography of the Honourable Robert Boyle.* Oxford: Clarendon Press, 1961.

Gabbey, Alan. "Henry More and the Limits of Mechanism." In *Henry More (1614–1687) Tercentenary Studies,* edited by Sarah Hutton, 19–35. Dordrecht: Kluwer Academic Publishers, 1990.

Garbers, Karl, and Jost Weyer. *Quellengeschichtliches Lesebuch zur Chemie und Alchemie der Araber im Mittelalter.* Hamburg: Helmut Buske Verlag, 1980.

Glaser, Christophle. *Traite de la chymie.* Paris, 1663.

Glauber, Johann Rudolph. *De auri tinctura sive auro potabile vero.* Amsterdam, 1646.

Gmelin, Johann Friedrich. *Geschichte der Chemie.* 3 vols. Göttingen, 1797. Reprint, Hildesheim: Georg Olms, 1965.

Golinski, Jan V. "Chemistry in the Scientific Revolution: Problems of Language and Communication." In *Reappraisals of the Scientific Revolution,* edited by David C. Lindberg and Robert S. Westman, 367–96. Cambridge: Cambridge University Press, 1990.

———. "A Noble Spectacle: Phosphorus and the Public Culture of Science in the Early Royal Society." *Isis* 80 (1989): 11–39.

———. "Robert Boyle: Scepticism and Authority in Seventeenth-Century Chemical Discourse." In *The Figural and the Literal: Problems of Language and Communication,* edited by Andrew E. Benjamin, Geoffroy N. Cantor, and John R. R. Christie, 58–82. Manchester: Manchester University Press, 1987.

Greene, R. A. "Henry More and Robert Boyle on the Spirit of Nature." *Journal of the History of Ideas* 23 (1962): 451–74.

Grew, Nehemiah. *Musaeum Regalis Societatis*. London, 1681.

Guibert, Nicholas. *Alchymia ratione et experientia ita demum viriliter impugnata & expugnata, una cum suis fallaciis et deliramentis, quibus homines imbobinarat, ut numquam in posterum se erigere valeat*. Strasbourg, 1603.

———. *De interitu alchimiae metallorum transmutatoriae*. Toul, 1614.

Guichenon, Samuel. *Histoire généalogique de la Royale Maison du Savoie*. 5 vols. Turin, 1778–1780.

Güldenfalk, Siegmund Heinrich. *Sammlung von mehr als hundert wahrhaftigen Transmutationgeschichten*. Frankfurt and Leipzig, 1784.

Gunnoe, Charles D., Jr. "Thomas Erastus and His Circle of Anti-Paracelsians." In *Analecta paracelsica*, edited by Joachim Telle, 127–48. Stuttgart: Franz Steiner Verlag, 1994.

Hall, Marie Boas. "Acid and Alkali in Seventeenth-Century Chemistry." *Archives Internationales d'Histoire des Sciences* 9 (1956): 13–28.

———. "An Early Version of Boyle's *Sceptical Chymist*." *Isis* 45 (1954): 153–68.

———. "Frederic Slare, F.R.S. (1648–1727)." *NRRS* 46 (1992): 23–41.

———. "Henry Miles, F.R.S. (1698–1763) and Thomas Birch (1705–1766)." *NRRS* 18 (1963): 39–44.

———. "Newton's Voyage in the Strange Seas of Alchemy." In *Reason, Experiment, and Mysticism*, edited by M. L. Righini Bonelli and William R. Shea, 239–46. New York: Science History Publications, 1975.

———. *Promoting Experimental Learning: Experiment and the Royal Society 1660–1727*. Cambridge: Cambridge University Press, 1991.

———. *Robert Boyle and Seventeenth-Century Chemistry*. Cambridge: Cambridge University Press, 1958.

———. *Robert Boyle on Natural Philosophy*. Bloomington: Indiana University Press, 1965.

Hall, Marie Boas, and A. Rupert Hall. "Newton's Chemical Experiments." *Archives internationales d'histoire des sciences* 11 (1958): 113–52.

Halleux, Robert. "Le mythe de Nicolas Flamel, ou les méchanismes de la pseudépigraphie alchimique." *Archives internationales d'histoire des sciences* 33 (1983): 234–55.

———. *Les textes alchimiques*. Turnhout, Belgium: Brepols, 1979.

Hannaway, Owen. *The Chemists and the Word: The Didactic Origins of Chemistry*. Baltimore: Johns Hopkins University Press, 1975.

Harkness, Deborah. *Talking with Angels: John Dee and the End of Nature*. Cambridge: Cambridge University Press, 1998.

Harwood, John T. *The Early Essays and Ethics of Robert Boyle*. Carbondale: Southern Illinois University Press, 1991.

———. "Science Writing and Writing Science: Boyle and Rhetorical Theory." In *RBR*, 37–56.

Helvetius, Johann Friedrich. *Vitulus aureus*. In *MH*, 815–63.

Henry, John. "Boyle and Cosmical Qualities." In *RBR*, 119–38.

———. "Henry More versus Robert Boyle: The Spirit of Nature and the Spirit of Providence." In *Henry More (1614–1687) Tercentenary Studies*, edited by Sarah Hutton, 55–76. Dordrecht: Kluwer Academic Publishers, 1990.

Heyd, Michael. *"Be Sober and Reasonable": The Critique of Enthusiasm in the Seventeenth and Early Eighteenth Century.* Leiden: Brill, 1995.

Hoefer, J. C. Ferdinand. *Histoire de la chimie.* 2 vols. Paris, 1842–1843.

Hoffmann, Johann Maurice. *Acta laboratorii chemici Altdorfini.* Nuremberg and Altdorf, 1719.

Holmyard, E. J. *Chemistry to the Time of Dalton.* Oxford: Oxford University Press, 1925.

Hooykaas, Reijer. "Die Elementenlehre der Iatrochemiker." *Janus* 41 (1937): 1–28.

———. "Die Elementenlehre des Paracelsus," *Janus* 39 (1935): 175–87.

Hoppen, K. T. "The Nature of the Early Royal Society." *British Journal for the History of Science* 9 (1976): 1–24, 243–73.

Hornius, Christopher. *De auro medico philosophorum.* In *TC,* 5:869–912.

Hunter, Michael. "Alchemy, Magic and Moralism in the Thought of Robert Boyle." *British Journal of the History of Science* 23 (1990): 387–410.

———. "Boyle versus the Galenists: A Suppressed Critique of Seventeenth-Century Medical Practice and Its Significance." *Medical History* 41 (1997): 322–61.

———. "The Conscience of Robert Boyle: Functionalism, Dysfunctionalism, and the Task of Historical Understanding." In *Renaissance and Revolution: Humanists, Scholars, Craftsmen, and Natural Philosophers in Early Modern Europe,* edited by J. V. Field and F.A.J.L. James, 147–59. Cambridge: Cambridge University Press, 1993.

———. *The Early Royal Society and the Shape of Knowledge.* Dordrecht: Kluwer Academic, 1991.

———. *Establishing the New Science.* Rochester: Boydell Press, 1989.

———. "How Boyle Became a Scientist." *History of Science* 33 (1995): 59–103.

———. *Letters and Papers of Robert Boyle: A Guide to the Manuscripts and Microfilm.* Bethesda, MD: University Publications of America, 1992.

———. "A New Boyle Find." *British Society for the History of Science Newsletter* 45 (October 1994): 20–21.

———. "The Reluctant Philanthropist: Robert Boyle and the 'Communication of Secrets and Receits in Physick.'" In *Religio Medici: Medicine and Religion in Seventeenth-Century England,* edited by Ole Peter Grell and Andrew Cunningham, 247–72. London: Scholar Press, 1996.

———. "Robert Boyle and the Dilemma of Biography in the Age of the Scientific Revolution." In *Telling Lives in Science: Studies in Scientific Biography,* edited by Michael Shortland and Richard Yeo, 115–37. Cambridge: Cambridge University Press, 1996.

———. *Robert Boyle by Himself and His Friends.* London: William Pickering, 1994.

———. *The Royal Society and Its Fellows 1660–1700: The Morphology of an Early Scientific Institution.* 2d ed. Oxford: British Society for the History of Science, 1994.

———. "Science and Heterodoxy: An Early Modern Problem Reconsidered." In *Reappraisals of the Scientific Revolution,* edited by David C. Lindberg and Robert S. Westman, 437–60. Cambridge: Cambridge University Press, 1990.

———. *Science and Society in Restoration England.* Cambridge: Cambridge University Press, 1981.

———, ed. *Robert Boyle Reconsidered.* Cambridge: Cambridge University Press, 1994.

Hunter, Michael, and Edward B. Davis. "The Making of Robert Boyle's *Free En-*

quiry into the Vulgarly Receiv'd Notion of Nature (1686)." *Early Science and Medicine* 1 (1996): 204–71.

Hunter, Michael, and Lawrence M. Principe. "The Lost Papers of Robert Boyle." In preparation.

Hunter, Michael, and David Wootton, eds. *Atheism from the Reformation to the Enlightenment.* Oxford: Clarendon Press, 1992.

Hunter, R. A., and Macalpine, Ida. "Robert Boyle—Poet." *Journal of the History of Medicine* 12 (1957): 390–92.

———. "Robert Boyle's Poem—An Addendum." *Journal of the History of Medicine* 27 (1972): 85–88.

Ihde, Aaron. "Alchemy in Reverse: Robert Boyle on the Degradation of Gold." *Chymia* 9 (1964): 47–57.

Ince, Joseph. "Ambrose Godfrey Hanckwitz." *Pharmaceutical Journal,* ser. 1, 18 (1858): 126–30, 157–62, 215–22.

———. "The Old Firm of Godfrey." *Pharmaceutical Journal,* ser. 4, 2 (1896): 166–69, 205–7, 245–48.

Jacob, James R. *Robert Boyle and the English Revolution.* New York: Franklin, 1977.

Johnson, L. W., and M. L. Wolbarsht. "Mercury Poisoning: A Probable Cause of Isaac Newton's Physical and Mental Ills." *NRRS* 34 (1970): 1–9.

Josten, C. H. *Elias Ashmole (1617–1692).* 5 vols. Oxford: Clarendon Press, 1966.

Jung, Carl G. "Die Erloesungsvorstellungen in der Alchemie." In *Eranos-Jahrbuch 1936,* 13–111. Zurich: Rhein Verlag, 1937.

———. "The Idea of Redemption in Alchemy." In *The Integration of the Personality,* edited by Stanley Dell. New York: Farrar & Rinehart, 1939.

Kaplan, Barbara B. *"Divulging Useful Truths in Physick": The Medical Agenda of Robert Boyle.* Baltimore: Johns Hopkins University Press, 1993.

Karcher, Johannes. "Thomas Erastus (1524–1583), der unversöhnliche Gegner des Theophrastus Paracelsus." *Gesnerus* 14 (1957): 1–13.

Kargon, Robert. *Atomism in England from Hariot to Newton.* Oxford: Clarendon Press, 1966.

———. "The Testimony of Nature: Boyle, Hooke and Experimental Philosophy." *Albion* 3 (1971): 72–81.

Karpenko, V. "Coins and Medals Made of Alchemical Metal." *Ambix* 35 (1988): 65–76.

Kent, Andrew, and Owen Hannaway. "Some New Considerations on Beguin and Libavius." *Annals of Science* 16 (1960): 241–50.

Kerckring, Theodore. *Commentarius in currum triumphalem Basilii Valentini.* Amsterdam, 1671.

Kopp, Hermann. *Die Alchemie in Älterer und Neuerer Zeit.* Heidelberg, 1886. Reprint, Hildesheim: Georg Olms Verlag, 1971.

Kraus, Paul. *Jabir ibn Hayyan: Contribution à la histoire des idées scientifiques dans l'Islam.* In *Memoires présentés à l'Institut d'Egypte,* 44. 2 vols. Cairo, 1943.

Kuhn, Thomas S. "Robert Boyle and Structural Chemistry in the Seventeenth Century." *Isis* 43 (1952): 12–36.

Kunckel, Johann. *Chymische Brille contra non-entia chymica.* Wittenberg, 1677.

———. *Collegium Physico-Chymicum.* Hamburg and Leipzig, 1716.

———. *Perspicilium chymicum contra non-entia chymica.* Amsterdam, 1694.

Langelott, Joel. *Epistola ad praecellentissimos Naturae Curiosos de quibusdam in chymia praetermissis.* Hamburg, 1672.
Levi della Vida, G. "Something More about Artefius and the *Clavis sapientiae.*" *Speculum* 13 (1938): 80–85.
L[ongueville], T[homas]. *The Life of Sir Kenelm Digby.* London: Longmans, Green and Co., 1896.
Lull(pseudo-). *Clavicula.* In *TC,* 3:295–303.
———. *Testamentum.* In *BCC,* 1:707–77.
Lux, David S. *Patronage and Royal Science in Seventeenth-Century France.* Ithaca: Cornell University Press, 1989.
MacIntosh, J. J. "Locke and Boyle on Miracles and God's Existence." In *RBR,* 193–214.
———. "Robert Boyle on Epicurean Atheism and Atomism." In *Atoms, Pneuma, and Tranquillity: Epicurean and Stoic Themes in European Thought,* edited by Margaret J. Osler, 197–219. Cambridge: Cambridge University Press, 1991.
Maddison, R.E.W. "The Earliest Published Writing of Robert Boyle." *Annals of Science* 17 (1961): 165–73.
———. "The First Edition of Robert Boyle's Medicinal Experiments." *Annals of Science* 18 (1962): 43–47.
———. *The Life and Works of the Honorable Robert Boyle F.R.S.* London: Taylor & Francis, 1969.
———. "Robert Boyle and the Irish Bible." *Bulletin of the John Rylands Library* 41 (1958): 81–101
———. "Studies in the Life of Robert Boyle, F.R.S.: Part I, Robert Boyle and Some of His Foreign Visitors." *NRRS* 9 (1951): 1–35.
———. "Studies in the Life of Robert Boyle, F.R.S.: Part IV, Robert Boyle and Some of His Foreign Visitors." *NRRS* 11 (1954): 38–53.
———. "Studies in the Life of Robert Boyle, F.R.S.: Part V, Boyle's Operator: Ambrose Godfrey Hanckwitz, F.R.S." *NRRS* 11 (1955): 159–88.
———. "A Summary of Former Accounts of the Life and Work of Robert Boyle." *Annals of Science* 13 (1957): 90–108.
———. "A Tentative Index of the Correspondence of the Honourable Robert Boyle, F.R.S." *NRRS* 13 (1958): 128–201.
Martin, Luther H. "A History of the Psychological Interpretation of Alchemy." *Ambix* 22 (1975): 10–20.
McGuire, J. E. "Boyle's Conception of Nature." *Journal of the History of Ideas* 33 (1972): 523–42.
McGuire, J. E., and P. M. Rattansi. "Newton and the Pipes of Pan." *NRRS* 21 (1961): 108–43.
Mertens, Michele. *Les alchimistes grecs.* Paris: Les belles lettres, 1995.
———. "Project for a New Edition of Zosimus of Panopolis." In *Alchemy Revisited,* edited by Z.R.W.M. van Martels, 121–26. Leiden: Brill, 1990.
Metzger, Hélène. *Les doctrines chimiques en France du début du XVIIe à la fin du XVIIIe siècle.* Paris: Les Presses Universitaires de France, 1923.
Meynell, Guy. "Locke, Boyle and Peter Stahl." *NRRS* 49 (1995): 185–92.
Micreris. *Tractatus suo discipulo mirnefindo.* In *TC,* 5:90–101.
Mintz, Samuel I. *The Hunting of Leviathan.* Cambridge: Cambridge University Press, 1962.
More, Henry. *The Immortality of the Soul.* London, 1662.

More, Louis Trenchard. "Boyle as Alchemist." *Journal of the History of Ideas* 2 (1941): 61–76.

———. *The Life and Works of the Honourable Robert Boyle.* New York: Oxford University Press, 1944.

Morhof, Daniel Georg. *Epistola ad Joelum Langelottum de transmutatione metallorum.* In *BCC,* 1:168–92.

Moufet, Thomas. *De iure et praestantia chemicorum medicamentorum dialogus.* In *TC,* 1:64–89.

Müller-Jahncke, Wolf-Dieter, and Joachim Telle. "Numismatik und Alchemie: Mitteilungen zu Münzen und Medaillen des 17. und 18. Jahrhunderts." In *Die Alchemie in der europäischen Kultur- und Wissenschaftgeschichte,* edited by Christoph Meinel, 229–75. Wiesbaden: Harrassowitz, 1986.

Musaeum hermeticum reformatum et amplificatum. Frankfurt, 1678. Reprint, Graz: Akademische Druck, 1970.

Museum criticum. Cambridge, 1826.

Newman, William R. "The Alchemical Sources of Robert Boyle's Corpuscular Philosophy." *Annals of Science* 53 (1996): 567–85.

———. "Alchemy, Assaying, and Experiment." In *Instruments and Experimentation in the History of Chemistry.* Chicago: University of Chicago Press, 1998.

———. "The Authorship of the *Introitus apertus ad occlusum regis palatium.*" In *Alchemy Revisited,* edited by Z.R.W.M. van Martels, 139–44. Leiden: Brill, 1990.

———. "Boyle's Debt to Corpuscular Alchemy." In *RBR,* 107–18.

———. "The Corpuscular Theory of J. B. van Helmont and Its Medieval Sources." *Vivarium* 31 (1993): 161–91.

———. "The Corpuscular Transmutational Theory of Eirenaeus Philalethes." In *Alchemy and Chemistry,* 161–82.

———. *Gehennical Fire: The Lives of George Starkey, an American Alchemist in the Scientific Revolution.* Cambridge: Harvard University Press, 1994.

———. "Newton's *Clavis* as Starkey's *Key.*" *Isis* 78 (1987): 564–74.

———. *The Summa Perfectionis of the Pseudo-Geber.* Leiden: Brill, 1991.

———. "Technology and Alchemical Debate in the Late Middle Ages." *Isis* 80 (1989): 423–45.

Newman, William R., and Lawrence Principe. "Alchemy vs. Chemistry: The Etymological Origins of a Historiographic Mistake." *Early Science and Medicine* 3 (1998).

Newton, Isaac. *The Correspondence of Isaac Newton.* Edited by H. W. Turnbull. 7 vols. Cambridge: Cambridge University Press, 1960.

Noll, Richard. *The Jung Cult.* Princeton: Princeton University Press, 1994

Norton, Thomas. *Ordinall of Alchymie.* In *TCB,* 1–106.

Osler, Margaret J. *Divine Will and Mechanical Philosophy.* Cambridge: Cambridge University Press, 1994.

———. "Fortune, Fate, and Divination: Gassendi's Voluntarist Theology and the Baptism of Epicureanism." In *Atoms, Pneuma, and Tranquillity: Epicureans and Stoic Themes in European Thought,* edited by Margaret J. Osler. Cambridge: Cambridge University Press, 1991.

———. "The Intellectual Sources of Boyle's Philosophy of Nature: Gassendi's Voluntarism and Boyle's Physico-Theological Project." In *Philosophy, Science, and Religion in England 1640–1700,* edited by Richard Kroll, Richard Ashcroft, and Perez Zagorin, 178–98. Cambridge: Cambridge University Press, 1992.

Paracelsus. *Sämtliche Werke.* Edited by Karl Sudhoff. Munich and Berlin, 1922–1923.

Partington, J. R. *A History of Chemistry.* 3 vols. London: Macmillan, 1961.

Pasmore, Stephen. "Thomas Henshaw, F.R.S." *NRRS* 36 (1982): 177–82.

Patterson, T. S. "Jean Beguin and His *Tyrocinium chymicum.*" *Annals of Science* 2 (1937): 243–98.

Penotus, Bernard. *Chrysorrhoas, sive de arte chemica dialogus.* In *TC,* 2:139–50.

Pereira, Michela. *The Alchemical Corpus Attributed to Raymond Lull.* London: Warburg Institute, 1989.

Perrault, Francois. *The Devill of Mascon.* Translated by Peter Du Moulin. Oxford, 1658.

Petersson, R. T. *Sir Kenelm Digby: The Ornament of England.* Cambridge: Harvard University Press, 1956.

Philalethes, Eirenaeus Philoponus. *Introitus apertus ad occlusum regis palatium.* In *MH,* 649–99.

———. *The Marrow of Alchemy.* London, 1654–55.

Pighetti, Clelia. *L'Influsso scientifico di Robert Boyle nel tardo '600 italiano.* Milan: Franco Angeli, 1988.

Plato (pseudo-). *Quartorum cum commento Hebuhabes Hamed.* In *TC,* 5:101–9.

Priesner, Claus. "Johann Thoelde und die Schriften des Basilius Valentinus." In *Die Alchemie in der europäischen Kultur- und Wissenschaftgeschichte,* edited by Christoph Meinel, Wolfenbütteler Forschungen 32, 107–18. Wiesbaden: Verlag Otto Harrassowitz, 1986.

Principe, Lawrence M. "The Alchemies of Robert Boyle and Isaac Newton: Alternate Approaches and Divergent Deployments." In *Canonical Imperatives: Rethinking the Scientific Revolution,* edited by Margaret J. Osler. Cambridge: Cambridge University Press, 1998.

———. "Apparatus and Reproducibility in Alchemy." In *Instruments and Experimentation in the History of Chemistry.* Chicago: University of Chicago Press, 1998.

———. "Boyle's Alchemical Pursuits." In *RBR,* 91–105.

———. "'Chemical Translation' and the Role of Impurities in Alchemy: Examples from Basil Valentine's *Triumph-Wagen.*" *Ambix* 34 (1987): 21–30.

———. "Diversity in Alchemy: The Case of Gaston 'Claveus' DuClo, a Scholastic Mercurialist Chrysopoeian." In *Reading the Book of Nature: The Other Side of the Scientific Revolution,* edited by Allen G. Debus and Michael Walton, 169–85. Kirksville, MO: Sixteenth Century Press, 1997.

———. "The Gold Process: Directions in the Study of Robert Boyle's Alchemy." In *Alchemy Revisited,* edited by Z.R.W.M. von Martels, 200–205. Leiden: Brill, 1990.

———. "Newly-Discovered Boyle Documents in the Royal Society Archive: Alchemical Tracts and His Student Notebook." *NRRS* 49 (1995): 57–70.

———. "Robert Boyle's Alchemical Secrecy: Codes, Ciphers, and Concealments." *Ambix* 39 (1992): 63–74.

———. "Style and Thought of Early Boyle: Discovery of the 1648 Manuscript of *Seraphic Love.*" *Isis* 85 (1994): 247–60.

———. "Virtuous Romance and Romantic Virtuoso: The Shaping of Robert Boyle's Literary Style." *Journal of the History of Ideas* 56 (1995): 377–97.

Principe, Lawrence M., and William Newman. "The Historiography of Alchemy." *Archimedes* (forthcoming 1998).

Rattansi, P.M. "The Helmontian-Galenist Controversy in Restoration England." *Ambix* 12 (1964): 1–23.

Reti, Ladislao. "Van Helmont, Boyle and the Alkahest." In *Some Aspects of Seventeenth Century Medicine and Science.* Berkeley and Los Angeles: University of California Press, 1969.

Reyher, Samuel. *Dissertatio de nummis quibusdam ex chymico metallo factis.* Kiel, 1692.

Ripley, George. *The Bosom Book.* In *Collectanea chemica,* 121–47. London, 1684. Reprint, London, 1898.

———. *Compound of Alchymie.* In *TCB,* 107–93.

Rocke, Alan J. "Agricola, Paracelsus, and 'Chymia.'" *Ambix* 32 (1985): 37–45.

Rolfinck, Werner. *Chimia in artis formam redacta.* Jena, 1661.

Rolleston, Humphry. "Edmund Dickinson." *Annals of Medical History,* ser. 3, 4 (1942): 175–80.

Der Römisch Kaiserlichen Akademie der Naturforscher auserlesene Medicinisch-Chirurgisch- Anatomisch- Chymisch- und Botanische Abhandlungen. Nuremberg, 1762.

Rosarium philosophorum: ein alchemisches Florilegium des Spatmittelalters. Weinheim: VCH, 1992.

Rowbottom, M. E. "The Earliest Published Writing of Robert Boyle." *Annals of Science* 6 (1950): 376–89.

———. "Some Huguenot Friends and Acquaintances of Robert Boyle (1627–91)." *Proceedings of the Huguenot Society of London* 20 (1959–1960): 177–94.

Sailor, Danton B. "Newton's Debt to Cudworth." *Journal of the History of Ideas* 49 (1988): 511–18.

Sala, Angelus. *Processus de auro potabili.* Strasbourg, 1630.

Sargent, Rose-Mary. *The Diffident Naturalist.* Chicago: University of Chicago Press, 1995.

———. "Scientific Experiment and Legal Expertise: The Way of Experience in Seventeenth-Century England." *Studies in the History and Philosophy of Science* 20 (1989): 19–45.

Sarton, George. "Boyle and Bayle, the Sceptical Chemist and the Sceptical Historian." *Chymia* 3 (1950): 155–89.

Schmieder, Karl Christoph. *Geschichte der Alchemie.* Halle, 1832. Reprint, Ulm: Arkana-Verlag, 1958.

Schneider, Wolfgang. "Chemiatry and Iatrochemistry." In *Science, Medicine, and Society in the Renaissance,* edited by Allen G. Debus, 141–50. New York: Science History Publications, 1972.

Schröder, Gerald. "Neuere Ergebnisse der Beguin-Forschung." In *Die Vorträge der Hauptversammelung der Internationalen Gesellschaft für Geschichte der Pharmazie,* edited by George E. Dann, 227–33. Stuttgart: Wissenschaftliche Verlag, 1966.

Schröder, Wilhelm von. *Unterricht vom Goldmachen.* In *Deutsches Theatrum Chemicum,* edited by Friedrich Roth-Scholtz,, 1:219–88. Nuremberg, 1727. Reprint, Hildesheim: Georg Olms Verlag, 1976).

Schuler, Robert M. "Some Spiritual Alchemies of Seventeenth-Century England." *Journal of the History of Ideas* 41 (1980): 293–318.

Screta, Heinrich. "De mercurio cum auro incalescente." *Miscellanea curiosa sive Ephemeridum Medico-Physicarum Germanicarum Academiae Naturae Curiosiorum* 1682, Decuria II, Annus primus (Nuremberg, 1683): 83–93.

Sendivogius, Michael. *Dialogus mercurii, alchemistae et naturae*. In *TC*, 4:448–56, and *MH*, 590–600.

———. *Novum lumen chemicum*. In *MH*, 545–600.

Shapin, Steven. "The House of Experiment in Seventeenth Century England." *Isis* 79 (1988): 373–404.

———. "Pump and Circumstance: Robert Boyle's Literary Technology." *Social Studies of Science* 14 (1984): 481–520.

———. *Social History of Truth*. Chicago: University of Chicago Press, 1994.

Shapin, Steven, and Simon Schaffer. *Leviathan and the Air-Pump*. Princeton: Princeton University Press, 1985.

Shapiro, Barbara. "The Concept 'Fact': Legal Origins and Cultural Diffusion." *Albion* 26 (1994): 227–52.

Shaw, Peter. *The Philosophical Works of the Honourable Robert Boyle Esq*. 3 vols. London, 1725.

Smith, Pamela. "Alchemy as a Language of Mediation in the Habsburg Court." *Isis* 85 (1994): 1–25.

———. *The Business of Alchemy: Science and Culture in the Holy Roman Empire*. Princeton: Princeton University Press, 1994.

Spargo, P. E., and C. A. Pounds. "Newton's 'Derangement of the Intellect': New Light on an Old Problem." *NRRS* 34 (1970): 11–32.

Stahl, George Ernst. *Philosophical Principles of Universal Chemistry*. Translated by Peter Shaw. London, 1730.

Starkey, George. See also Philalethes, Eirenaeus.

———. *Dr. Georg Starkeys Chymie*. Nuremburg, 1722.

———. *Liquor Alchahest*. London, 1675.

Stillman, John Maxson. *The Story of Early Chemistry*. 1924. Reprint, New York: Dover Publications, 1960.

Streibinger, R., and W. Reif. "Das alchemistische Medaillion Kaisers Leopold I." *Mitteilungen der Numismatischen Gesellschaft in Wien* 16 (1932): 209–13.

Szydlo, Zbigniew. *Water Which Does Not Wet Hands: The Alchemy of Michael Sendivogius*. Warsaw: Polish Academy of Sciences, 1994.

Tachenius, Otto. *Epistola de famoso liquore alkahest*. Venice, 1655.

———. *Hippocrates chymicus . . . Translated into English by J[ohn] W[arr, Sr.]*. London, 1677.

Tenison, Thomas. *The Creed of Hobbes Examined*. London, 1670.

Theatrum chemicum. Edited by Lazarus Zetzner. 6 vols. Strasbourg, 1659–1661. Reprint, Turin: Bottega d'Erasmo, 1981.

Theatrum chemicum britannicum. Edited by Elias Ashmole. London, 1652. Reprint, New York: Johnson Reprint Co., 1967.

Thompson, Charles O. "Robert Boyle." *Proceedings of the American Antiquarian Society*, n.s., 2 (1882): 54–79.

Thorndike, Lynn. *A History of Magic and Experimental Science*. 8 vols. New York: Columbia University Press, 1958.

Thorpe, T. E. *Essays in Historical Chemistry*. London: Macmillan and Co., 1894.

Tractatus de lapide, Manna benedicto. In *Aurifontina chymica*, edited by John Fredrick Houpreght, 107–43. London, 1680.

Trommsdorf, Johann B. *Versuch einer allgemeinen Geschichte der Chemie*. Erfurt, 1806. Reprint, Leipzig: Zentral-Antiquariat der DDR, 1965.

Turnbull, George. *Hartlib, Dury and Comenius: Gleanings from Hartlib's Papers.* London: University Press of Liverpool, 1947.

Uffenbach, Z. C. von. *Merkwürdige Reisen durch Neidersachsen, Holland und Engelland.* 3 vols. Ulm and Memmingen, 1753–1754.

Vanel, Gabriel. *Une grande ville aux XVIIe et XVIIIe siècles.* 3 vols. Caen, 1910.

van Helmont, Jan Baptista. *Opuscula medica inaudita.* Amsterdam, 1648. Reprint, Brussels: Culture et Civilisation, 1966.

———. *Ortus medicinae.* Amsterdam, 1648. Reprint, Brussels: Culture et Civilisation, 1966.

van Suchten, Alexander. *Of the Secrets of Antimony.* London, 1670.

Vickers, Brian. "Alchemie als verbale Kunst." In *Chemie und Geisteswissenschaften: Versuch einer Annäherung,* edited by Jürgen Mittelstrass and Günter Stock, 17–34. Berlin: Akademie Verlag, 1992.

"Vom Gold machen des Dr. Price (ein Auszug des Hrn. Prof. Gmelin aus des Doctors Schrift)." *Göttingisches Magazin der Wissenschaften und Litteratur* 3 (1783): 410–52. Reprint, Osnabrück: Otto Zeller Verlag, 1977.

Wagner, Henry. *Pedigrees of the Du Moulin and De L'Angle Families.* London, 1883.

Walton, Michael T. "Boyle and Newton on the Transmutation of Water and Air, from the Root of Helmont's Tree." *Ambix* 27 (1980): 11–18.

Ward, Seth. *In Thomae Hobbesii philosophiam exercitatio epistolica.* Oxford, 1656.

Webster, Charles. *The Great Instauration: Science, Medicine, and Reform 1626–60.* London: Duckworth, 1975.

———. *Samuel Hartlib and the Advancement of Learning.* Cambridge: Cambridge University Press, 1970.

———. "Water as the Ultimate Principle of Nature: The Background to Boyle's Sceptical Chymist." *Ambix* 13 (1966): 96–107.

Webster, John. *The Displaying of Supposed Witchcraft.* London, 1677.

West, Muriel. "Notes on the Importance of Alchemy to Modern Science in the Writings of Francis Bacon and Robert Boyle." *Ambix* 9 (1961): 102–14.

Westfall, Richard S. "Alchemy in Newton's Library." *Ambix* 31 (1984): 97–101.

———. "Isaac Newton's *Index chemicus.*" *Ambix* 22 (1975): 174–85.

———. "Newton and the Hermetic Tradition." In *Science, Medicine, and Society in the Renaissance,* edited by Allen G. Debus, 183–92. New York: Science History Publications, 1972.

———. "Unpublished Boyle Papers Relating to Scientific Method." *Annals of Science* 12 (1956): 63–73, 103–17.

Weyer, Jost. "The Image of Alchemy in Nineteenth and Twentieth Century Histories of Chemistry." *Ambix* 23 (1976): 65–70.

Whalen, Kathleen. "Robert Boyle, Experimental Reports, and Agricultural Literature." Forthcoming.

Whitby, C. L. "John Dee and Renaissance Scrying." *Bulletin of the Society for Renaissance Studies* 3 (1985): 25–36.

Wojcik, Jan W. "Pursuing Knowledge: Robert Boyle and Isaac Newton." In *Canonical Imperatives: Rethinking the Scientific Revolution,* edited by Margaret J. Osler. Cambridge: Cambridge University Press, 1998.

———. *Robert Boyle and the Limits of Reason.* Cambridge: Cambridge University Press, 1997.

———. "The Theological Context of *Things Above Reason*." In *RBR*, 139–55.
Wood, Alfred C. *A History of the Levant Company.* London: Oxford University Press, 1935.
Wood, Anthony a. *Athenae oxonienses.* 2 vols. London, 1692.
Yates, Francis A. *Theatre of the World.* Chicago: University of Chicago Press, 1969.

INDEX

ABOUT THE AUTHOR

Lawrence M. Principe is Assistant Professor in the Department of Chemistry and the Institute for the History of Science, Medicine, and Technology at The Johns Hopkins University.